Molecular Farming of Plants and Animals for Human and Veterinary Medicine

Molecular Farming of Plants and Animals for Human and Veterinary Medicine

Edited by

L. Erickson
University of Guelph,
Department of Plant Agriculture,
Geulph, Ontario, Canada

W.-J. Yu
Syngenta Biotechnology Inc.,
Research Triangle Park,
North Carolina, U.S.A.

J. Brandle
Research Branch,
Agriculture and Agri-Food Canada,
London, Ontario, Canada

and

R. Rymerson
Research Branch,
Agriculture and Agri-Food Canada,
London, Ontario, Canada

KLUWER ACADEMIC PUBLISHERS
DORDRECHT / BOSTON / LONDON

A C.I.P. Catalogue record for this book is available from the Library of Congress.

ISBN 1-4020-0835-X

Published by Kluwer Academic Publishers,
P.O. Box 17, 3300 AA Dordrecht, The Netherlands.

Sold and distributed in North, Central and South America
by Kluwer Academic Publishers,
101 Philip Drive, Norwell, MA 02061, U.S.A.

In all other countries, sold and distributed
by Kluwer Academic Publishers,
P.O. Box 322, 3300 AH Dordrecht, The Netherlands.

Printed on acid-free paper

Printed in the Netherlands.

Contents

List of Tables

List of Figures

Preface

Molecular farming has been hailed as the "third wave" of genetically-modified organisms produced through biotechnology for the bio-based economy of the future. Unlike products of the first wave, such as herbicide resistant crop plants, which were perceived to benefit only the farmers who used them and the agrochemical companies who developed them, products of molecular farming are designed specifically for the benefit of the consumer. Such products could be purified from food or non-food organisms for a range of applications in industry, as well as animal and human health. Alternatively, the products of this technology could be consumed more directly in some edible format, such as milk, eggs, fruits or vegetables.

There is a rapidly-growing interest on the part of the public as well as in the medical community in the role food plays in health, especially in the immunophysiological impact of food over and above the role of basic nutrition. As a result, there is an expanding opportunity and need for those working in plant and animal biology to conduct research in a broader scientific and social context than in the past. Such research requires a multidisciplinary approach and, despite the potential for discovery and wider relevance of such research, there are not many examples of plant or animal scientists straying far beyond the boundaries of their disciplines to collaborate with colleagues, for example, in the medical community. Yet, the significance of food and feed products far exceeds that of providing the biochemical building blocks for humans and their animals. Many constituents of food have both a beneficial and harmful impact on health through effects on the immune, endocrine, nervous, circulatory and digestive systems. The rapidly expanding bodies of knowledge in the various branches of biological science now enable us to combine and utilize this knowledge in new ways to solve problems old and new in human and animal health.

The multidisciplinary approach required for development of the products of molecular farming is amply illustrated in the chapters of this book.

The problems addressed by the authors range from HIV-Aids in humans to diseases in swine. The approaches to these problems range from oral vaccines produced in plants to antibodies produced in eggs. The array of platforms and strategies being developed for the products of molecular farming demonstrates not only the creativity of the scientists involved and the flexibility of biological systems as biosynthetic factories, but also the potential complexities for the regulation and commercialization of these products.

L. Erickson,
Principal Editor

Acknowledgements

The editors are indebted to the infinitely patient and superbly competent assistance of Jennifer Kingswell and Angela Hill in the Department of Plant Agriculture, University of Guelph, for the completion of this work.

Chapter 1

TOBACCO, A PLATFORM FOR THE PRODUCTION OF RECOMBINANT PROTEINS

R.T. Rymerson, R. Menassa and J.E. Brandle
Agriculture and Agrifood Canada, 1391 Sandford St, London, ON, Canada, N5V 4T3

Keywords: recombinant protein, tobacco, molecular farming, biopharmaceutical

Abstract Tobacco is readily amenable to genetic engineering and has many desirable agronomic attributes, like high biomass yield and high soluble protein levels that are essential for crops used to produce recombinant proteins. It is a non-food crop, making containment in an agricultural setting feasible. Most production systems are based on the accumulation of proteins in leaves, eliminating the need for flowering and pollen production. These attributes make tobacco an ideal bioreactor for the large-scale production of biopharmaceutical recombinant proteins. As a consequence, a wide variety of the recombinant proteins, from simple peptides to complicated multimeric molecules like hemoglobin or secretory antibodies, have been produced successfully in tobacco. Many of these proteins have therapeutic or industrial uses.

1. INTRODUCTION

Although plants have long been sources of pharmacologically active products, the advent of plant genetic engineering has created a myriad of new possibilities. It has also become abundantly clear that a wide array of high-value biopharmaceuticals, antibodies and vaccines can be produced in crop plants (Giddings et al., 2000). Biopharmaceuticals represent a global market in excess of $12 billion and growth is expected to continue at a brisk pace, especially when discoveries based on the human genome begin to reach the clinic (Walsh, 2000). Bottlenecks that are the result of limited manufacturing capacity are creating even more opportunity for plant-based manufacturing of recombinant

L. Erickson et al. (eds.),
Molecular Farming of Plants and Animals for Human and Veterinary Medicine, 1–31.

proteins (Garber, 2001). The ability to provide low-cost recombinant biopharmaceuticals free from human pathogens and nearly unlimited scalability are the strengths of plant-based production systems. However, low costs will only be realized if regulatory restrictions are minimized and if recombinant protein accumulation is maximized. Tobacco is one of the few platforms that can address the technical and regulatory considerations of the biopharmaceutical industry. This plant is easily transformed, and for some time has been the model system for plant transformation (Horsch et al., 1989). Because it is a non-food crop, tobacco provides inherent containment. This containment is further enhanced by production systems based on leaves that do not require flowering or pollination. With these attributes, there is a major role for tobacco as a vehicle for the production of recombinant proteins. In this review we will examine the use of tobacco as a bioreactor species, evaluate what has already been accomplished and address what remains to be done.

2. AGRICULTURAL PRODUCTION OF BIOMASS

Tobacco is a highly self-pollinated annual species that, although there is no record of the species in the wild state, is thought to have originated in the border area of Argentina and Bolivia (Goodspeed, 1954). It is an allotetraploid (2n=48) that probably arose from hybridization between *N. sylvestris* and *N. otophora*. There are many types of tobacco cultivars, ranging in size from the small oriental types like the common lab tobacco "SR1" to the large flue-cured types like 81V9 (Maliga et al., 1973; Menassa et al., 2001). Most of the research conducted to date has been with model system cultivars like SR1. However, the larger tobacco varieties will be more suitable for commercial production simply because they are designed for high biomass production and because they are bred for adaptation to agricultural production conditions and resistance to pests. Scale-up with tobacco will be rapid because it is a prolific seed producer (3000 seeds/capsule), so sufficient seed for large commercial acreage could easily be produced in a contained setting like a greenhouse or phytotron. Furthermore, there is a wide range of germplasm available, so the genetic backgrounds used for molecular farming could be customized to remove metabolites like nicotine that may impact end-use, or those like phenolics that may interfere with processing (Miele, 1997; Menassa et al., 2001).

Unlike conventional crop biotechnology that focuses on crop improvement, the purpose of making bio-pharmaceuticals in plants is to produce molecules that are biologically active. Therefore, issues like safety and containment are of paramount importance and need to be considered. The effect of chronic exposure to biopharmaceuticals through the food chain cannot be predicted, so there is a real need to ensure containment. Tobacco is clearly a non-food crop for mammals, which is an advantage for molecular farming as it effectively excludes the material from the food chain. Rootstalks from harvested tobacco plants will not overwinter in Canada, and with the exception of sporadic occurrences of *N. attenuata* in Southern British Columbia, there are no wild relatives in Canada (Goodspeed, 1954). These characteristics help to reduce the potential for gene leakage into the environment. We have developed a production system based on the use of male-sterile hybrids that further enhances containment (Menassa et al., 2001). The crop is harvested prior to flowering, thereby preventing the production of pollen or seeds. In addition, the crop is grown from transplants that are produced in contained facilities, thus eliminating the release of seed that could remain dormant in the soil.

Although tobacco was originally developed for other applications, production systems have been established for biomass production and protein extraction (Woodleif et al., 1981). Those production systems have been based on high density (100,000 plants/ha) field production of transplanted seedlings and multiple harvests from re-growth following the first cutting (Woodleif et al., 1981). Under these conditions, tobacco is capable of producing substantial amounts of biomass. For example, Woodleif and co-workers (1981) reported fresh weight yields in excess of 50,000 kg/ha from tobacco harvested 40 days following transplanting. It should also be possible to use multiple plantings to achieve the same level of biomass production and at the same time buffer against the uncertainty associated with crop re-growth.

Soluble protein levels in tobacco biomass ranged between 2.3 and 2.8% of total biomass and when stems are excluded concentrations as high as 9.2% were possible (Woodleif et al., 1981). Yields of extractable protein were dependent on the year of production, and varied between 155 and 228 kg/ha. The concentration of protein was also dependent on the time of harvest, but there were no differences between the three cultivars used in the trial. A more exhaustive screening of cultivars may, however, reveal useful variation in protein content.

The soluble proteins of tobacco have been classified into fractions based on solubility at a range of pH values. The most abundant fraction is fraction I. Pilot systems have been developed to purify fraction I protein (Knuckles et al., 1979; Montanari et al., 1993). Fraction I protein is largely made up of the photosynthetic enzyme ribulose bisphophate carboxylase and accounts for about 50% of the total soluble protein (Ershoff et al., 1978). Biological evaluations of the fraction I protein from tobacco show that it is suitable for use in both food and feed (Ershoff et al., 1978; Knuckles et al., 1979; Montanari et al., 1993).

Recent controlled environment studies conducted with tobacco plants expressing two chains of an IgG1 antibody showed that antibody concentration in leaves varied with stage of development, growth temperature and light levels (Stevens et al., 2000). These results demonstrate that research needs to be focused on both the molecular and whole plant levels if maximal yields of recombinant proteins are to be achieved. Stevens and co-workers (2000) also found that the concentration of antibody was closely correlated with total soluble protein levels, demonstrating that total soluble protein can be used as a marker for agronomic studies aimed at improving recombinant protein yields.

3. RECOMBINANT PROTEIN EXPRESSION IN TOBACCO

A large variety of recombinant proteins have been expressed in tobacco. However, recombinant protein yields are frequently low. In an effort to improve yields, recombinant proteins have been targeted to different organs or to various subcellular locations. As well, transient expression systems are being explored for the high level expression of some proteins. In this section, we will describe several methods used in molecular farming for optimizing recombinant protein yields.

3.1 Production in Seeds

Recombinant proteins have been produced in tobacco seeds through the use of both constitutive and tissue-specific expression systems. GM-CSF and the cytomegalovirus glycoprotein B were targeted to seeds using the rice glutelin promoter (Ganz et al., 1996; Tackaberry et al., 1999). scFv antibodies were targeted to seed with the seed-specific promoters of legumin 4 or the unknown seed protein of *Vicia faba* (Fiedler and Conrad, 1995; Fiedler et al., 1997). Dieryck et al. (1997) relied on the constitutive expression of the CaMV 35S promoter for seed expression of human hemoglobin. The seed is a good environment for the accumulation and storage of proteins because it is dry and devoid of enzyme activity prior to germination. However, the use of seeds as a repository for recombinant proteins raises issues of containment, since seed is easily dispersed by wind, birds, or animals. Tobacco seed yields are also quite low, so seed-based systems would be practical for only the most high-value recombinant proteins. Seed-specific expression does not utilize tobacco's full potential as an excellent producer of leaf biomass.

3.2 Organelle Targeting

In the plant cell, proteins synthesized in the cytosol are shuffled to various organelle compartments, among which are the ER, mitochondria, chloroplasts, vacuoles and the apoplast. Transit peptides can target proteins to specific compartments. These transit peptides are cleaved proteolytically once the protein reaches its destination, giving rise to an intact mature protein that can then fold and assemble into its oligomeric form (Glaser et al., 1998).

The endoplasmic reticulum (ER) is the largest membrane-bound organelle in a typical eukaryotic cell (Pelham, 1989). Its surface area is at least six times that of the plasma membrane. Membrane and secretory proteins containing signal peptides are synthesized on the ribosomes, which stud the rough ER, and then enter the organelle. Such proteins remain in the ER for only a short time before being transported to the Golgi and subsequently to the cell surface (Pfeffer, 1987). The ER also contains a population of resident proteins that must avoid export. It is thought that membrane proteins may be retained in the ER because of their tendency to form aggregates. However, the retention of luminal proteins results from a signal in the form of a C-terminal tetrapeptide, KDEL or HDEL (Pelham, 1989 and references cited therein). When the KDEL-containing protein leaves the ER, it binds to a membrane receptor which then recycles it back into the ER.

In a number of cases, the addition of a KDEL retention signal to recombinant proteins has been found to increase their levels of accumulation. For example, Schouten et al. (1996) demonstrated a 100-fold increase in the expression of a single-chain antibody in transgenic tobacco. Likewise, recent experiments in our lab on the expression of human interleukin-10 in tobacco demonstrated that addition of the KDEL signal increased recombinant protein levels by approximately 70 times (Menassa et al., 2001). Clearly, the addition of an ER retention signal is beneficial for increasing expression levels of recombinant proteins.

The pathway for general import of proteins into the chloroplast is a multi-step process that starts with synthesis, in the cytosol, of the precursor protein containing a N-terminal leader peptide (Dobberstein et al., 1977). The precursor protein then binds to the chloroplast surface, inserts into a proteinaceous outer envelope translocation complex and then proceeds across the outer and then inner envelope membranes in an extended conformation (Waegemann and Soll, 1991). Import into the plastid is only half of the localization process because proteins must then be routed into their proper suborganellar compartment. Precursor proteins must be specifically and faithfully routed to one of six subcompartments: the outer and inner envelopes, the inter-envelope space, the stroma, the thylakoid membrane and the thylakoid lumen. Access to the general import pathway is governed by stroma-targeting domains of transit peptides. Stromal protein precursors possess transit peptides that contain only a stroma-targeting domain, whereas thylakoid lumenal protein precursors have an additional targeting domain in their transit peptides (De Boer and Weisbeek, 1991).

There are a few examples of proteins targeted to chloroplasts. Human hemoglobin, for which chloroplast localization was essential for proper assembly in the presence of heme, accumulated in tobacco seeds (Dieryck et al., 1997). Human somatotropin expressed at 0.025% TSP when targeted to the chloroplast and accumulated at approximately 6 times higher levels than when the protein was targeted for secretion, but 300 times less than when the gene was incorporated into the chloroplast genome (Staub et al., 2000).

3.3 Chloroplast Transformation

Plant plastids contain a circular genome of about 290 kb that is present in very high copy number. About 60 copies of the chloroplast genome are present in each plastid and 50-60 chloroplasts are present in a single leaf cell (Bogorad, 2000). Plastids have an active homologous recombination system, which facilitates the precise, targeted integration of cloned DNA, and its propagation throughout the pool of plastid genomes present in an organelle. Therefore, any gene introduced into the chloroplast genome will be present at thousands of copies per cell, creating the potential for very high amounts of recombinant protein expression. Cloned DNA fragments have been integrated into the plastid genome in a number of plant species. So far, the regeneration of fertile transplastomic plants has been limited to tobacco and potato (Svab et al., 1990; Sidorov et al., 1999).

The chief advantage of chloroplast transformation over nuclear transformation is the increased yields of recombinant protein. For example, somatotropin accumulated in the chloroplast to 7% TSP (Staub et al., 2000) and Bt toxin reached 3-5% TSP (McBride et al., 1994). However, chloroplasts are not always the best location to express proteins. For example, a synthetic polymer did not accumulate to high levels despite very high transcript levels (Guda et al., 2000). Nevertheless, transplastomic technology holds other advantages than protein yields. The strictly maternal inheritance of chloroplast genes eliminates pollen dispersal of the transgenes and provides better containment. The ability to introduce groups of genes as operons for transcription as a unit in the plastid, or as a series of independently transcribed DNA sequences constitutes an advantage in the engineering of new biochemical pathways in transgenic plants. Also, the location of transgene insertion is predictable and gene expression is uniform in all independently transformed lines.

3.4 Secretion to the Apoplast

The large majority of recombinant proteins expressed in tobacco have been targeted for secretion to the apoplast with variable success (Tables 1 and 2). Some proteins accumulate to very high levels in the apoplast of plant cells such as the *Aspergillus* phytase at 14% TSP (Pen et al., 1993), while others are not able to accumulate to high levels such as erythropoietin at 0.0025% TSP (Matsumoto et al., 1995). The reasons for such variations in expression are not yet clear. One aspect that was investigated is the efficiency of secretion of the SIgA/G. In tobacco transformed with SIgA/G, only 10% of newly synthesized and assembled molecules were secreted after 24 hours, with the bulk remaining in the ER. A proportion of this multimeric protein was also delivered to the vacuole where it became fragmented, presumably by protease activity (Frigerio et al., 2000). Thus, a portion of the proteins targeted for secretion may be delivered and degraded in the vacuole.

Regardless of accumulation levels, secretion to the cell exterior should facilitate the purification of recombinant proteins when methods are developed allowing the easy extraction of extracellular proteins from plants. Recently, the group of Ilya Raskin in Rutgers has developed two methods for the collection of secreted proteins, guttation and rhizosecretion (Komarnytsky et al., 2000; Borisjuk et al., 1999).

Guttation is the loss of water and dissolved material from uninjured plant organs, which commonly occurs in higher plants. Guttation fluid (GF) is released from the ends of the tracheids through hydathodes into the apoplastic space, usually under conditions of high humidity. It may also be released through the cuticle or stomata. Some proteins are naturally found in GF. This phenomenon was used to express recombinant proteins targeted for secretion. Komarnytsky et al, (2000) produced alkaline phosphatase, green fluorescent protein (GFP) and xylanase which were targeted for secretion by the native alkaline phosphatase signal peptide, the calreticulin ER targeting signal peptide, or the proteinase inhibitor II signal peptide, respectively. All three proteins were detected in the GF. Alkaline phosphatase was produced at 0.3-2.8% TSP of the guttation fluid.

There are two benefits to this system. The method of collection is non-destructive and GF may be collected over the life of the plant. Also, GF is a much less complex mixture of proteins than the usual total soluble protein extract, making purification of expressed recombinant proteins simpler. However, unless it is developed for large-scale extraction, this method is largely impractical.

Another interesting method for collecting recombinant proteins is rhizosecretion. With this method, the protein of interest is targeted for expression and secretion in root tissue. Plants are grown hydroponically and the protein of interest is secreted into the media. Borisjuk et al. (1999) used this method to produce xylanase, human placental secreted alkaline phosphatase (SEAP) and GFP in tobacco. These proteins were secreted into the media and were all biologically active. GFP was expressed at levels up to 900 ng GFP/g dry root weight, while SEAP was secreted at 20 μg/g dry root weight/day. This system allows for continuous, non-destructive protein production and should permit easy purification of the protein from the simple hydroponic fluid.

3.5 Transient Expression

Heterologous proteins are expressed transiently when a foreign gene is introduced into plant cells without stable integration into the host genome. Genes are expressed only for a short time, until the host cell recognizes and destroys the foreign DNA. The advantages of transient gene expression include the rapidity of protein production, the lack of position effects in the host genome, and the ability to use species that are recalcitrant to tissue culture and regeneration. Several methods of obtaining transient expression have been developed, including electroporation, particle bombardment, microinjection, viral vectors and *Agrobacterium* infiltration. The latter two methods have been used for molecular farming applications and are described below.

Plant viruses such as tobacco mosaic virus (TMV), alfalfa mosaic virus (AMV), or potato virus X (PVX) have been engineered for the expression of recombinant proteins. Proteins of interest may replace the viral coat protein, but these infections are not systemic (Takamatsu et al., 1987). A gene may be introduced into the genome under the control of an additional copy of a coat protein subgenomic promoter, but these sequences are often rapidly deleted from the virus (Dawson et al., 1989). However, recombinant viruses containing heterologous coat protein and subgenomic promoters appear to be much more stable (Yusibov et al., 1999). The phenomenon of leaky termination, exhibited by some plant viruses (Skuzeski et al., 1991), may be employed to express foreign genes as C-terminal fusions to the coat protein. This system is advantageous as only 5% of the coat proteins produced are recombinant, allowing the virus to form particles properly and move systemically (Turpen et al., 1995; Hamamoto et al., 1993). Specific epitopes of proteins may be engineered into the viral coat protein at locations resulting in their presentation at the surface of the virus particle. There are size restrictions on the introduced peptide in some viral systems. For example, the TMV coat protein can only accomodate peptides of less than 25 amino acids (Yusibov et al., 1997), while the coat protein of AMVcan accomodate larger peptides. This approach has been used for the production of several different subunit vaccines (Belanger et al., 2000; Yusibov et al., 1997; Turpen et al., 1995).

The chief advantages of transient viral expression are the rapid production of protein-expressing plant material (2 weeks versus 3 to 4 months for stable transformation) and the high levels of protein produced (in the range of mg/g leaf tissue). A significant disadvantage to this system is the potential for viral spread because of the broad host range of the viruses employed. The stability of recombinant viruses containing different foreign genes can vary significantly, but it appears that the inserted genes are lost eventually (Yusibov et al., 1999). This inherent instability may be advantageous for field production, as the hybrid viruses do not survive long.

Another method to transiently express a recombinant protein in tobacco is the infiltration of tobacco leaves with *Agrobacterium* harbouring the heterologous gene. Upon infection of a plant cell with *Agrobacterium*, the T-DNA is transferred to the nucleus. Some copies of the T-DNA integrate into the plant genome while many other copies remain transient in the nucleus. Genes present on the unincorporated T-DNAs can be transcribed and translated, leading to the rapid accumulation of the recombinant protein. Vaquero et al. (1999) used the vacuum infiltration of *Agrobacterium* to evaluate the transient expression of a tumor-specific antibody in tobacco. Interestingly, the full-size antibody was assembled in tobacco upon the simultaneous infiltration of tobacco leaves with two independent strains of *Agrobacterium* containing the heavy or light chain genes. Neither chain alone was able to accumulate, while the assembled antibody accumulated to 1mg/kg fresh leaf tissue.

4. EXAMPLES

4.1 Antibodies

In the first report of an antibody expressed in plants (Hiatt et al., 1989), the γ and κ chains of a murine hybridoma-derived antibody were expressed at low levels in individual plant lines. When the plant lines were crossed, functional assembled antibody was expressed at 1.3% of leaf protein. The native murine secretory signals of the two antibody chains enhanced the expression levels, as plants containing leaderless constructs produced very low levels of γ or κ chains. Since this study, there have been numerous reports on the expression of whole or single chain variable fragment (scFv) antibodies in tobacco against a wide range of antigens. The scFv is a synthetic antibody derivative containing the heavy and light chain variable domains of an immunoglobulin joined by a flexible peptide linker. While the scFv retains full binding capabilities, it does not need to pass through the endomembrane system to be assembled. A further advance in that field is the production of bispecific scFvs, formed by joining two different scFvs with the cellobiohydrolase linker I from *Trichoderma reesi*. These bispecific scFvs, with their abilities to cross-link two different antigens, have potential as unique therapeutic agents (Fischer et al., 1999). Some of the research, which employs both transient and transgenic expression systems for producing antibodies, is described below.

4.1.1 Antibodies Against Tumour Epitopes

Verch et al. (1998) used a TMV-based transient expression system to produce tumour-specific, full-length monoclonal antibodies. The genes for the heavy and light chains were engineered into independent TMV vectors and were used to co-infect tobacco. This resulted in the assembly of a full-length antibody *in planta*. Such a product has great potential for cancer immunotherapy. Similarly, Vaquero et al. (1999) used the transient *Agrobacterium* system to express a full-size antibody against a human carcinoembryonic antigen in tobacco by the simultaneous infiltration of tobacco leaves with two independent strains of *Agrobacterium* containing the heavy or light chain genes. This antigen has great therapeutic and diagnostic potential, since it is found in most colon cancers, 50% of breast cancers and in tumours of epithelial origin. The transient TMV-based system was also used for the production of tumour-derived, single-chain Fv epitopes in tobacco (McCormick et al., 1999). The affinity-purified scFv elicited an immune response in mice, which protected the animals from an injected, lethal dose of tumour cells.

4.1.2 Antibodies Against Herbicides

The production of antibodies against herbicides may be useful for developing immunoassay-based detection methods for these organic pollutants and the subsequent bioremediation of contaminated soil or water. Single chain variable fragments against two organic herbicides, atrazine and paraquat, were produced in transgenic tobacco at levels up to 0.014% of soluble leaf protein (Longstaff et al., 1998). The binding profile of the anti-atrazine scFv was similar to that produced in *E. coli*. These antibodies also recognized related herbicides. Using the transient PVX system, scFvs against the herbicide diuron were expressed at an estimated 0.1 - 0.25% of fresh tissue weight (Smolenska et al., 1998).

4.1.3 Antibodies Against *S. Mutans*

A recombinant murine monoclonal antibody against streptococcal antigen I/II cell surface adhesion molecules (Guy's 13) was expressed and assembled correctly in transgenic tobacco (Ma et al., 1994). A secretory version (SIgA/G) of this antibody was produced by expressing the four required genes in separate plant lines and then crossing the plants (Ma et al., 1995). This multimeric protein, expressed at up to 500 μg/g leaf, was assembled correctly and was fully functional. A detailed study of the N-glycosylation found that there were many glycoforms produced and 60% of the glycans contained β(1,2) xylose and α(1,3) fucose residues not found on mammalian proteins (Cabanes-Macheteau et al., 1999). Because these plant glycans may elicit an immune response, the immunogenicity of plant-produced Guy's 13 was tested in a murine system. Antibodies against the protein or the glycans were not detected (Chargelegue et al., 2000). No immune response was observed in human clinical topical applications of this antibody to prevent dental caries caused by *Streptococcus mutans* (Ma et al., 1998). However, fucose and xylose residues are known to be targets of the immune response against three plant glycoproteins (van Ree et al., 2000). Altered glycosylation may also affect the biological properties of proteins *in vivo*. For example, tobacco-produced erythropoietin exhibited biological activity in cell assays, but no activity when injected into rats (Matsumoto et al., 1995). The authors attribute this lack of activity to the rapid clearance of administered erythropoietin from the circulation, likely a result of the altered glycosylation. Because N-glycosylation patterns on proteins differ between plants and animals, efforts have been made to "humanize" the glycosylation of plant-produced therapeutic proteins. A murine antibody expressed in tobacco was partially galactosylated when this plant line was crossed with one expressing the human β1,4 galactosyltransferase (Bakker et al., 2001). This approach may be useful for resolving differences in glycosylation that may exist between the organism of gene origin and the bioreactor of choice.

4.2 Biopharmaceuticals

The expression of therapeutic proteins in plants is gaining in popularity because plants have the ability to properly express and assemble eukaryotic proteins. Plants lack endotoxins that may contaminate bacterially produced proteins and do not harbour mammalian pathogens, two important considerations when producing recombinant proteins for therapeutic purposes. Many therapeutic proteins have been expressed in tobacco, including several cytokines, blood proteins, milk proteins, and autoantigens (Table 1).

4.2.1 Cytokines

Cytokines act as biological regulators of the functional activities of cells and tissues. Most cytokines are glycoproteins that are secreted through normal pathways. Since cytokines are involved in a vast array of physiological processes, including the pathogenesis of many diseases, they have great potential for therapeutic use. Several cytokines, including GM-CSF, erythropoietin, interferon (α, β), IL-2 and IL-11 (Meager, 1998; Walsh, 2000), have established or emerging clinical applications. Cytokines are often multimeric, disulfide bonded and/or glycosylated proteins, so their production in prokaryotic systems can be difficult. Although effective production systems based on mammalian, yeast or insect cells exist, they all have limitations related to cost and scale-up. Several cytokine genes have been expressed in tobacco, all of them by *Agrobacterium*-mediated transformation. In most cases, the native signal peptide was used to target the protein through the secretory system to the apoplast. Expression of these genes and levels of protein accumulation have been lower than is necessary for a thorough characterization of the recombinant protein and for clinical trials to be initiated. However, recombinant cytokines were biologically active when tested in *in vitro* cell assays. The low expression levels may be due to the short half-lives of these proteins *in vivo*. It is also possible, however, that a different subcellular compartment may be more suitable to protect the recombinant protein from proteolysis.

We have produced the human interleukin-10 cytokine (hIL-10) in low nicotine tobacco plants (cv. "81V9"). Targeting hIL-10 to the apoplast resulted in very low protein accumulation (0.8 ng/mg protein; 0.00008% TSP). Retention of IL-10 in the lumen of the ER resulted in a 70-fold increase in accumulation (55 ng/mg protein or 0.0055% TSP). The highest expressing plant contained a single copy of the transgene and produced biologically active human IL-10 (Menassa et al., submitted). This plant was selfed and homozygous progeny were used as pollen donors in a cross with male-sterile low nicotine tobacco (81V9). The resulting progeny were hemizygous for IL-10 and male sterile. A field test was conducted with homozygous fertile and hemizygous sterile tobacco plants, and plants were examined for nicotine content and IL-10 levels. No differences were observed between the male sterile and fertile plants (Menassa et al., 2001). This system allows containment of the transgene by growing in the field only male sterile plants that cannot spread pollen to other plants, and containment of the recombinant protein by producing it in a non-food crop. The level of nicotine in 81V9 is 10 times lower than commercial tobacco varieties, making the oral administration of plant tissue or crude protein extracts possible while reducing nicotine side effects.

4.2.2 Autoantigens

The oral administration of autoantigens can induce immune tolerance in autoimmune diseases such as insulin-dependent diabetes mellitus (IDDM). Glutamic acid decarboxylase (GAD) has been implicated as a major autoantigen in IDDM (De Aizpurura et al., 1992). There are two isoforms of GAD in mammals, GAD65 and GAD67. GAD65 is the major form in human pancreatic islets, while GAD67 is the major form in mouse islets. However, large amounts of low cost autoantigen are needed before such treatments become possible. To assess the effect of oral administration of GAD-containing plant material on the onset of diabetes in NOD mice, Ma et al. (1997) expressed mouse GAD67 in transgenic tobacco and potato. At a level of expression of 0.4% TSP, GAD67 was found to protect against the development of diabetes in NOD mice when administered orally at a dose of 1 mg/day.

The expression of human GAD65 in tobacco and carrot taproot was recently reported (Porceddu et al., 1999). The human GAD65 was immuno-reactive with IDDM-associated antibodies and also retained its enzymatic activity. Surprisingly, immunogold labelling showed that GAD65 was localized to chloroplast thylakoids and to mitochondria, with very little labeling in the cytoplasm. This is curious, given the lack of a signal sequence; however, the authors reasoned that the amino terminal region of GAD65 is anchored in the membrane by protein-protein interactions and a similar interaction in the membranes of chloroplasts and mitochondria may be occurring.

4.2.3 Blood Proteins

Perhaps the most striking example of the ability of plants to assemble complex multimeric proteins was the production of human hemoglobin in tobacco. Hemoglobin is a tetrameric protein composed of two α chains and two β globin chains, with each globin subunit containing one heme molecule. The integration of the heme cofactor is necessary for stability and for activity of the assembled protein (Theisen, 1999). A plant expression vector was engineered with the α and the β globin genes. Each gene was fused to the transit peptide of the Rubisco small subunit gene from pea to direct the polypeptide to the chloroplast where assembly would take place in the presence of heme. Functional recombinant human hemoglobin accumulated in seeds, and Western blots showed globin chains of similar mass to the native protein, indicating that the transit peptide had been properly cleaved (Dieryck et al., 1997).

Human serum albumin is synthesized as a prepro-protein, which is translocated to the ER where the prepeptide is removed. The protein then leaves the ER and is further processed in the Golgi where the propeptide is cleaved off by a serine protease. Finally, the mature protein is secreted (Judah et al., 1973). Human serum albumin fused to a plant signal peptide (PR-S) was properly processed and secreted in tobacco cell suspension culture and in potato tubers (Sijmons et al., 1990). The native prepro-albumin protein was also secreted, however, only its presequence was cleaved. The propeptide was not processed, presumably because serine proteases are rare in plants. In any case, removal of the prosequence was not necessary for secretion (Sijmons et al., 1990).

Although expression levels for each of these proteins was low (Table 1), the ability of plants to process and assemble complex multimeric proteins has been established.

4.2.4 Growth Hormones and Factors

The human epidermal growth factor, the human growth hormone and human somatotropin genes have all been introduced into tobacco. A synthetic version of the 6 kDa human epidermal growth factor, codon-optimized for expression in *E. coli*, was introduced into tobacco. The coding sequence was devoid of signal or targeting sequences, implying that the recombinant protein would remain in the cytoplasm. Accumulation levels were extremely low: 60 pg/mg TSP (Higo et al., 1993). The human growth hormone, on the other hand, was targeted for secretion by the α-coixin signal peptide, under the control of a sorghum γ-kafirin seed-specific promoter. Interestingly, the monocot promoter retained its tissue specificity in tobacco, as human growth hormone was detected only in seeds, and not in leaves or roots (Leite et al., 2000). Alternatively, the human somatotropin gene was introduced into the chloroplast genome. Somatotropin accumulation reached 7% TSP as a ubiquitin fusion, and under the control of the strong constitutive chloroplast ribosomal RNA operon promoter (Prrn) with the ribosome binding site region of the bacteriophage T7 gene 10 leader (G10L). The somatotropin produced in the chloroplast was soluble, disulfide-linked and biologically active (Staub et al., 2000).

4.2.5 Milk Proteins

The human α lactalbumin and human lactoferrin genes have been introduced into tobacco via *Agrobacterium* transformation. The genes were expressed and in each case, the native signal peptide was properly cleaved off prior to secretion to the apoplast. Both proteins were found to be biologically active *in vitro* (Salmon et al., 1998; Takase and Hagiwara, 1998). However, the glycosylation patterns of lactoferrin in tobacco were different from those of the native protein. Indeed, the recombinant lactoferrin contained no N-acetyl neuraminic acid residues, and fewer galactose and N-acetylglucosamine residues than in human lactoferrin. In addition, it also contained xylose residues, which are found exclusively in plants (Salmon et al., 1998).

4.2.6 Other Proteins

Several other therapeutically useful proteins from various organisms have been expressed in tobacco (Table 1). The most informative examples will be discussed below.

Human glucocerebrosidase is a lysosomal glycoprotein that catalyses the degradation of complex glycosylceremide lipids. A deficiency in GC production in humans causes a condition called Gaucher's disease. Enzyme replacement therapy results in significant reduction in the symptoms of the disease. However, the high cost of this enzyme places it beyond the reach of many (Barranger et al., 1995). The GC enzyme was introduced into tobacco under the control of the wound-inducible MeGA promoter (CropTech, Virginia). The glycosylated hGC was found to be enzymatically active and represented a spectacular 10% TSP only 8 hours after induction (Cramer et al., 1996).

Ricin is a type II ribosome inactivating protein produced in the endosperm of castor seeds. The mature toxin consists of a disulfide-linked heterodimer of RTA, an RNA-specific N-glycosidase a chain, and RTB, a galactose-binding lectin b chain (Funatsu and Funatsu, 1977). Both RTA and RTB are synthesized as part of the same precursor 576 amino acid prepro-polypeptide. The signal peptide translocates the protein to the lumen of the ER where it is processed to the pro-ricin form. The protein is then transported to the storage vacuole where the 12 amino acid linker is processed endoproteolytically, giving rise to the mature heterodimeric toxin (Lamb et al., 1985). Ricin has applications in the treatment of cancer and AIDS. However, commonly used protein expression systems like insect cell baculovirus, frog oocytes or yeast have failed to produce ricin due to the susceptibility of the host's ribosomes to RTA. Ricin was successfully expressed in tobacco plants (Sehnke et al., 1994), in tobacco protoplasts (Frigerio et al., 1998) and in tobacco suspension culture (Sehnke and Ferl, 1999). Tobacco ribosomes are much less sensitive to RTA than mammalian ribosomes, although a large accumulation of RTA in the cytosol would compromise protein synthesis (Taylor et al., 1994). By testing various constructs containing RTA, Frigerio et al. (1998) found that only when RTA is expressed as part of the preproricin does it get delivered to the vacuoles where it accumulates with minimal toxicity. Sehnke et al. (1994) found that in suspension culture, the ricin heterodimer is secreted as a soluble protein at a concentration of 0.05% TSP.

Table 1. Pharmaceutical proteins expressed in tobacco

Gene	Method of expression / targeting	Protein size / levels	Function / uses	Reference
HSA	stable / secreted	66.5 kDa / 0.025%		Sijmons et al., 1990
Human hemoglobin	stable / chloroplasts	tetramer, one heme/globin subunit / 0.05% in seeds	blood substitute	Dieryck et al., 1997
Erythropoietin	stable / secreted	36 kDa / 0.0026%	regulation of erythrocyte mass / anemia	Matsumoto et al., 1995
Human IL-2	stable / secreted	15-18 kDa / 90 μg/l culture medium and 250-350 ng/g callus	treatment of renal cell carcinoma	Magnuson et al., 1998

Gene	Method of expression / targeting	Protein size / levels	Function / uses	Reference
Human IL-4	stable / secreted	19 kDa / 455 μg/l culture medium and 1.1 μg/g callus	modulation of immune and inflammatory responses	Magnuson et al., 1998
Human IL-6	stable / cytoplasmic	21-28 kDa / ND	immunological, hematological and antitumor effects	Kwon et al., 1995
Human interferon α	stable / secreted	18.5 kDa homodimer / ND	treatment of hairy cell leukemia, hepatitis B, C	Smirnov et al., 1990
Human interferon β	stable / secreted	21 kDa homodimer / ND	treatment of multiple sclerosis	Edelbaum et al., 1992
Human GM-CSF	stable / secreted	18-34 kDa / 0.5 % TSP in seeds and 258 ng/ml culture medium	treatment of neutropenia	James et al., 2000; Ganz et al., 1996
Human EGF	stable / cytoplasm	6 kDa synthetic gene codon-optimized for E. coli / 10^{-6}%	mitogenic stimulates proliferation of mesenchymal and epithelial cells	Higo et al., 1993
Human growth hormone	stable / secreted/seed	22 kDa / 0.16%		Leite et al., 2000
Human somatotropin	chloroplast / stroma	22 kDa / 7%	hypopituitary dwarfism in children	Staub et al., 2000
Human α lactalbumin	stable / secreted	14 kDa / 0.2%	regulates lactose biosynthesis / nutraceutical	Takase and Hagiwara 1998
Human lactoferrin	stable / secreted	80 kDa / 0.3%	antimicrobial	Salmon et al., 1998

Gene	Method of expression / targeting	Protein size / levels	Function / uses	Reference
Mouse GAD67	stable / cytoplasm	67 kDa / 0.4%	autoantigen for NOD mice	Ma et al., 1997
Human GAD65	stable / cytoplasm	65 kDa / 0.04%	autoantigen for IDDM in humans	Porceddu et al., 1999
α trichosanthin	TMV / secreted	27 kDa / 2%	inhibition of HIV replication	Kumagai et al., 1993
Birch pollen allergen	TMV / ND	17.5 kDa / 2.5%	tolerisation against the allergen	Krebitz et al., 2000
Ricin	stable / storage vacuole	0.25 % in leaves, 0.05% in culture medium	AIDS and cancer treatment	Frigerio et al., 1998; Sehnke et al., 1994; Sehnke and Ferl., 1999
ACEI	TMV / ND	12 amino acids / 100 μg/g fresh weight	antihypertensive	Hamamoto et al., 1993
Human protein C	stable / secreted	0.002%	anticoagulant	Cramer et al., 1996
Gluco-cerebrosidase	stable / inducible / ND	69 kDa lysosomal enzyme/ 10 %	treatment of Gaucher disease	Cramer et al., 1996
Human collagen	stable / secreted	120 kDa homotrimer / 0.1 mg/g fresh weight	natural biomaterial	Ruggiero et al., 2000

4.3 Industrial Proteins

The common feature shared by industrial proteins expressed in tobacco is the high level of expression achieved (Table 2). Indeed, bacterial xylanase accumulated to 4-5% TSP in the apoplast of tobacco leaves and in the culture medium of tobacco suspension culture (Herbers et al., 1995; Kimura et al.,

1997); bacterial phytase accumulated to 6-14.4% TSP in the apoplast of tobacco leaves (Ullah et al., 1999; Verwoerd et al., 1995) and to 1% in tobacco seeds (Pen et al., 1993). Rice alpha amylase, when expressed in a TMV vector, accumulated to 5% TSP, while bacterial alpha amylase accumulated to only 0.3% in stably transformed plants (Kumagai et al., 2000; Pen et al., 1992). It is possible that this bacterial gene does not contain optimal codons for expression in plants. Other industrial proteins such as bovine stomach lysozyme (Wilcox et al., 1997), bacterial levansucrase (Ebskamp et al., 1994), and rat Δ9 desaturase (Grayburn et al., 1992) were also stably expressed in tobacco. Expression levels, however, were not reported in percentage of total soluble proteins and although levansucrase appears to be expressed at high levels (8 mg/g fresh weight), it cannot be compared to other systems.

Table 2. Industrial proteins expressed in tobacco

Gene / organism	Method of expression / targeting	Expression protein size / % TSP	Uses	References
Xylanase / *Clostridium*	stable / secreted	37 kDa / 4-5% in leaves	pulp and paper industry	Herbers and Sonnewald 1995 ; Kimura et al., 1997
Phytase / *Aspergillus*	stable / secreted	67 kDa / 1% in seeds, 14.4% in leaves	phosphorus release, animal feed supplement	Pen et al., 1993; Verwoerd et al., 1995; Ullah et al., 1999
α amylase / *Bacillus*	stable / secreted	64 kDa / 0.3% in leaves	starch liquefaction	Pen et al., 1992
α amylase / rice	TMV / secreted	46 kDa / 5% in leaves	starch liquefaction	Kumagai et al., 2000
Lysozyme / bovine stomach	stable	15 kDa / 1.8% F2 proteins	antibacterial	Wilcox et al., 1997
Levansucrase / *Bacillus*	stable / vacuole	8 mg/g fresh weight	low calorie sweetener	Ebskamp et al., 1994
Δ9 desaturase / rat	stable / cytoplasm	ND	nutraceutical	Grayburn et al., 1992

Gene / organism	Method of expression / targeting	Expression protein size / % TSP	Uses	References
Protein-based polymer / synthetic	chloroplast / chloroplast	60 kDa / low levels	prevention of post-surgical adhesion	Guda et al., 2000
Cellulase / *Ruminococcus*	stable / cytoplasm	ND	animal feed supplement	Kimura et al., 1997

Guda et al. (1999) have introduced a synthetic sequence coding for repeated amino acid sequences of GVGVP into the chloroplast and the nuclear genomes of tobacco. This sequence is observed in elastin proteins, and would constitute a protein-based polymer. Although transcript levels were very high in transplastomic lines, protein levels were lower than in nuclear transgenic lines, indicating that chloroplasts may not be always efficient at synthesizing heterologous proteins.

4.4 Vaccines

There are several reports on the production of antigens in tobacco for use as vaccines. Some of these vaccines are designed for oral administration, while others are intended for injection and thus require purification. A common feature of many of these studies is the production of large, multiple subunit molecules, such as intact recombinant plant viruses or self-assembled virus-like particles. Immunologically, these large particles are ideal for oral administration because of their greater immunogenicity that may arise from their slower breakdown in the gastro-intestinal tract. M cells in the gut-associated lymphoid tissue sample antigens from the gut lumen, especially large particulate antigens including bacteria and viruses which bind to the lectin-like receptors on the cell surface (Giannasca et al., 1994; Mayer, 2000). Particularly for diseases of the respiratory and gastro-intestinal tracts, oral vaccination is desirable because it mimics the route of entry of the causal agent. Producing subunit vaccines is preferred, since there is little risk of the recombination of subunits to make an infectious virus.

The production of vaccines in plants has been accomplished by using modified plant viruses to express antigenic epitopes on the viral coat protein for transient expression, or by expressing whole proteins in transgenic plants. Plant viruses like TMV or AMV have had antigenic epitopes added to their coat proteins such that the epitope is exposed in the assembled virus particle.

4.4.1 Transient Viral Expression

Chimaeric plant virus particles expressing antigenic peptides from the rabies virus and from HIV-1 as fusions with the alfalfa mosaic virus (AMV) coat protein were produced (Yusibov et al., 1997). The purified viral particles were used to immunize mice, resulting in an antigen-specific production of virus-neutralizing antibodies. Using the same transient viral system, Belanger et al. (2000) expressed two peptides from the human respiratory syncytical virus (RSV) in tobacco. The purified virus particles elicited the production of RSV-specific serum antibodies when injected into mice. Most of the immunized mice were protected when challenged with the RSV.

Turpen et al. (1995) produced malarial epitopes identified from B cells using the tobacco mosaic virus (TMV). The epitopes were introduced at the surface loop or C-terminus of the TMV coat protein. Two other peptide epitopes from the influenza virus hemagglutinin and one epitope from the principal neutralization determinant of the HIV-1 virus were expressed using the TMV transient system (Sugiyama et al., 1995). The production of these fusion proteins was confirmed by their reactivity with anti-peptide antibodies.

An antigenic peptide epitope of the outer membrane protein F of *Pseudomonas aeruginosa* was engineered into the TMV coat protein and expressed in tobacco (Staczek et al., 2000). The immunization of mice with purified chimeric viruses elicited the production of anti-peptide antibodies that were protective against *P. aeruginosa* infection. This vaccine has potential use to prevent infections by *P. aeruginosa* in immuno-compromised individuals. The same method was used to establish protective immunity against the murine hepatitis virus (MHV) by immunizing mice with chimeric, plant-produced TMV expressing an antigenic epitope from the MHV spike protein (Koo et al., 1999).

4.4.2 Stable Expression

There are also several examples of antigen production in transgenic tobacco, some of which were aimed at increasing the levels of expression of the recombinant proteins. The binding subunit B (LT-B) of the *E. coli* heat labile enterotoxin was produced in tobacco as a potential subunit vaccine against acute diarrhea (Haq et al., 1995). The leaves of transgenic tobacco expressed the LT-

B at 0.014% TSP and this material was used to immunize mice by gavage. The mice made antibodies against this antigen that were found to neutralize the activity of LT-B in cell assays.

The capsid protein of the Norwalk virus, a causal agent of epidemic acute gastroenteritis in humans, was expressed in transgenic tobacco and self-assembled into virus-like particles *in planta* (Mason et al., 1996). The capsid protein, expressed in leaves at 0.23% TSP, was partially purified and administered orally to mice. An immune response was detected, but was higher when cholera toxin was included as an adjuvant.

The spike protein from the transmissible gastroenteritis virus was expressed in transgenic tobacco at 0.1-0.2% TSP (Tuboly et al., 2000). Injection of piglets with a protein extract from the plant tissue resulted resulted in an immune response that was capable of viral neutralization. Tackaberry et al. (1999) produced the large cytomegalovirus glycoprotein B in tobacco seeds. This protein was expressed at about 0.015% of extracted seed protein and was demonstrated to inhibit an immunofluorescence assay against virally infected fibroblasts.

The hepatitis B surface antigen (HBSAg) was successfully expressed in transgenic tobacco at approximately 0.01% TSP (Mason et al. 1992). This protein, which self-assembled into virus-like particles *in planta*, was recognized by monoclonal antibodies against HBSAg. The level of expression was noted to be too low for use as an oral vaccine. In this case, tobacco was used as a preliminary system, since these authors have recently reported high levels of accumulation of HBSAg in potato tubers, suitable for use as an oral vaccine (Richter et al., 2000).

The measles virus hemagglutinin, a surface-exposed glycoprotein, was produced in tobacco leaves (Huang et al., 2001). Mice were treated IP or orally with crude soluble leaf extract. Both treatments resulted in the production of neutralizing antibodies against the measles virus. The nucleocapsid protein of the Puulama virus, a hantavirus that causes nephropathia epidemica, was expressed at about 0.03% dry weight in tobacco leaves (Kehm et al., 2001). This level of protein was sufficient to induce an immune response directed against the nucleocapsid protein.

The ultimate proof of these plant-expressed proteins as vaccines is the production of neutralizing antibodies against the infectious agent in question. Levels of expression are also important, as the immune response was higher when greater amounts of antigen were administered. Particularly for orally administered vaccines, higher levels of antigen may be required to compensate for proteolysis in the digestive tract.

5. CONCLUSION

The wide array of biopharmaceutical and industrial proteins described in this chapter clearly demonstrates the utility of tobacco as the production platform of choice for recombinant proteins. This plant is easily transformed using a variety of methods, including the commonly used *Agrobacterium*-mediated plant transformation. Tobacco is a non-food crop, and because most production systems are based on leaf biomass, the production of pollen is circumvented. These facts minimize the risks to the environment and the food chain. Proof of principle for the safe and efficaceous production of recombinant proteins in tobacco is firmly established. The next challenge will be to optimize recombinant protein production and to develop efficient and low-cost handling, extraction and purification systems.

REFERENCES

Bakker H, Bardor M, Molthoff JW Gomord V, Elbers I , Stevens LH, Wjordi W, Lommen A, Faye L, Lerouge P and Bosch D, 2001. Galactose-extended glycans of antibodies produced by transgenic plants. Proc Natl Acad Sci USA, 98:2899-2904

Barranger JA, Tomich J, Weiler S, Sakallah S, Sansieri C, Mifflin T, Bahnson A, Wei FS, Wei JF and Vallor M, 1995. Molecular biology of glucocerebrosidase and the treatment of Gaucher disease. Cytokines Mol Ther, 1:149-63

Belanger H, Fleysh N, Cox S, Bartman G, Deka D, Trudel M, Koprowski H and Yusibov V, 2000. Human respiratory syncytial virus vaccine antigen produced in plants. FASEB J, 14:2323-2328

Bogorad L, 2000. Engineering chloroplasts: an alternative site for foreign genes, proteins, reactions and products. Trends Biotechnol, 18:257-263

Borisjuk NV, Borisjuk LG, Logendra S, Petersen F, Gleba Y and Raskin I, 1999. Production of recombinant proteins in plant root exudates. Nat Biotechnol, 17:466-469

Cabanes-Macheteau M, Fitchette-Laine AC, Loutelier-Bourhis C, Lange C, Vine ND, Ma JK, Lerouge P and Faye L, 1999. N-Glycosylation of a mouse IgG expressed in transgenic tobacco plants. Glycobiology, 9:365-372

Chargelègue D, Vine ND, van Dolleweerd CJ, Drake PM and Ma JK, 2000. A murine monoclonal antibody produced in transgenic plants with plant-specific glycans is not immunogenic in mice. Transgenic Res, 9:187-194

Cramer CL, Weissenborn DL, Oishi KK, Grabau EA, Bennett S, Ponce E, Grabowski GA and Radin DN, 1996. Bioproduction of human enzymes in transgenic tobacco. Ann N Y Acad Sci, 792:62-71

Dawson WO, Lewandowski DJ, Hilf ME, Bubrick P, Raffo AJ, Shaw JJ, Grantham GL and Desjardins PR, 1989. A tobacco mosaic virus-hybrid expresses and loses an added gene. Virology, 173:285-292

De Aizpura HJ, Wilson YM and Harrison LC, 1992. Glutamic acid decarboxylase autoantibodies in preclinical insulin-dependent diabetes. Proc Natl Acad Sci USA ,89:9841-9845

de Boer A and Weisbeek PJ, 1991. Chloroplast protein topogenesis: import, sorting and assembly. Biochim Biophys Acta 1071:221-253

Dieryck W, Pagnier J, Poyart C, Marden MC, Gruber V, Bournat P, Baudino S and Merot B, 1997. Human haemoglobin from transgenic tobacco. Nature, 386:29-30

Dobberstein B, Blobel G and Chua NH, 1977. In vitro synthesis and processing of a putative precursor for the small subunit of ribulose-1,5-bisphosphate carboxylase of *Chlamydomonas reinhardtii*. Proc Natl Acad Sci USA ,74:1082-1085

Ebskamp MJ, van der Meer IM, Spronk BA, Weisbeek PJ and Smeekens SC, 1994. Accumulation of fructose polymers in transgenic tobacco. Biotechnology (N Y), 12:272-275

Edelbaum O, Stein D, Holland N, Gafni Y, Livneh O, Novick D, Rubinstein M, and Sela I, 1992. Expression of active human interferon-beta in transgenic plants. J Interferon Res, 12:449-453

Ershoff BH, Wildman SG and Kwanyuen P, 1978. Biological evolution of fraction I protein from tobacco. Proc Soc Exp Med Biol, 157: 66-630

Fiedler U, and Conrad U, 1995. High-level production and long-term storage of engineered antibodies in transgenic tobacco seeds. Biotechnology (N Y), 13:1090-1093

Fiedler U, Phillips J, Artsaenko O and Conrad U, 1997. Optimization of scFv antibody production in transgenic plants. Immunotechnology, 3:205-216

Fischer R, Schumann D, Zimmermann S, Drossard J, Sack M and Schillberg S, 1999. Expression and characterization of bispecific single-chain Fv fragments produced in transgenic plants. Eur J Biochem, 262:810-816

Frigerio L, Vine ND, Pedrazzini E, Hein MB, Wang F, Ma JK and Vitale A, 2000. Assembly, secretion, and vacuolar delivery of a hybrid immunoglobulin in plants. Plant Physiol, 123:1483-1494

Frigerio L, Vitale A, Lord JM, Ceriotti A and Roberts LM, 1998. Free ricin A chain, proricin, and native toxin have different cellular fates when expressed in tobacco protoplasts. J Biol Chem, 273:14194-14199

Funatsu G and Funatsu M, 1977. Separation of the two constituent polypeptide chains of ricin D [toxic protein of the castorbean]. Agric Biol Chem, 41:1211-1215

Ganz PR, Dudani AK, Tackaberry ES, Sardana R, Sauder C, Cheng XY and Altosaar I, 1996. Expression of human blood proteins in transgenic plants: the cytokine GM-CSF as a model protein. In: *Transgenic plants: a production system for industrial and pharmaceutical proteins.* Eds. Owen MRL and Pen J, John Wiley and Sons, New York, pp. 281-297

Garber K, 2001. Biotech industry faces new bottleneck. Nat Biotechnol, 19:184-185

Giannasca PJ, Giannasca KT, Falk P, Gordon JI and Neutra MR, 1994. Regional differences in glycoconjugates of intestinal M cells in mice: potential targets for mucosal vaccines. Am J Physiol, 267:G1108-121

Giddings G, Allison G, Brokks D and Carter A, 2001. Transgenic plants as factories for biopharmaceuticals. Nat Biotechnol, 18:1151-1155

Glaser E, Sjoling S, Tanudji M and Whelan J, 1998. Mitochondrial protein import in plants. Signals, sorting, targeting, processing and regulation. Plant Mol Biol, 38:311-338

Goodspeed TH, 1954. *The Genus Nicotiana.* Chronica Botanica Company, Waltham, Mass. USA

Grayburn WS, Collins GB and Hildebrand DF, 1992. Fatty acid alteration by a delta 9 desaturase in transgenic tobacco tissue. Biotechnology (N Y), 10:675-678

Guda C, Lee SB and Daniell H, 1999. Stable expression of a biodegradable protein-based polymer in tobacco chloroplasts. Plant Cell Rep, 19:257-262

Hamamoto H, Sugiyama Y, Nakagawa N, Hashida E, Matsunaga Y, Takemoto S, Watanabe Y, and Okada Y, 1993. A new tobacco mosaic virus vector and its use for the systemic production of angiotensin-I-converting enzyme inhibitor in transgenic tobacco and tomato. Biotechnology (N Y), 11:930-932

Haq TA, Mason HS, Clements JD, and Arntzen CJ, 1995. Oral immunization with a recombinant bacterial antigen produced in transgenic plants. Science, 268:714-716

Herbers K, Wilke I and Sonnewald U, 1995. A thermostable xylanase from *Clostridium thermocellum* expressed at high levels in the apoplast of transgenic tobacco has no detrimental effects and is easily purified. Biotechnology, 13:63-66

Hiatt A, Cafferkey R and Bowdish K, 1989. Production of antibodies in transgenic plants. Nature, 342:76-78

Higo, K, Saito Y, and Higo H, 1993. Expression of a chemically synthesized gene for human epidermal growth factor under the control of cauliflower mosaic virus 35S promoter in transgenic tobacco. Biosci Biotechnol Biochem, 57:1477-1481

Horsch RB, Fry JE, Hoffman NL, Eicholtz D, Rogers SG, and Fraley RT, 1985. A simple and general method of transferring genes into plants. Science, 227:1229-1231

Huang Z, Dry I, Webster D, Strugnell R and Wesselingh S, 2001. Plant-derived measles virus hemagglutinin protein induces neutralizing antibodies in mice. Vaccine, 19:2163-2171

James EA, Wang C, Wang Z, Reeves R, Shin JH, Magnuson NS and Lee JM, 2000. Production and characterization of biologically active human GM-CSF secreted by genetically modified plant cells. Protein Expr Purif, 19:131-138

Judah JD, Gamble M and Steadman JH, 1973. Biosynthesis of serum albumin in rat liver, evidence for the existence of 'proalbumin'. Biochem J, 134:1083-1091

Kehm R, Jakob NJ, Welzel TM, Tobiasch E, Viczian O, Jock S, Geider K, Sule S, and Darai G, 2001. Expression of immunogenic Puumala virus nucleocapsid protein in transgenic tobacco and potato plants. Virus Genes, 22:73-83

Kimura T, Kawazu T, Sun JL, Karita S, Sakka K, and Ohmiya K, 1997. Expression of bacterial cellulase and xylanase genes in plant cell. Rec Res Develop Agric Biol Chem, 1:249-258

Knuckles BE, Kohler GO, and deFremery D, 1979. Processing of fresh tobacco leaves for protein fractions. J Agric Food Chem, 27:414-418

Komarnytsky S, Borisjuk NV, Borisjuk LG, Alam MZ and Raskin I, 2000. Production of recombinant proteins in tobacco guttation fluid. Plant Physiol, 124:927-934

Koo M, Bendahmane M, Lettieri GA, Paoletti AD, Lane TE, Fitchen JH, Buchmeier MJ and Beachy RN, 1999. Protective immunity against murine hepatitis virus (MHV) induced by intranasal or subcutaneous administration of hybrids of tobacco mosaic virus that carries an MHV epitope. Proc Natl Acad Sci USA, 96:7774-7779

Krebitz M, Wiedermann U, Essl D, Steinkellner H, Wagner B, Turpen TH, Ebner C, Scheiner O, and Breiteneder H, 2000. Rapid production of the major birch pollen allergen Bet v 1 in *Nicotiana benthamiana* plants and its immunological in vitro and in vivo characterization. FASEB J, 14:1279-288

Kumagai MH, Donson J, della-Cioppa G and Grill LK, 2000. Rapid, high-level expression of glycosylated rice alpha-amylase in transfected plants by an RNA viral vector. Gene, 245:169-174

Kumagai MH, Turpen TH, Weinzettl N, della-Cioppa G, Turpen AM, Donson J, Hilf ME, Grantham GL, Dawson WO and Chow TP, 1993. Rapid, high-level expression of biologically active alpha-trichosanthin in transfected plants by an RNA viral vector. Proc Natl Acad Sci USA, 90:427-430

Kwon SY, Yang Y, Hong CB and Pyun KH, 1995. Expression of active human interleukin-6 in transgenic tobacco. Mol Cells, 5:486-492

Lamb FI, Roberts LM and Lord JM, 1985. Nucleotide sequence of cloned cDNA coding for preproricin. Eur J Biochem, 148:265-270

Leite A, Kemper EL, da Silva MJ, Luchessi AD, Siloto RMP, Bonaccorsi ED, El-Dorry HF and Arruda P, 2000. Expression of correctly processed human growth hormone in seeds of transgenic tobacco plants. Mol Breed, 6:47-53

Longstaff M, Newell CA, Boonstra B, Strachan G, Learmonth D, Harris WJ, Porter AJ and Hamilton WD, 1998. Expression and characterisation of single-chain antibody fragments produced in transgenic plants against the organic herbicides atrazine and paraquat. Biochim Biophys Acta, 1381:147-160

Ma JK, Hiatt A, Hein M, Vine ND, Wang F, Stabila P, van Dolleweerd C, Mostov K and Lehner T, 1995. Generation and assembly of secretory antibodies in plants . Science, 268:716-719

Ma JK, Hikmat BY, Wycoff K, Vine ND, Chargelegue D, Yu L, Hein MB and Lehner T, 1998. Characterization of a recombinant plant monoclonal secretory antibody and preventive immunotherapy in humans. Nat Med, 4:601-606

Ma JK, Lehner T, Stabila P, Fux CI and Hiatt A, 1994. Assembly of monoclonal antibodies with IgG1 and IgA heavy chain domains in transgenic tobacco plants. Eur J Immunol, 24:131-138

Ma SW, Zhao DL, Yin ZQ, Mukherjee R, Singh B, Qin HY, Stiller CR and Jevnikar AM, 1997. Transgenic plants expressing autoantigens fed to mice to induce oral immune tolerance. Nat Med, 3:793-796

Magnuson NS, Linzmaier PM, Reeves R, An G, HayGlass K and Lee JM, 1998. Secretion of biologically active human interleukin-2 and interleukin-4 from genetically modified tobacco cells in suspension culture. Protein Expr Purif, 13:45-52

Maliga P, Sz-Breznovits A and Marton L, 1973. Streptomycin resistant plants from callus culture of haploid tobacco. Nat New Biol, 244:29-30

Mason HS, Ball JM, Shi JJ, Jiang X, Estes MK and Arntzen CJ, 1996. Expression of Norwalk virus capsid protein in transgenic tobacco and potato and its oral immunogenicity in mice. Proc Natl Acad Sci USA, 93:5335-5340

Matsumoto S, Ikura K, Ueda M and Sasaki K, 1995. Characterization of a human glycoprotein (erythropoietin) produced in cultured tobacco cells. Plant Mol Biol, 27:1163-1172

Mayer L, 2000. Mucosal immunity and gastrointestinal antigen processing. J Pediatr Gastroenterol Nutr 30 Suppl:S4-12

McCormick AA, Kumagai MH, Hanley K, Turpen TH, Hakim I, Grill LK, Tuse D, Levy S and Levy R, 1999. Rapid production of specific vaccines for lymphoma by expression of the tumor-derived single-chain Fv epitopes in tobacco plants. Proc Natl Acad Sci USA, 96:703-708

Meager A, 1998. *The molecular biology of cytokines. Molecular Medical Science Series.* Eds. James K and Morris A, John Wiley & Sons, London

Menassa R, Jevnikar A and Brandle J, 2001. A contained system for the field production of plant recombinant proteins. In *Molecular Farming.* Eds. Toutant JP and Balazs E, INRA Editions, Versailles, France, pp 197-205

Miele L, 1997. Plant bioreactors as hosts for biopharmaceuticals: Regulatory considerations. TIBTECH, 15:45-50

Montanari L., Fantozzi P and Pedone S, 1993. Tobacco fraction 1 (F1P) utilization for oral and enteral feeding of patients. 1. Heavy metal evaluation. Food Sci Tech 26:259-263.

Pelham HR, 1989. Control of protein exit from the endoplasmic reticulum. Ann Rev Cell Biol, 5:1-23

Pen J, Molendijk L, Quax WJ, Sijmons PC, van Ooyen AJ, van den Elzen PJ, Rietveld K and Hoekema A, 1992. Production of active *Bacillus licheniformis* alpha-amylase in tobacco and its application in starch liquefaction. Biotechnology (N Y), 10:292-296

Pen J, Verwoerd TC, van Paridon PA, Beudeker RF, van den Elzen PJM, Geerse K, van der Klis JD, Versteegh HAJ, van Ooyen AJJ, Hoekema A, Van Paridon PA, Van den Elzen PJM, Van der Klis JD and Van Ooyen AJJ, 1993. Phytase-containing transgenic seeds as a novel feed additive for improved phosphorus utilization. Bio\Technology, 11:811-814

Porceddu A, Falorni A, Ferradini N, Cosentino A, Calcinaro F, Faleri C, Cresti M, Lorenzetti F, Brunetti P and Pezzotti M, 1999. Transgenic plants expressing human glutamic acid decarboxylase (GAD65), a major autoantigen in insulin-dependent diabetes mellitus. Mol Breed, 5:553-560

Richter LJ, Thanavala Y, Arntzen CJ and Mason HS, 2000. Production of hepatitis B surface antigen in transgenic plants for oral immunization. Nat Biotechnol, 18:1167-1171

Ruggiero F, Exposito JY, Bournat P, Gruber V, Perret S, Comte J, Olagnier B, Garrone R and Theisen M, 2000. Triple helix assembly and processing of human collagen produced in transgenic tobacco plants. FEBS Lett, 469:132-136

Salmon V, Legrand D, Slomianny MC, el Yazidi I, Spik G, Gruber V, Bournat P, Olagnier B, Mison D, Theisen M and Merot B, 1998. Production of human lactoferrin in transgenic tobacco plants. Protein Expr Purif, 13:127-135

Schouten A, Roosien J, de Boer JM, Wilmink A, Rosso MN, Bosch D, Stiekema WJ, Gommers FJ, Bakker J and Schots A. 1997. Improving scFv antibody expression levels in the plant cytosol. FEBS Lett, 415:235-241

Schouten A, Roosien J, van Engelen FA, de Jong GA, Borst-Vrenssen AW, Zilverentant JF, Bosch D, Stiekema WJ, Gommers FJ, Schots A and Bakker J, 1996. The C-terminal KDEL sequence increases the expression level of a single-chain antibody designed to be targeted to both the cytosol and the secretory pathway in transgenic tobacco. Plant Mol Biol, 30:781-793

Sehnke PC and Ferl RJ, 1999. Processing of preproricin in transgenic tobacco. Protein Expr Purif, 15:188-195

Sehnke PC, Pedrosa L, Paul AL, Frankel AE and Ferl RJ, 1994. Expression of active, processed ricin in transgenic tobacco. J Biol Chem, 269:22473-22476

Shuzeski JM, Nichols LM, Gesteland RF and Atkins F, 1991. The signal for a leaky AUG codon in several plant viruses includes the two downstream codons. J Mol Biol, 218: 365-373

Sidorov VA, Kasten D, Pang SZ, Hajdukiewicz PT, Staub JM and Nehra NS, 1999. Stable chloroplast transformation in potato: use of green fluorescent protein as a plastid marker. Plant J, 19:209-216

Sijmons PC, Dekker BM, Schrammeijer B, Verwoerd TC, van den Elzen PJ and Hoekema A,. 1990. Production of correctly processed human serum albumin in transgenic plants. Bio/Technology, 8:217-221

Smirnov SP, Teverovskaia EKH, Krasheninnikova LV and Pukhal'ski VA, 1990. Design of an expression integrative vector and its application for introducing the human recombinant alfa interferon gene into plants. Genetika, 26:2111-2121

Smolenska L, Roberts IM, Learmonth D, Porter AJ, Harris WJ, Wilson TM and Santa Cruz S, 1998. Production of a functional single chain antibody attached to the surface of a plant virus. FEBS Lett, 441:379-382

Staczek J, Bendahmane M, Gilleland LB, Beachy RN and Gilleland HE, 2000. Immunization with a chimeric tobacco mosaic virus containing an epitope of outer membrane protein F of *Pseudomonas aeruginosa* provides protection against challenge with *P. aeruginosa*. Vaccine, 18:2266-2274

Staub JM, Garcia B, Graves J, Hajdukiewicz PT, Hunter P, Nehra N, ParadkarV, Schlittler M, Carroll JA, Spatola L, Ward D, Ye G and Russell DA, 2000. High-yield production of a human therapeutic protein in tobacco chloroplasts. Nat Biotechnol, 18:333-338

Stevens LH, Stoopen GM, Elbers IJ, Molthoff JW, Bakker HA, Lommen A, Bosch D and Jordi W, 2000. Effect of climate conditions and plant developmental stage on the stability of antibodies expressed in transgenic tobacco. Plant Physiol, 124:173-182

Sugiyama Y, Hamamoto H, Takemoto S, Watanabe Y and Okada Y, 1995. Systemic production of foreign peptides on the particle surface of tobacco mosaic virus. FEBS Lett, 359:247-250

Svab Z, Hajdukiewicz P and Maliga P, 1990. Stable transformation of plastids in higher plants. Proc Natl Acad Sci USA, 87: 8526-8530

Tackaberry ES, Dudani AK, Prior F, Tocchi M, Sardana R, Altosaar I and Ganz PR, 1999. Development of biopharmaceuticals in plant expression systems: cloning, expression and immunological reactivity of human cytomegalovirus glycoprotein B (UL55) in seeds of transgenic tobacco. Vaccine, 17:3020-3029

Takamatsu N, Ishikawa N, Meshi T and Okada Y, 1987. Expression of bacterial chloramphenicol acetyltransferase gene in tobacco plants mediated by TMV-RNA. EMBO J, 6:307

Takase K and Hagiwara K, 1998. Expression of human alpha-lactalbumin in transgenic tobacco. J Biochem (Tokyo), 123:440-44

Taylor S, Massiah A, Lomonossoff G, Roberts LM, Lord JM and Hartley M, 1994. Correlation between the activities of five ribosome-inactivating proteins in depurination of tobacco ribosomes and inhibition of tobacco mosaic virus infection. Plant J, 5:827-835

Theisen M, 1999. Production of recombinant blood factors in transgenic plants. Adv Exp Med Biol, 464:211-220

Tuboly T, Yu W, Bailey A, Degrandis S, Du S, Erickson L and Nagy E, 2000. Immunogenicity of porcine transmissible gastroenteritis virus spike protein expressed in plants. Vaccine, 18:2023-2028

Turpen TH, Reinl SJ, Charoenvit Y, Hoffman SL, Fallarme V and Grill LK, 1995. Malarial epitopes expressed on the surface of recombinant tobacco mosaic virus. Biotechnology (N Y), 13:53-57

Ullah AHJ, Sethumadhavan K, Mullaney EJ, Ziegelhoffer T and Austin-Phillips S, 1999. Charcterization of recombinant fungal phytase (phyA) expressed in tobacco leaves. Biochem Biophys Res Comm, 264:201-206

van Ree R, Cabanes-Macheteau M, Akkerdaas J, Milazzo J-P, Loutelier-Bourhis C, Rayon C, Villalba M, Koppelman S, Aalberse R, Rodriguez R, Faye L and Lerouge P, 2000. β(1,4)-xylose and α(1,3)-fucose residues have a strong contribution in IgE binding to plant glycoallergens. J Biol Chem, 275:11451-11458

Vaquero C, Sack M, Chandler J, Drossard J, Schuster F, Monecke M, Schillberg S and Fischer R, 1999. Transient expression of a tumor-specific single-chain fragment and a chimeric antibody in tobacco leaves. Proc Natl Acad Sci USA, 96:11128-11133

Verch T, Yusibov V and Koprowski H, 1998. Expression and assembly of a full-length monoclonal antibody in plants using a plant virus vector. J Immunol Methods, 220:69-75

Verwoerd TC, van Paridon PA, van Ooyen AJ, van Lent JW, Hoekema A and Pen J, 1995. Stable accumulation of *Aspergillus niger* phytase in transgenic tobacco leaves. Plant Physiol, 109:1199-1205

Waegemann K and Soll J, 1991. Characterization of the protein import apparatus in isolated outer envelopes of chloroplasts. Plant J, 1:149-158

Walsh G, 2000. Biopharmaceutical benchmarks. Nat Biotechnol, 18:831-833

Wilcox CP, Weissinger AK, Long RC, Fitzmaurice LC, Mirkov TE and Swaisgood HE, 1997. Production and purification of an active bovine lysozyme in tobacco (*Nicotiana tabacum*): Utilization of value-added crop plants traditionally grown under intensive agriculture. J Agric Food Chem, 45:2793-797

Woodleif WG, Chaplin JF, Campbell CR and DeJong DW, 1981. Effect of variety and harvest treatments on protein yield of close-grown tobacco. Tob Sci, 25: 83-86

Yusibov VA, Modelska K, Steplewski M, Agadjanyan D, Weiner D, Hooper DC and Koprowski H, 1997. Antigens produced in plants by infection with chimeric plant viruses immunize against rabies virus and HIV-1. Proc Natl Acad Sci USA, 94:5784-5788

Yusibov VA, Shivprasad S, Turpen RH, Dawson W and Koprowski H, 1999. Plant viral vectors based on tobamoviruses. Curr Top Microbiol Immunol, 240:81-94

Chapter 2

ALFALFA: AN EFFICIENT BIOREACTOR FOR CONTINUOUS RECOMBINANT PROTEIN PRODUCTION

M.-A. D'Aoust[1], U.Busse[1], M. Martel[1], L. Faye[2], D. Levesque[3] and L.-P. Vezina[1]

[1]Medicago Inc., Ste-Foy, Québec, Canada, G1K 7P4; [2] Laboratoire des transports intracellulaires, CNRS, Université de Rouen, Mont Saint-Aignan cédex, France; [3] Avenir Luzerne, Aulnay-les-Planches, France.

Keywords: alfalfa recombinant protein, molecular farming, bioreactor

Abstract Alfalfa has many desirable agronomic properties including perennial growth, low secondary metabolite content and symbiotic association with nitrogen-fixing bacteria. These attributes make alfalfa a good bioreactor for the production of recombinant proteins. The generation of transgenic alfalfa, the ramp-up of the population, and the extraction and purification processes are outlined in this chapter. The utility of this bioreactor is illustrated by describing its use in the production of an antibody.

1. INTRODUCTION

In the early 90s, and throughout 1997, when Medicago Inc. was founded, the selection and development of alfalfa (*Medicago sativa* L.) as a bioreactor for recombinant protein production was not only based on its attractive agronomic characteristics but also on the advantageous physiological and biochemical aspects of the plant. Alfalfa is a perennial forage crop that is productive for up to 5 years in the field and 10 years when grown in a greenhouse. It is easy to grow on a large scale and its symbiotic association with root-invading rhizobia gives the plant the capacity to fix atmospheric nitrogen, hence limiting the need for fertilization. Alfalfa can be propagated by stem cuttings and shows a strong

L. Erickson et al. (eds.),
Molecular Farming of Plants and Animals for Human and Veterinary Medicine, 33–47.

regenerative capacity, giving the possibility to produce large clonal populations within a limited time period. Finally, the low secondary metabolite content of its leaves has contributed to the choice of alfalfa as one of the few among the best-suited plants for molecular farming.

As for any other plant-based production system, the process leading to the production of recombinant proteins using alfalfa can be divided into five major steps. The cDNA encoding the protein of interest is inserted into an efficient expression cassette (1), this cassette is transferred into a plant cell, and a first transgenic plant is regenerated from this cell (2). From the first transgenic plant(s), a whole population of plants is produced to obtain sufficient biomass (3) for the extraction and purification (4) of the desired amount of recombinant protein, while the co-products of extraction are destroyed or recycled (5).

This chapter presents the actual state of the art for each of these five major steps for molecular farming using alfalfa. Emphasis is given on the specific needs and advantages of using alfalfa for recombinant protein production taking into account the agronomy, in vitro biology and molecular biology of the plant.

2. EXPRESSION CASSETTES

In the first reports on transgenic alfalfa production, transgene expression was controlled using commercially available expression cassettes, predominantly based on the cauliflower mosaic virus (CaMV) 35S promoter (Odell et al., 1985; Jefferson et al., 1987) or its derivatives. It was later found that this promoter drives relatively weak transcription in alfalfa. An example of this low activity comes from experiments by Narváez-Vásquez and co-workers (1992) who showed that the capacity of the 35S promoter to promote the expression of the tomato proteinase inhibitor I gene is 3- to 4-fold lower in alfalfa than in tobacco.

The development of promoters specifically designed for alfalfa started with a study from Khoudi (1997) who cloned a genomic fragment of the alfalfa Rubisco small subunit (RbcSK-1A), and tested the activity of the upstream sequences for their capacity to drive the expression of the ß-glucuronidase (GUS) reporter gene in transgenic alfalfa. The RbcSK-1A promoter directed more variable expression levels of the transgene than the 35S promoter, but the mean activity of the RbcSK-1A promoter was 10-fold higher than that of the 35S promoter (Khoudi et al., 1997).

This was an impressive result, but the intellectual property context surrounding the use of Rubisco promoters encouraged us to push forward the finding of new candidates for strong constitutive and inducible expression in transgenic alfalfa. The beacons guiding this search were based on two main concerns, biosafety and yield. We wanted promoters to be derived from alfalfa genomic sequences, and we wanted them to drive high levels of expression in alfalfa leaves. Using these guidelines, three families of promoters have since been developed including an inducible promoter, a leaf-specific promoter and one constitutive promoter.

Since most recombinant molecules to be produced in molecular farming are bioactive molecules, it can be expected that a certain number of them will have detrimental effects on normal plant growth and development. For this reason, an inducible promoter family was developed that can be induced by a cheap and safe inorganic compound (patent pending). In a situation where the bioactive protein to be produced shows a deleterious effect on plant growth, the inducible promoter is used to withhold overexpression of the transgene until the plant reaches a satisfying biomass. Then, upon application of the inducer, maximal expression of the protein of interest rapidly raises leaf bioactive protein content to the desired level and mature plants are harvested with the molecule of interest. As shown in Figure 1, in its induced state, the promoter Med1302T is almost ten-fold stronger than the 35S promoter when placed upstream of the GUS coding sequences in transgenic tobacco.

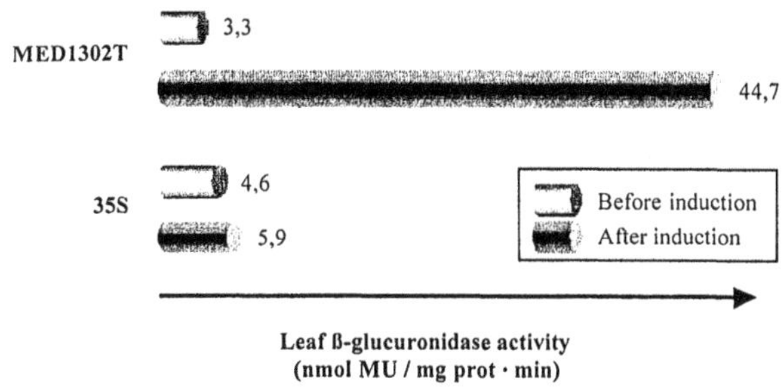

Figure 1. Inducibility of GUS expression with Med-1302T expression cassette in tobacco leaves.

Recent results obtained in transgenic alfalfa not only show that the promoter is inducible in alfalfa, but also that it is silent before induction. The leaf specific promoter (Med2101, patent pending) was tested in transgenic tobacco. Figure 2 presents the results obtained with this promoter in comparison with the 35S promoter. In the leaves of Med2101-GUS- transformed plants, GUS activity was 25-fold higher than in the 35S-GUS plants.

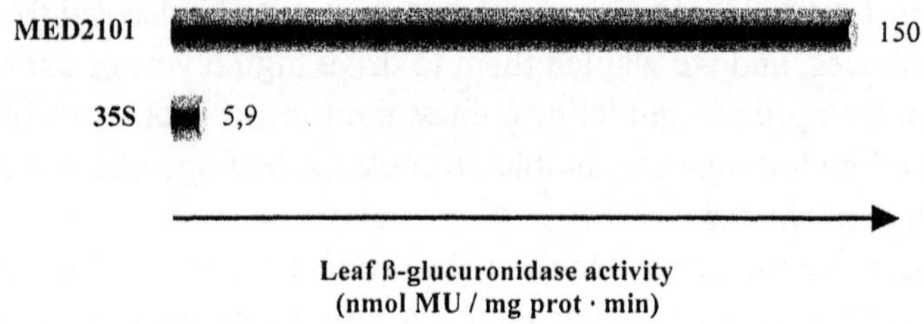

Figure 2. Comparison of Med-2101 and 35S promoters for leaf GUS expression capacity in transgenic tobacco.

Since the accumulation of a given protein is the result of its production level minus its degradation level, the next challenge is to improve the stability of recombinant proteins in the plant cell and during extraction. Past experience has shown that the sub-cellular targeting of proteins in transgenic alfalfa has a dramatic effect on protein stability (Bagga et al., 1992; Wandelt et al., 1992; Tabe et al., 1995; Benmoussa, 1999), and we are exploring this avenue as a means of improving recombinant protein accumulation.

3. TRANSFORMATION METHODS

3.1 In vitro Culture

The main limits for the in vitro regeneration of alfalfa are ascribable to its inability to undergo organogenesis. Hence, since 1980 protocols for the regeneration of alfalfa through somatic embryogenesis were developed (Kao and Michayluk, 1980; Johnson et al., 1981; Groose and Bingham, 1984; Atanassov and Brown, 1984; Brown and Atanassov, 1985). Early during this period, it was demonstrated that the regeneration capacity was highly genotype dependent, and that a *Medicago falcata* background was largely the determinant of the in vitro regeneration capacity of the plants (Brown and Atanassov, 1985). The most highly regenerative material was obtained through selection of alfalfa populations mostly derived from *M. falcata* (Brown and Atanassov, 1985; Wan et al., 1988; Wandelt et al., 1992; Micallef et al., 1995). Unfortunately, high *M. falcata* background also results in undesirable agronomic characteristics like tiny leaves, low leaf-to-stem ratio and high dormancy.

This situation prompted the development of regenerative genotypes which also present the agronomic characteristics desired for a commercial crop (Nowak et al., 1992; Desgagnés et al., 1996). Transgenics obtained from these plants have the advantage of being identical to cultivated alfalfa except for the characteristics inherent to the inserted trait.

Regeneration of alfalfa from protoplasts has also been achieved (Kao and Michayluk, 1980; Johnson et al., 1981; Atanassov and Brown, 1984; Song et al., 1990, Holbrook et al., 1985). These studies show that embryogenesis from protoplasts is possible under appropriate conditions. However, the genotype used for embryogenesis from protoplasts is determinant of the capacity to obtain cell division. It appears that only the most embryogenic material is capable of dividing and forming healthy embryos.

Except for pollen electroporation, all transformation techniques developed inevitably use in vitro regeneration of plants from treated cells or tissue. The success of a plant transformation technique is highly dependent on the regenerability of the material used and on its compatibility with the constraints inherent to the transformation method.

3.2 Transformation with *Agrobacterium*

Successful transformation of alfalfa through *Agrobacterium tumefaciens* infection dates back to 1986 (Deak et al., 1986; Shahin et al., 1986). Thanks to the recent progress in somatic embryogenesis, protocols for the production of transgenic plants from single transformed cells have been developed. However, it was rapidly found that regenerative potential was not the only character needed for transformation. Albeit they are highly regenerative, some genotypes should be avoided for transformation because they do not survive the co-cultivation step with *Agrobacterium*. A study from Desgagnés and coworkers (1995) has shown that within regenerative and *Agrobacterium*-resistant genotypes, susceptibility to *Agrobacterium* transfection varies widely, ranging from 1-2 to 50 independent transformation events per 50 leaf disks.

Agrobacterium-mediated transformation requires the selection of a few transformed cells from an organized tissue. With the actual tools available, it is performed by joining an antibiotic resistance gene to the gene of interest in the construct to be transfected, and this requirement diminishes the interest in

Agrobacterium-based transformation methods for molecular farming in open-field conditions.

3.3 Direct DNA Transfer

3.3.1 Particle Bombardment

Particle bombardment first appeared as an ideal alternative to *Agrobacterium*-mediated transformation because it allowed the transformation of *Agrobacterium*-insensitive plants. The problem of insensitivity was most obvious for the monocotyledoneous plants because this class comprises the world's most cultivated crops like wheat, rice and maize. Soon after the development of the technique, particle bombardment of plant tissue was also seen as an option for the transformation of *Agrobacterium*-sensitive species for commercial uses because of the intellectual property issue about the use of *Agrobacterium* and the T-DNA borders for gene transfer into plant cells. In alfalfa, transformed plants with stable integration were obtained by bombarding calli from either petiole and stem section (Pereira and Erickson, 1992) or excised immature embryos (Buising and Tomes, 1995). Another attempt at particle bombardment of suspension-cultured cells showed that bombardment was only efficient in integrating DNA transiently into cells (Brown et al., 1994). It proved impossible to reproducibly obtain regeneration of transgenic material from these cells. Bombardment was also performed on alfalfa pollen grains. The method was described as highly efficient, but the transformed plants derived from fertile seeds gradually lost the inserted transgene, even if multiplied by stem cuttings, indicating that the transgene was not integrated into the chromosomes and was lost during cell division (Ramaiah and Skinner, 1997).

Particle bombardment is also advantageous because, in contrast to *Agrobacterium*, the capacity to incorporate DNA into the cells is not genotype dependent. Any embryogenic genotype can be bombarded to produce transgenic alfalfa, even though special attention should be given to the type of tissue to be bombarded. The explant should be selected based on its capacity to generate somatic embryos and to regenerate transgenic plants with stable integration of the transgene.

3.4 Other Methods

The most original and eclectic work to develop cell transfection methods comes from animal and human research because, in contrast to plants cell transformation, a natural DNA transfer and integration mechanism has not been controlled yet. Electroporation and DNA-complexing polymers or lipids are among the most popular techniques developed.

Because of the success of these methods for producing stable transgenic animals, plant scientists took inspiration from their colleagues in animal science and adapted direct transformation methods for plant cells. For example, electroporation was performed on alfalfa protoplasts and resulted in transient expression of the transgene (Harrison et al., 1991). Likewise, pollen electroporation was used with success to introduce DNA into alfalfa pollen grains, but no transgenic plants could be recovered from fertilization with these pollen grains (J.A. Saunder, personal communication). The microinjection of naked DNA into protoplasts from highly regenerative alfalfa has been used with more success and led to the production of stable transgenics (Reich et al., 1986). This study was published the same year as the first *Agrobacterium* transformation of alfalfa, but there is no more recent report of the use of microinjection for alfalfa transformation whereas *Agrobacterium* has been routinely used since that date. At the time the microinjection method was published, it was laborious work, but the method should not be left aside because both microinjection and protoplast regeneration are by far more efficient than at that time.

The production of stable transgenic maize plants was obtained through polybrene- or lipofectin-based transfection of protoplasts (Antonelli and Stadler, 1990). In rice, poly-L-ornithine was successfully used to introduce DNA into protoplasts (Tsugawa et al., 1998). Even though few, these results exemplify the capacity of complexing molecules to favor the introduction of DNA into plants cells, and it is to be expected that the number of plants transformed with polymers or lipids – hopefully including alfalfa – will increase during the next decade.

4. POPULATION RAMP-UP

The procedure of plant transformation ends with an individual transgenic plant bearing the transgene of interest. Starting with this first plant, a whole population of plants must be obtained to produce the biomass needed for the extraction of sufficient recombinant proteins. The producing population can be created by stem cutting, somatic embryogenesis or by the production of seeds, and each method responds to a different need in terms of the amount of recombinant protein to be produced (Table 3).

Stem cutting is the easiest method for alfalfa multiplication. It can be performed with very simple facilities and the skills required are taught rapidly to non-specialized workers. Using stem cuttings, up to 200 g of recombinant protein can be produced in alfalfa within 10 months and the population created will supply 2 kg of protein in each following harvest for the next 10 years.

Table 3. Methods and conditions for increasing the population of alfalfa plants to production scales of recombinant protein

Yield	Interval to first delivery (months)	Number of plants	Yield/ harvest (5-6 weeks)	Propagation method	Growing premises
10 mg	4	70/1 m^2	50 mg	Stem cutting	Greenhouse
100 mg	6	900/9 m^2	1 g	Stem cutting	Greenhouse
10 g	8	11000/110 m^2	40 g	Stem cutting	Greenhouse
200g	10	140000/ 1.4 ha	2 kg	Stem cutting	Greenhouse
5 kg	20	2 ha	10-20 kg/ season	Seeds	Field
100 kg	30	36 ha	200-400 kg/season	Seeds	Field
1 t	30	360 ha	2-4 t/ season	Seeds	Field

These calculations assume an expression level of 0.5% of total soluble protein and purification yields of 10%(10 mg), 60% (10 g), and 90% (2kg-1 tonne).

In the field, tonnes of recombinant proteins can be produced using alfalfa. In order to produce the largest populations, a first population of transgenic plants, created by stem cutting, is backcrossed to an elite population adapted to the local growth conditions. The backcrossing manipulations lead to the creation of simplex (Axxx) transgenic segregants for the transgene, and larger seed populations are obtained by intercrossing these simplex segregants in the field.

5. EXTRACTION AND PURIFICATION

5.1 Extraction and Purification

The original use of alfalfa as a source for animal feed products encouraged the development of extraction methods for alfalfa leaf constituents. The narrow net operating margin associated with the commercialization of these products forced the elaboration of cheap and efficient extraction methods from massive amounts of alfalfa leaves. Nowadays, molecular farming benefits from these earlier developments. These methods have been developed with the aim of concentrating proteins and now represent excellent first steps for the purification of recombinant proteins from large amounts of biomass. These facilities are already available in most locations appropriate for alfalfa production. Figure 3 presents the bases of a wet fractionation process optimized for alfalfa leaf constituents. The wet fractionation process was developed by Viridis (France) to purify Rubisco from alfalfa leaves, and co-products of this process are further separated into a protein and xanthophyll rich fraction for monogastrics, and into a fibre-rich fraction destined for ruminants.

Previous developments in extraction methods for animal feed bring another benefit to molecular farming. Recombinant molecules developed for oral administration to animals can be produced in alfalfa and recovered in specific fractions destined to animal feed. For example, based on the knowledge of the chemical and physical conditions used for the separation of leaf constituents using the wet fractionation method, it is possible to alter specific properties of the recombinant molecule so that it will follow the route to the desired fraction. In other cases, the separation parameters can be modified to favor the fractionation of the molecule of interest towards a specific fraction.

Using these fractionation tools, it is possible to commercialize new therapeutic agents not only in fresh plants (as well as in bailed forage and silage) destined for animal consumption, but also in specialized animal food products which are already on the market.

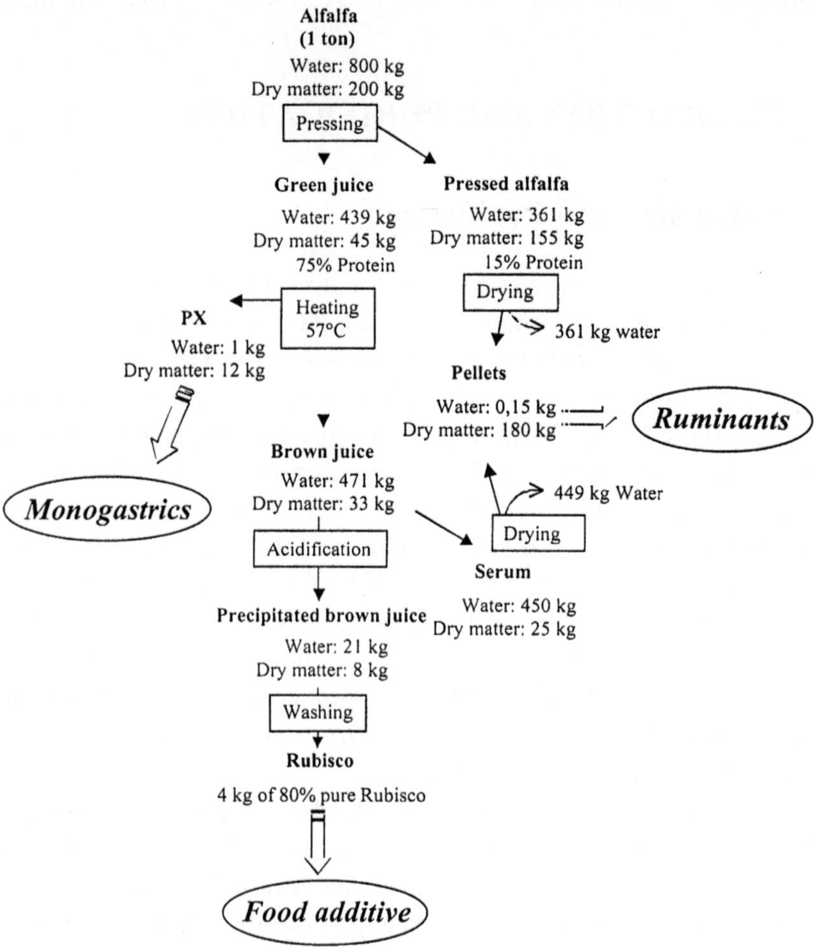

Figure 3. Steps of alfalfa processing toward the purification of Rubisco

Even though the wet fractionation method is well suited to introduce recombinant proteins in animal feed, most of the recombinant proteins to be produced in molecular farming will have to be purified. In these cases, alfalfa leaves offers the advantage of containing very few alkaloids and phenols (Jones et al., 1973; Goplen et al., 1980). These compounds interfere with the purification processes and additional purification steps are required to remove them when producing purified therapeutics.

6. N-GLYCOSYLATION OF ALFALFA PROTEINS

Most plant N-glycans found on mature proteins differ from their mammalian counterparts (Lerouge et al., 1998). Because N-glycosylation of some mammalian proteins is essential for their activity, and because some structures of plant N-glycans are associated with allergenic reactions, specific attention must be given to this specific part of mature protein structure.

The actual knowledge on the incidence of each glycan residue of mature protein on their biological activity remains incomplete, and studies are underway to determine the structure of N-glycans found on alfalfa endogenous proteins as well as on exogenous proteins produced in alfalfa. Another set of experiments aims at resolving the structure to function relationship of mammalian N-glycans and at finding ways to engineer plant glycolysation pathways to produce humanized glycans.

The structure of the major N-glycans found in alfalfa leaves is presented in Figure 4. These glycans were characterized by mass spectrometry (MALDI-TOF MS) and confirmed by enzymatic sequencing. Additional minor complex oligosaccharides lacking one fucose residue and high-mannose-type glycans were also found in the pool of N-linked glycans.

7. CASE STUDY

7.1 Characteristics of an Anti-human IgG Produced in Alfalfa

Immunoglobulin (IgG) C5-1, a monoclonal antibody employed by blood banks for the detection of non-agglutinating antibodies, was produced in transgenic alfalfa (Khoudi et al., 1999). C5-1, being a glycosylated heterotetrameric protein composed of two different subunits held together by disulfide bonds, represented an ideal candidate for the demonstration of the capacity of alfalfa to produce complex recombinant proteins.

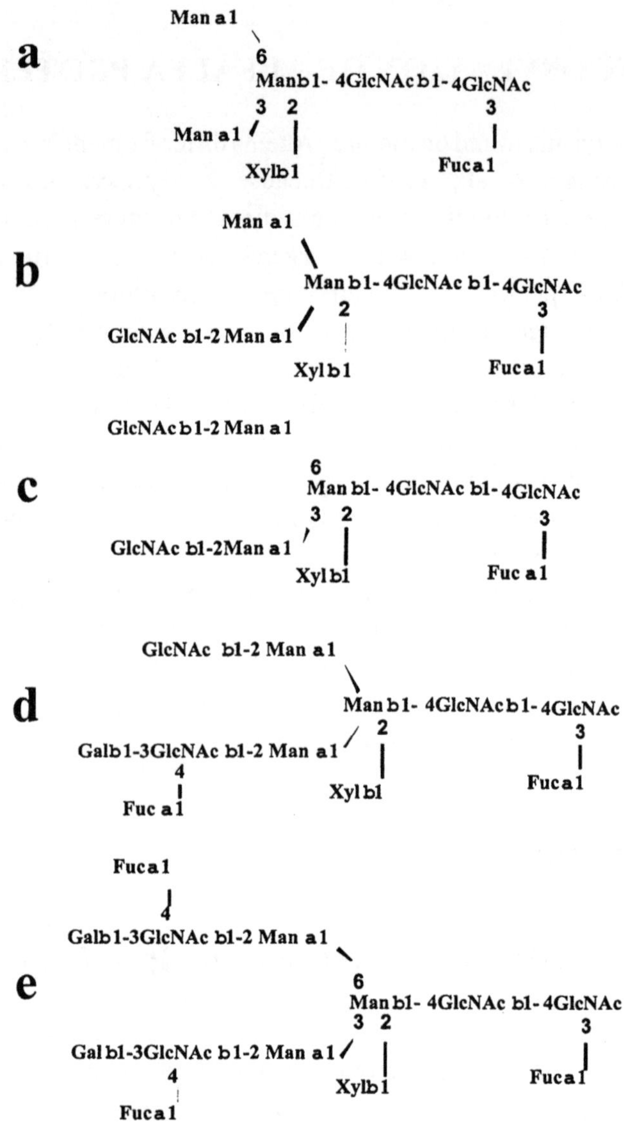

Figure 4. Major N-glycans found on alfalfa proteins

The characteristics of the plant-derived antibody were compared to those of the hybridoma. The specific activity of the plant derived C5-1 against human IgGs was demonstrated to be similar to that of C5-1 isolated from hybridoma cells. Furthermore, plant C5-1 showed comparable hemagglutination capacity to its hybridoma-derived counterpart against anti-D-sensitized human red blood cells.

After confirming the similarity of the plant-derived antibody with the original antibody produced by hybridoma cells, we demonstrated that the level of C5-1 production was similar after several steps of stem cutting multiplication and, noteworthy, that the antibody was stable in dry hay within 12 weeks of harvest. It was furthermore demonstrated that the proteolytic machinery of alfalfa, after grinding of the tissue, was much less aggressive than that of tobacco against either the plant- and the hybridoma-derived C5-1. Effectively, C5-1 (from alfalfa or hybridoma) was stable in alfalfa extract for at least 120 min whereas, in tobacco leaf extract, it was completely degraded within this same time period.

Finally, the demonstration of the stability of alfalfa C5-1 in the blood stream of mice to a level comparable to that of the original C5-1 further underscored the therapeutic potential of alfalfa-derived antibodies.

REFERENCES

Antonelli NM and Stadler J, 1990. Genomic DNA can be used with cationic methods for highly efficient transformation of maize protoplasts. Theor Appl Genet, 80:395-401

Atanassov A and Brown DCW, 1984. Plant regeneration from suspension culture and mesophyll protoplasts of *Medicago sativa* L. Plant Cell Tiss Org Cult, 3:149-162

Bagga S, Sutton D, Kemp JD and Sengupta-Gopalan C, 1992. Constitutive expression of the beta-phaseolin gene in different tissues of transgenic alfalfa does not ensure phaseolin accumulation in non-seed tissue. Plant Mol Biol, 19:951-958

Benmoussa M, 1999. Production d'une glutéine à faible poids moléculaire dans les feuilles de la luzerne et les tubercules de la pomme de terre. PhD thesis. Laval University, Québec, Canada

Brown DC and Atanassov A, 1985. Role of genetic background in somatic embryogenesis in *Medicago*. Plant Cell Tiss Org Cult, 4:111-122

Brown DCW, Tian L, Buckley DJ, Lefebvre M, McGrath A and Webb J, 1994. Development of a simple particle bombardment device for gene transfer into plant cells. Plant Cell Tiss. Org Cult, 37:47-53

Buising CM and Tomes D, 1995. Methods of regeneration of *Medicago sativa* and expressing foreign DNA in same. US Pat. # 5,994,626

Deak M, Kiss GB, Koncz C and Dudits D, 1986. Transformation of *Medicago* by *Agrobacterium* mediated gene transfer. Plant Cell Rep, 5:97-100

Desgagnés R, Laberge S, Allard G, Khoudi H, Castonguay Y, Lapointe J, Michaud R and Vézina LP, 1995. Genetic transformation of commercial breeding lines of alfalfa (*Medicago sativa*). Plant Cell Tiss Org Cult, 42:129-140

Goplen BP, Howarth RE, Sarkar SK and Lesins K, 1980. A search for condensed tannins in annual and perennial species of *Medicago*, *Trigonella* and *Onobrychis*. Crop Sci, 20:801-804

Groose RW and Bingham ET, 1984. Variation in plants regenerated from tissue culture of tetraploid alfalfa (*Medicago sativa*) heterozygous for several traits. Crop Sci, 24:655-658

Harrison MJ, Choudhary AD, Dubery I, Lamb CJ and Dixon RA, 1991. Stress responses in alfalfa (*Medicago sativa* L. 8. *Cis*-elements and *trans*-acting factors for the quantitative expression of a bean chalcone synthase gene promoter in electroporated alfalfa protoplasts. Plant Mol Biol, 16:877-890

Holbrook LA, Reich TJ, Iyer VN, Haffner M and Miki BL, 1985. Induction of efficient cell division in alfalfa protoplasts. Plant Cell Rep, 4:229-232

Jefferson RA, Kavanagh TA and Bevan MW, 1987. GUS fusions: ß-glucuronidase as a sensitive and versatile gene fusion marker in higher plants. EMBO J, 6:3901-3907

Johnson LB, Stuteville DL, Higgins RK and Skinner DZ, 1981. Regeneration of alfalfa (*Medicago sativa* cultivar Saranac) plants from protoplasts of selected Regen S clones. Plant Sci Lett, 20:297-304

Jones WT, Anderson LB and Ross MD, 1973. Bloat in cattle XXIX. Detection of protein precipitant (flavolans) in legumes. New Zeal J Agr Res, 16:441-446

Kao KN and Michayluk MR, 1980. Plant regeneration from mesophyll protoplasts of alfalfa. Z Pflanzenphysiol Bd, 96 S:135-141

Khoudi H, Laberge S, Ferullo JM, Bazin R, Darveau A, Castonguay Y, Allard G, Lemieux R and Vezina LP, 1999. Production of a diagnostic monoclonal antibody in perennial alfalfa plants. Biotechnol Bioeng, 64:135-143

Khoudi H, Vézina LP, Mercier J, Castonguay Y, Allard G and Laberge S, 1997. An alfalfa rubisco small subunit homologue shares cis-acting elements with the regulatory sequences of the RbcS-3A gene from pea. Gene, 197:343-351

Lerouge P, Cabanes-Macheteau M, Rayon C, Fischette-Lainé AC, Gomord V and Faye L, 1998. Plant Mol Biol, 38:31-48

Micallef MC, Austin S and Bingham ET, 1995. Improvement of transgenic alfalfa by backcrossing. In Vitro Cell Dev Biol-Plant, 31:187-192

Narvàez-Vàsquez J, Orozco-Càrdenas ML and Ryan CA, 1992. Differential expression of a chimeric CaMV-tomato proteinase Inhibitor I gene in leaves of transformed nightshade, tobacco and alfalfa plants. Plant Mol Biol, 20:1149-1157

Nowak J, Matheson S, Maclean NL and Havard P, 1992. Regenerative trait and cold hardiness in highly productive cultivars of alfalfa and red clover. Euphytica, 59:189-196

Odell JT, Nagy F and Chua NH, 1985. Identification of DNA sequences required for activity of the cauliflower mosaic virus 35S promoter. Nature, 313:810-812

Pereira LF and Erickson L, 1992. Stable transformation of alfalfa (*Medicago sativa L.*) by particle bombardment. Plant Cell Rep, 14:290-293

Ramaiah SM and Skinner DZ, 1997. Particle bombardment: A simple and efficient method of alfalfa (*Medicago sativa L.*) pollen transformation. Cur Sci (Bangalore), 73:674-682

Reich TJ, Iyer VN and Miki BL, 1986. Efficient transformation of alfalfa protoplasts by the intranuclear microinjections of Ti plasmids. Biotechnology, 4:1001-1004

Shahin EA, Spielmann A, Sukhapinda K, Simpson RB and Yashar M, 1986. Transformation of cultivated alfalfa using disarmed *Agrobacterium tumefaciens*. Crop Sci, 26:1935-1939

Song J, Sorensen EL and Liang GH, 1990. Direct embryogenesis from single mesophyll protoplats in alfalfa (*Medicago sativa* L. Plant Cell Rep, 9:21-25

Tabe LM, Wardley-Richardson T, Ceriotti A, Aryan A, McNabb WC, Moore A and Higgins TJV, 1995. A biotechnological approach to improving the nutritive value of alfalfa. J Plant Sci, 73:2752-2759

Tsugawa H, Otsuky Y and Susuky M, 1998. Efficient transformation of rice protoplasts mediated by a synthetic polycationic amino polymer. Theor Appl Genet, 97:1019-1026

Wan Y, Sorensen EL and Liang GH, 1988. Genetic control of in vitro regeneration in alfalfa (*Medicago sativa* L. Euphytica, 39:3-9.

Wandelt CI, Khan MRI, Craig S, Schroeder HE, Spencer D and Higgins TJV, 1992. Vicilin with carboxy-terminal KDEL is retained in the endoplasmic reticulum and accumulates to high levels in the leaves of transgenic plants. Plant J, 2:181-192

Chapter 3

THE USE OF VIRAL VECTORS FOR THE PRODUCTION OF RECOMBINANT PROTEINS IN PLANTS

G. P. Lomonossoff
John Innes Centre, Colney Lane, Norwich NR4 7UH; U.K.

Keywords: recombinant proteins, viral vectors

Abstract The use of genetically engineered plant viruses to express foreign proteins or polypeptides is well established. The advantages of this method are the ease of manipulation of the small viral genome, the quick and simple procedure for inoculation of plants, and the production of high levels of recombinant proteins due to multiplication of the viruses within infected cells. This article will discuss the progress which has been made in the use of virus-based vectors and will also consider limitations to this technology.

1. INTRODUCTION

There are currently two types of systems for expressing recombinant proteins in plants: stable genetic transformation and the use of plant virus-based vectors. The transgenic approach, which is currently widely used, involves integration of heterologous genes into the chromosomes of a host plant, expression of the integrated gene being under the control of a promoter positioned immediately upstream of the coding region. This method of foreign gene expression has the advantage that the integrated sequence is heritable and true-breeding lines of plants expressing the foreign gene can be created. There are, however, a number of disadvantages with this approach: the process of regenerating transformed plants can be difficult and time-consuming, the levels

L. Erickson et al. (eds.),
Molecular Farming of Plants and Animals for Human and Veterinary Medicine, 49–71.

of expression reached are often relatively low and undesirable phenomena, such as transgene silencing, often occur.

One of the main attractions of the use of virus-based approach for protein expression in plants is that viruses multiply within infected cells. Thus any foreign gene incorporated within the viral genome should be concomitantly amplified potentially leading to very high levels of protein expression. There are, however, potential additional advantages to this approach: viral genomes are small and therefore relatively easy to genetically manipulate and the infection process is simpler than transformation/regeneration. Inevitably, there are also potential downsides to this technology: The foreign gene is not heritable in the normal sense, there are likely to be limitations on the size and complexity of the sequences which can be expressed while maintaining virus viability and there are concerns about the ability of modified viruses to spread in the environment. It is the intention of this article to review the progress which has been made in the use of virus-based vectors for the expression of heterologous proteins in plants and to discuss those problems which remain to be solved.

2. CHOICE OF VIRUS FOR VECTOR DEVELOPMENT

The first viruses to be proposed as potential gene vectors were those whose genomes consist of DNA (members of the *Caulimo-* and *Geminiviridae*; Szeto et al., 1977; Hull, 1978). Though these make up only a minority of plant viruses, at the time they were the only viruses for which genetic manipulation was possible. However, because of their modes of genome replication, the DNA viruses have proved difficult to develop into practical vectors for the expression of foreign proteins. These DNA virus-based vectors have, nonetheless, proved very useful in providing information about cellular processes such as transcription, transactivation of expression and RNA splicing. However, these uses are outside the scope of the current review, and the reader is referred to Porta and Lomonossoff (1996; 2001) for a discussion of the development and use of DNA virus-based vectors.

The vast majority of plant viruses have genomes that consist of one or more strands of positive-sense RNA. These viruses can grow in a wide range of hosts and, in a number of instances, reach extremely high titres. They also have a variety of strategies for gene expression (e.g. the use of subgenomic promoters, polyprotein processing) which can be exploited to develop vectors for different purposes. The development of RNA virus-based vectors was initially hampered by the difficulties in manipulating a RNA, as opposed to a DNA, genome. The advent of efficient *in vitro* transcription systems dramatically changed the situation and the construction of full-length cDNA clones of brome mosaic virus (BMV) from which infectious transcripts could be obtained was reported in 1984 (Ahlquist and Janda, 1984; Ahlquist et al., 1984). Subsequently, full-length cDNA clones of many other RNA plant viruses have been produced and shown to be infectious either after transcription with the appropriate RNA polymerase, when directly applied as DNA or when introduced by "agroinoculation". In addition to the technical difficulties in developing RNA virus-based vectors, there was some concern expressed initially as to whether such vectors would be very useful due to the high rate of mutation of RNA viruses (van Vloten-Doting et al., 1985). This view was disputed at the time (Siegel, 1985) and the genetic stability of RNA viruses has actually proved to be quite adequate (see, for example, Kearney et al., 1993).

3. TYPES OF EXPRESSION SYSTEMS

With the advent of infectious cDNA clones, RNA viruses from numerous families have been developed as vectors. Many of the initial vectors were based on gene replacement in which a sequence encoding a non-essential viral function was replaced by the sequence of interest (e.g. French et al., 1986; Mori et al., 1993; Takamatsu et al., 1987). Many of these early constructs could multiply in protoplasts but were debilitated for the infection of whole plants. Thus, with a few exceptions noted below, gene replacement vectors have not proved suitable for the high-level expression of foreign proteins in whole plants. The more successful approach has been based on gene addition in which a foreign sequence is added to the viral genome, no viral function being deleted.

The gene addition approach to vector development has led to the development of two basic types of expression system. In the first type, often termed "epitope-presentation" systems, a foreign sequence, most usually encoding a short antigenic peptide, is fused to the viral coat protein gene. This results in expression of the inserted peptide on the surface of assembled virus particles. Such modified particles are particularly attractive as potential novel vaccines since they present multiple copies of the inserted peptide. Such multiple presentation can significantly increase the immunogenicity of the expressed peptide (Lomonossoff and Johnson, 1996). In the second type of expression system, a sequence encoding a whole gene is inserted into the viral genome and expressed in infected cells as an individual polypeptide. These systems are often referred to as "polypeptide expression" systems.

4. EXAMPLES OF THE EXPRESSION OF PEPTIDES AND PROTEINS USING VIRAL VECTORS

In the following section, specific examples of the use of virus vector systems to express both proteins and peptides in plants will be described. Though the examples given cover the main viral systems which are currently in use, it should be noted that there are additional instances where marker genes, the green fluorescent protein (GFP) or b-glucuronidase (GUS) have been introduced into a viral genome as an aid to following the infection process. Except where such studies with marker genes were carried out with the specific aim of developing generic vector systems, these types of studies will not be discussed further in this review.

4.1 Tobacco Mosaic Virus

Tobacco mosaic virus (TMV) has a number of properties which make it ideal for development as a gene vector. It is one of the best-studied plant viruses, it grows to extremely high titres in susceptible plants and, since it has rod-shaped particles, there are, in theory, no constraints on the size of RNA which can be packaged. The genome of TMV consists of a single molecule of RNA of 6.4 KB, which contains separate open reading frames (ORFs) for the polymerase functions, the viral movement protein (MP) and coat protein (CP). The latter two proteins are synthesized from subgenomic (sg) RNAs (see top section of Figure 5). Particles of TMV consist of a single molecule of genomic RNA encapsidated by 2130 copies of the 17.6 kDa coat protein arranged with helical symmetry (lower section of Figure 5). Because of the large amount of information which is available about both structure of the virions and its replication cycle, TMV has been investigated both as an epitope-presentation and polypeptide expression system.

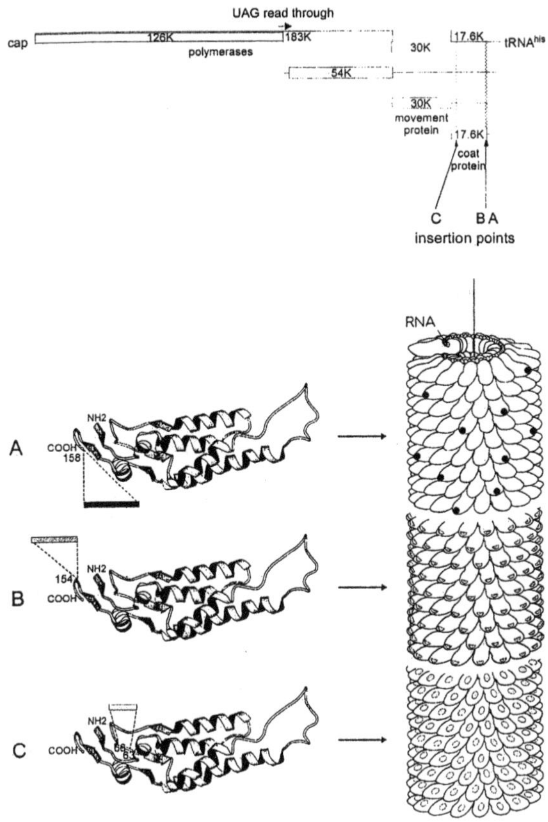

Figure 5. Epitope presentation using tobacco mosaic virus (TMV). The top section of the figure shows the genome organization of the virus and indicates the sites in the viral CP gene (A, B and C) into which oligonucleotides encoding peptides have been inserted. The resulting recombinant CP molecules are shown as ribbon diagrams with the sites of insertion indicated. The modified CP molecules are assembled into virus particles, the number of coat protein subunits carrying the modification depending on the site of insertion. In the case of insertion at site A, only 5% of the subunits are modified whereas when sites B and C are used, 100% of subunits in particles carry the heterologous peptide. The inserted epitopes are shown as black or shaded dots on the assembled virus particles. Reproduced from Porta and Lomonossoff (1998).

4.1.1 Epitope Presentation

TMV was the first plant virus whose coat protein was exploited as a means of presenting heterologous peptides (Haynes et al., 1986). However, in this initial case a modified version of the viral coat protein bearing an eight amino acid poliovirus epitope at its C-terminus was expressed in *E. coli* rather than in plants. The first attempt to produce modified TMV particles in plants was reported by Takamatsu et al. (1990). In these experiments, a sequence encoding Leu-enkephalin was fused to the C-terminus of the viral coat protein. Though the construct multiplied well in protoplasts, it gave only local lesions on tobacco plants and it was concluded that the modified coat protein was not competent for virion assembly. To address this problem, Hamamoto et al. (1993) developed a new vector which permitted the synthesis of both native and C-terminally modified versions of the coat protein from the same viral RNA. This was achieved by engineering a leaky termination codon at the C-terminus of the coat protein gene. This system produced particles in plants in which up to 5% of the coat protein subunits were modified at their C-termini (Figure 5A) to express either an enzyme inhibitor (Hamamoto et al., 1993), epitopes from animal viruses (Sugiyama et al., 1995) or an epitope from the malarial parasite, *Plasmodium yoelii* (Turpen et al., 1995). In each case, the inserted peptide could be detected on the surface of assembled virions. Subsequently, by modifying the site of peptide insertion, TMV-based systems were developed in which all the coat protein subunits could be modified to express foreign peptides without abolishing virus viability (Turpen et al., 1995; Fitchen et al., 1995; Figure 5B,C). Using the vector developed by Fitchen et al. (1995), Koo et al. (1999) were able to produce chimaeric TMV particles presenting epitopes from the Coronavirus, mouse hepatitis virus (MHV) and demonstrated that purified particles could protect mice against challenge by a normally lethal dose of MHV. One of the principal limitations on the use of TMV as an epitope-presentation system currently appears to lie in the size and charge of inserts that can be tolerated (Bendahmane et al., 1999).

4.1.2 Polypeptide Expression

The first attempt to use TMV as a vector for the expression of whole proteins involved substitution of the CP ORF by a sequence encoding chloramphenicol acetyl transferase (CAT; Takamatsu et al., 1987). The resulting construct was capable of causing local lesions when inoculated on to tobacco but the infection was unable to spread systemically. CAT expression could be detected in extracts from the inoculated leaves. In view of the problems of spread in the absence of CP, subsequent work on the development of TMV-based vectors concentrated on the gene addition approach. Dawson et al. (1989) inserted a copy of the CAT gene complete with its own TMV CP sg promoter between the MP and CP genes. The construct replicated well, CAT activity could be detected and virions of an appropriately increased length were produced. However during systemic infection, the insert was precisely deleted and wild-type RNA accumulated as a consequence of homologous recombination between the two copies of CP sg promoter. To address this problem, a vector was developed in which a cartridge consisting of the CP sg promoter and CP from the related Tobamovirus, *Odontoglossum* ringspot virus (ORSV; Donson et al., 1991) was inserted into the TMV genome in place of the TMV CP coding region (Figure 6A). The relatively low sequence identity between the CP sg promoters of TMV and ORSV effectively abolished homologous recombination and allowed the stable systemic expression of two bacterial genes (DHFR and NPT-II) in *N. benthamiana*. The success in developing genetically stable TMV vectors based on the strategy of using duplicated non-homologous sg promoters has led to such vectors being used to express several valuable proteins in plants. The first example of this was the high level expression (2% of soluble proteins) of a eukaryotic ribosome-inactivating protein (RIP), α-trichosanthin, in *N. benthamiana* (Kumagai et al., 1993). Subsequently it has proved possible to use TMV-based vectors to produce functional antibodies in plants. These can be either single-chain variable fragments (ScFvs; McCormick et al., 1999) or full-length monoclonal antibodies (Verch et al., 1998). In the latter case the heavy and light chains of the antibody were separately inserted into a TMV vector and the two constructs used to co-inoculate *N. benthamiana*. Assembly of the separately expressed chains in plants could be detected. Expression of a rice α-amylase gene in *N. benthamiana* using a TMV vector has demonstrated that it is possible to synthesize glycosylated proteins using viral vectors (Kumagai et al., 2000). Analysis of the purified protein revealed that the TMV-expressed α-amylase is less heavily glycosylated than when the protein is expressed in yeast. The TMV vector system has also been used to produce allergens for diagnostic purposes and possible therapy. Krebitz et al. (2000) expressed a major birch pollen antigen (Betv1) in *N. benthamiana* and demonstrated that the B-cell epitopes from natural Betv1 were preserved in the plant-expressed protein. Mice immunized with crude leaf extracts from *N. benthamiana* expressing Bet v 1 generated immunological responses comparable to those induced by the protein expressed in *E. coli* or extracted from birch pollen.

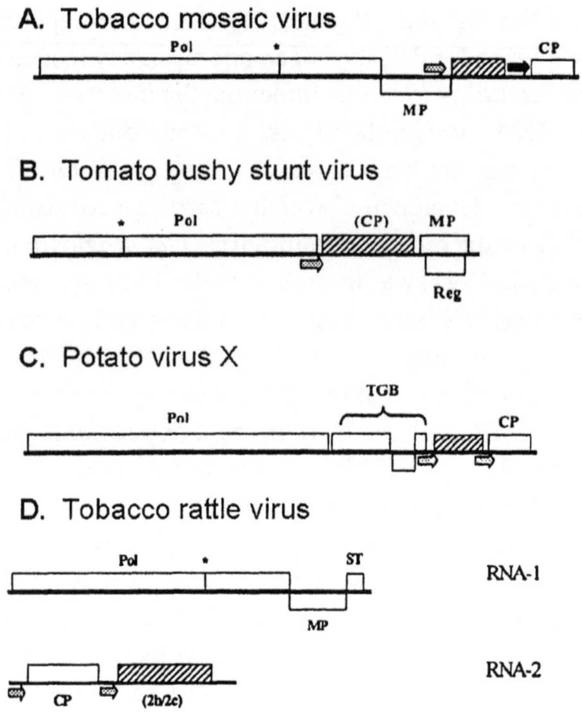

Figure 6. Genome organization of viruses in which the CP sg promoter (stippled arrow) is used to drive expression of the heterologous protein (hatched). In the case of TMV, CP expression is controlled by the sg promoter from a related virus (black arrow). Pol -polymerase; MP - movement protein; CP – coat protein; TGB - triple gene block; Reg – regulatory protein; ST – seed transmission; 2b/2c – proteins involved in nematode transmissions. Those genes deleted during vector construction are shown in brackets

In an approach combining TMV as a gene vector and the coat protein of alfalfa mosaic virus (AMV) as an epitope-presentation system, Yusibov et al. (1997) expressed modified versions of the AMV coat protein bearing epitopes from rabies virus and human immunodeficiency virus Type 1 (HIV-1) in tobacco. In infected leaf tissue the modified AMV CP subunits assembled into ellipsoid particles which expressed multiple copies of the antigenic insert. When purified and injected into mice, these particles elicited the production of appropriate neutralizing antibodies. It was subsequently shown that mice immunized with the Rabies virus construct were protected against a normally lethal challenge with the virus (Modelska et al., 1998).

4.2 Cowpea Mosaic Virus

Cowpea mosaic virus (CPMV) is the type member of the Genus *Comovirus* in the family *Comoviridae*. It infects a number of legume species and grows to particularly high titres in its natural host, cowpea (*Vigna unguiculata*). The genome of CPMV consists of two separately encapsidated positive-strand RNA molecules of 5889 (RNA-1) and 3481 (RNA-2) nucleotides. The RNAs each contain a single open reading frame and are expressed through the synthesis and subsequent processing of precursor polyproteins (Figure 7, upper section). The processing is mediated by the RNA-1-encoded 24K proteinase, which recognizes a number of specific cleavage sites. CPMV capsids contain 60 copies each of a large (L) and a small (S) CP arranged with icosahedral symmetry.

4.2.1 Epitope-presentation

The use of CPMV particle to express foreign sequences was first reported by Usha et al. (1993) and Porta et al. (1994). Since that time a large variety of epitopes have been expressed on the surface of CPMV particles (for examples, see Lomonossoff and Hamilton, 1999). In the majority of cases the foreign sequence was inserted into the most exposed loop of the S protein (the βB-βC loop; Figure 7, middle section). Recently, however, other sites, such as the βE-αB loop of the L protein and the βC'-βC" loop of the S protein, have also been used (Brennan et al., 1999; Taylor et al., 2000).

Generally, provided the inserted sequence was less than 40 amino acids long, the yields of modified particles were similar to those obtained with wild-type CPMV (approximately 1mg of particles per gram of infected leaf tissue). The resulting particles present 60 copies of the inserted peptide (Figure 7, lower section). A number of CPMV-based chimaeras have been subjected to detailed immunological analysis (e.g. McLain et al., 1995) and, in the case of a chimaera expressing an epitope from mink enteritis virus (MEV) have been shown to be capable of stimulating protective immunity (Dalsgaard et al., 1997). The immunological properties of various chimaeras have been summarized recently in Porta and Lomonossoff (1998) and Lomonossoff and Hamilton (1999) and the reader is referred to these reviews for further details.

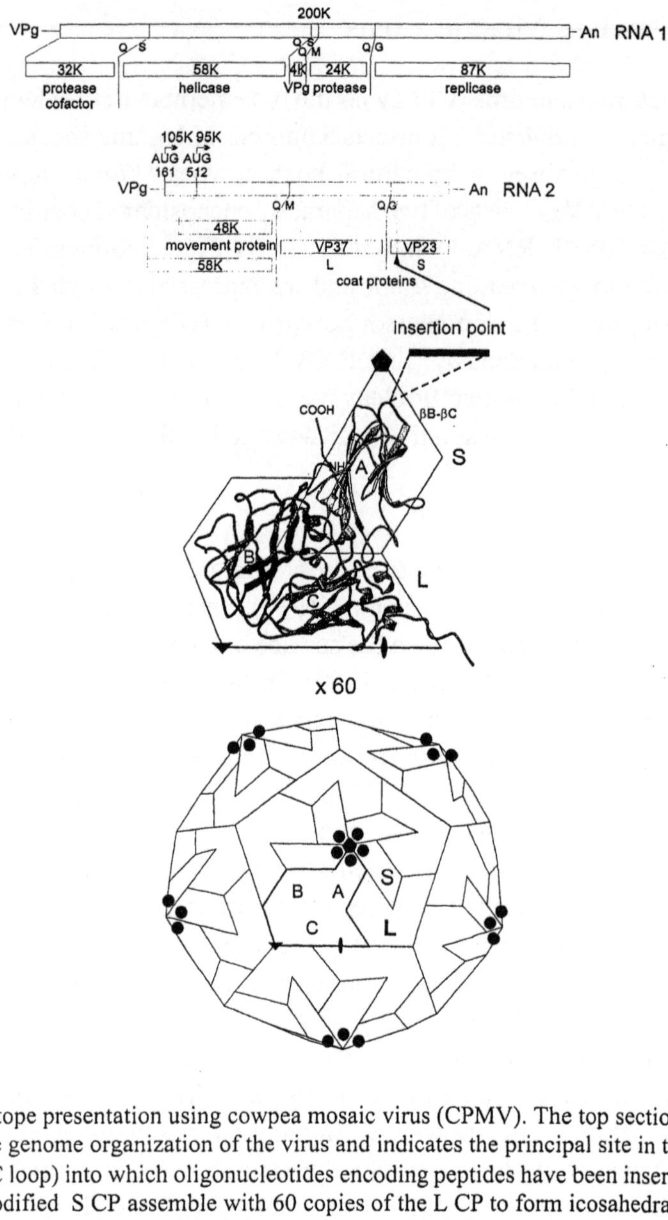

Figure 7. Epitope presentation using cowpea mosaic virus (CPMV). The top section of the figure shows the genome organization of the virus and indicates the principal site in the viral S CP gene (βB-βC loop) into which oligonucleotides encoding peptides have been inserted. Sixty copies of the modified S CP assemble with 60 copies of the L CP to form icosahedral capsids. The inserted epitopes are shown as black dots on the assembled virus particle. Reproduced from Porta and Lomonossoff (1998).

One curious feature of CPMV-based chimaeras is that proteolytic cleavage almost invariably occurs at or near the carboxy-terminus of the inserted sequence. This cleavage appears to be position- rather than sequence- dependent and occurs regardless of the site of insertion (Taylor et al., 1999; 2000). The cleavage does not result in the loss of the epitope from the surface of the virion but leads to it being anchored only at its N-terminus rather than being presented as a closed loop (Lin et al., 1996). To investigate the effects of this cleavage on the immunological properties of CPMV chimaeras, a series of constructs was produced all members of which contained the same epitope (residues 85-98 from VP1 of HRV-14) inserted at different positions in the S protein (Taylor et al., 2000). In all except one case (CPMV/HRV-L1), the epitope underwent efficient cleavage near its carboxy-terminus. Crystallographic analyses (Lin et al., 1996; Taylor et al., 2000) showed that CPMV/HRV-14 chimaeras in which cleavage occurs present the inserted peptides in a variety of different conformations, none of which resemble that adopted by the same sequence on the surface of HRV-14. These chimaeras raised antisera which, while reacting very well with denatured HRV-14 VP1, bound only very weakly to intact HRV-14 particles. By contrast, when the same epitope was presented as a closed loop in CPMV/HRV-L1, antisera which bound efficiently to HRV-14 particles were produced, demonstrating the importance of mode of display on the immunological properties of chimaeras (Taylor et al., 2000).

4.2.2 Polypeptide Expression

Verver et al. (1998) demonstrated that it was possible to insert a foreign gene, GFP, between the regions of CPMV RNA-2 encoding the MP and L CP (Figure 8A). Release of GFP was achieved by flanking it by extensive duplicated MP-L proteinase cleavage sites. Though the modified RNA-2 was as infectious as wild-type RNA 2 and expressed GFP, the construct tended to lose the insert by homologous recombination. When the length of repeated sequence was minimized, constructs with increased genetic stability were obtained (Gopinath et al., 2000). In an alternative approach to achieving genetic stability, Gopinath et al. (2000) showed that it was possible to replace either one of the two duplicated proteinase sites with a sequence encoding the 2A catalytic peptide from foot-and-mouth disease virus (FMDV). When 2A was inserted between GFP and L (Figure 8A), cleavage was only partial, resulting in some GFP-2A-L fusion protein being produced. This fusion protein could be incorporated into virus particles which, as a result, displayed green fluorescence under ultraviolet illumination. Gopinath et al. (2000) also fused GFP to the C-terminus of S via a 2A sequence (Figure 8A). The resulting construct multiplied well in plants, was genetically stable and expressed levels of GFP up to 1% of total soluble protein.

4.3 Tomato Bushy Stunt Virus

The genome of tomato bushy stunt virus (TBSV) consists of a single molecule of RNA of 4.7 kb which is encapsidated in isometric particles made up of 180 copies of a single coat protein of 41kDa. The CP, which is synthesized from a subgenomic mRNA (Figure 2B), is not essential for infection of certain *Nicotiana* species, though its presence does enhance systemic movement.

4.3.1 Epitope Presentation

Sequences derived from gp120 of HIV-1 have been fused to the C-terminus of the TBSV coat protein and the effect on particle formation determined (Joelson et al., 1997). When a sequence encoding 162 amino acids was expressed at this site, a large proportion of the inserted sequenced was lost on serial passaging. By contrast, when a 13 amino acid sequence, corresponding to the V3 loop of gp41, was expressed at the same location, the construct was genetically stable and the inserted epitope could be detected immunologically. Though the modified virions stimulated only a weak response when injected into mice (Sjö`lander et al., 1996), plates coated with particles could detect anti-V3 antibodies in HIV-positive individuals (Joelson et al., 1997).

4.3.2 Polypeptide Expression

The fact that the TBSV coat protein is not essential for infectivity was exploited by Scholtof et al. (1993) to produce constructs in which most of the region encoding the coat protein was replaced with that encoding either GUS or CAT (Figure 6B). Both constructs produced high levels of the appropriate protein in inoculated leaves, though the level of RNA accumulation was substantially reduced in the case of the GUS-containing virus. A refined version of the TBSV vector was subsequently produced in which the CP was replaced with a polylinker (Scholtof, 1999). This, coupled with improvements to the infection process, permitted the facile expression of heterologous sequences in the inoculated leaves of plants.

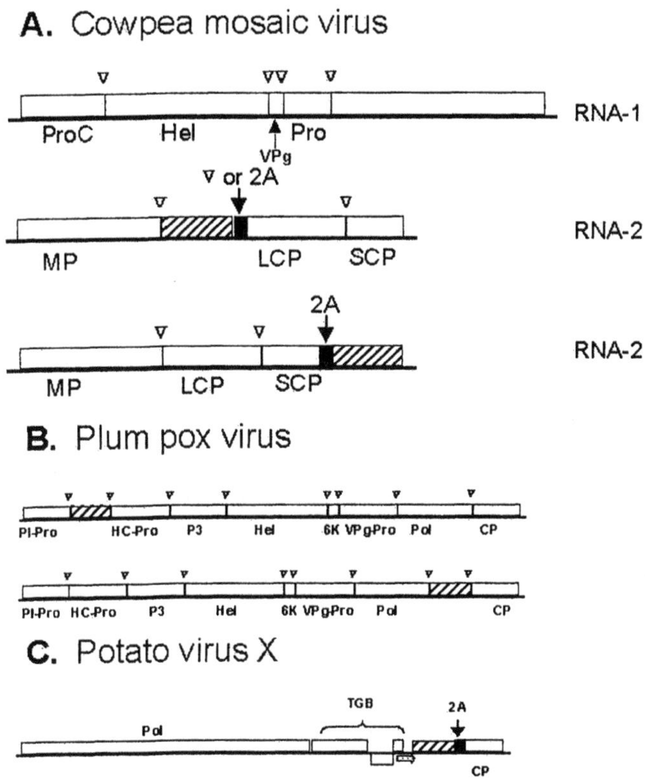

Figure 8. Genome organization of viruses in which protein processing is used to obtain expression of the heterologous protein (hatched). ∇ indicates processing by a virus-encoded proteinase while 2A indicates where processing is mediated by the FMDV 2A catalytic peptide. The functions of various virus genes are indicated: ProC – proteinase co-factor; Hel – helicase; VPg – virus protein genome-linked; Pol – polymerase; MP – movement protein; LCP – large coat protein; SCP – small coat protein; TGB – triple gene block; CP – coat protein; P1-Pro – P1-Proteinase; HC-Pro – helper component proteinase; P3 – protein P3; 6K – 6 kilodalton protein; VPg-Pro – VPg-Proteinase.

A similar approach has also been used to express the nucleocapsid protein p24 from HIV-1 as a fusion with the 5' terminal portion of the CP gene (Zhang et al., 2000). The modified TBSV RNA was shown to be capable of replicating in both protoplasts and in the inoculated leaves of whole plants. Accumulation of the CP-p24 fusion protein could be detected in inoculated leaves.

4.4 Plum Pox Virus

Plum pox virus (PPV) is a member of the genus *Potyvirus,* a genus which contains several hundred species. PPV, like other potyviruses, has a monopartite genome which is encapsidated in flexuous rod-shaped particles consisting of more than 2000 copies of a single coat protein. The RNA contains a single ORF which encodes a polyprotein which is self- processed by proteinase domains within it to produce the mature viral proteins, including the viral coat protein (Figure 8B). A potential advantage of vectors based on PPV is that as well as infecting experimental hosts, such as *Nicotiana* species, the virus can also infect perennial hosts.

4.4.1 Epitope Presentation

Though a detailed structure of a potyviral coat protein is not available, immunological analyses suggest that both the N- and C-termini are surface exposed. Making use of this information, Fernandez-Fernandez et al. (1998) fused a 15 amino acid epitope from VP2 of canine parvovirus (CPV) to the N-terminus of the PPV coat protein, either as a single copy or as a tandem duplication. Both constructs could be propagated in *N. clevelandii*, gave yields of virus particles similar to those obtained with wild-type PPV and the inserted epitope could be detected on the virion surface. Antisera raised against particles from either construct showed neutralizing activity in a monolayer-protection assay.

4.4.2 Polypeptide Expression

The first example of the expression of a foreign protein from any potyvirus was the expression of GUS from tobacco etch virus (TEV; Dolja et al., 1992). Though very useful for monitoring virus movement within a plant or analyzing the functions of genes from other plant viruses, such TEV-based constructs have not been used for the high-level expression of foreign proteins in plants. However the experience gained with TEV was instrumental in developing gene vectors based on PPV. The first attempt to develop a polypeptide expression vector based on PPV involved the insertion of the GUS gene between the regions of the genome encoding the proteins P1-Pro and HC-Pro (Guo et al., 1998; Figure 8B). Release of GUS was achieved by flanking it with duplicated proteinase cleavage sites. The resulting construct could multiply in *N. benthamiana* but gave a symptomless infection. After a single passage, deletions in the GUS gene could be detected, a situation similar to that found with the TEV-GUS construct described by Dolja et al. (1992). Subsequently, a PPV vector in which the foreign sequence is inserted between the polymerase (Pol) and CP genes (Figure 8B) was developed and used to express the VP60 structural protein of rabbit hemorrhagic disease virus (RHDV) in *N. clevelandii* (Fernandez-Fernandez et al., 2001). Inoculation with a crude preparation from infected leaf tissue protected rabbits against subsequent challenge with a lethal dose of RHDV.

4.5 Other Potyviruses

In addition to PPV, other poty- and closely related viruses have been explored as potential polypeptide expression vectors. These include clover yellow vein virus (ClYVV) as a vector for the production of proteins in legumes (Masuta et al., 2000) and wheat streak mosaic virus (WSMV) as a vector for monocots (Choi et al., 2000). In each case, release of the inserted sequences was obtained through the use of duplicated proteinase cleavage sites and the inserted protein could be detected in infected tissue.

4.6 Potato virus X

The genome of the potato virus X (PVX) consists of a single molecule of RNA of 6.4 kb encapsidated in flexuous rod-shaped particles consisting of approximately 1400 coat protein subunits of 25kDa. The viral coat protein is encoded by the 3' terminal cistron and is expressed from a sg RNA. Virus movement within infected plants is mediated by the "Triple gene block" (TGB) proteins whose precise functions have yet to be delineated.

4.6.1 Polypeptide Expression

Using an approach similar to that originally developed for vectors based on TMV, Chapman et al. (1992) showed that it was possible to express GUS from PVX when the sequence gene is inserted downstream of a duplicated copy of the CP sg promoter (Figure 6C). Similar results were subsequently obtained with PVX expressing GFP (Baulcombe et al., 1995). Though some loss of the inserted sequence was found, presumably as a result of homologous recombination across the duplicated CP sg promoter, the vector system has proved stable enough for many applications, such as the induction of virus-induced gene silencing (VIGS; Baulcombe, 1999), which require only transient expression.

Despite their general utility, PVX vectors with duplicated sg promoters have not been widely used to produce large amounts of valuable proteins in plants because of their tendency to delete all or part of the inserted sequence. However, there are some examples of such a use. O'Brien et al. (2000) expressed the major capsid protein, VP6, from a murine rotavirus and showed that although the protein retained its ability to form trimers, it tended to assemble into paracrystalline sheets and tubes rather than virus-like particles (VLPs). More recently, Saitoh et al. (2001) expressed the 51 amino acid defensin WT-1 from *Wisabia japonica* in *N. benthamiana* using a PVX vector. Though plants expressing WT-1 were not protected from fungal attack, the purified defensin did show considerable anti-fungal activity.

Though not extensively used for the production of large amounts of foreign proteins, PVX vectors have proved very useful in transient studies on gene constructs prior to their use for stable genetic transformation of plants. Examples of this include the expression of ScFv antibodies against proteins from either potato virus V (PVV; Hendy et al., 1999), tomato spotted wilt virus (TSWV; Franconi et al., 1999) or against granule-bound starch synthase I (Ziegler et al., 2000). These experiments also showed that it is possible to direct PVX-expressed proteins to the secretory pathway.

To overcome the problem of recombination, Santa-Cruz et al. (1996) developed a novel system in which a foreign gene, initially GFP, is fused to the N-terminus of the CP gene via a sequence encoding the 2A catalytic peptide from FMDV (Figure 8C). The 2A sequence promotes cleavage between the foreign gene insert and the CP though this is not 100% efficient, resulting in some CP subunits still bearing the inserted protein. These fusion proteins were found to retain their ability to be incorporated into virus capsids, resulting in particles which display the inserted polypeptide. Using this approach it is possible, using the same construct, to produce a protein of interest in both a free (unfused) state where cleavage by 2A has occurred and as a CP fusion where it is incorporated in PVX particles. The functionality of a foreign protein when incorporated into virions was demonstrated by the observation that a ScFv expressed as a CP fusion could still bind to its antigen, the herbicide diuron (Smolenska et al., 1998). When this system was used to express the rotavirus VP6 sequence, the uncleaved VP6-2A-CP assembled into PVX virions while the VP6-2A cleavage product formed typical VP6 VLPs (O'Brien et al., 2000).

4.7 Tobacco Rattle Virus

Tobacco rattle virus (TRV) is a bipartite rod-shaped virus. The larger RNA-1 (6.8 kb) encodes proteins involved in replication and movement of the virus while the smaller RNA-2 (1.8–4.5 kb) encodes the CP and a variable number (depending on virus isolate) of additional ORFs. These latter ORFS encode proteins that are non-essential for virus replication and movement within a plant but which are involved in nematode transmission of the virus. The CP and the other RNA-2-encoded proteins are synthesised from sg RNAs.

4.7.1 Polypeptide Expression

It has been shown that it is possible to replace the non-essential ORFs of RNA-2 of TRV, and related members of the genus *Tobravirus*, with GFP (under the control of a duplicated CP sg promoter) without affecting the ability of the virus to multiply and spread in plants (MacFarlane and Popovich, 2000). The construct was able to express GFP efficiently in both the leaves and roots of infected plants, the latter being a property particularly associated with tobraviruses. The ability of the TRV vector to express a protein which might be useful in protecting plants against nematodes was tested by inserting the sequence encoding a lectin (GNA) from snowdrop. Significant levels of the lectin could be detected in the roots of inoculated *N. benthamiana* plants.

4.8 Subviral Systems

In addition to the use of autonomously replicating plant viruses, there have also been some attempts to develop a number of subviral systems into usable vectors for polypeptide expression. Such systems include the use of both satellites and defective RNAs. Satellites are small RNA molecules associated with many genera of plant viruses. They are completely dependent on helper viruses, with which they have no sequence homology, for their replication. They have no role in the replication of their helper virus but can influence symptomatology. Defective RNAs (dRNAs) are small RNAs which, like satellites, are completely dependent on a helper virus. However, unlike satellites, they have sequence similarity with their helpers and, indeed, are derived from them through a process of sequence deletion. Because they encode no functions essential for replication, satellites and dRNAs have been considered potential candidates as vectors.

4.8.1 Polypeptide Expression

Bamboo mosaic virus (BaMV) is the only member of the genus *Potexvirus* (type member PVX) with a satellite (satBaMV). This is a linear molecule of 836 nucleotides which contains an ORF capable of encoding a 20kDa protein whose function is obscure. Lin et al. (1996) replaced the 20kDa ORF of satBaMV with the sequence encoding CAT and showed that the modified satellite could multiply in both barley protoplasts and *C. quinoa* plants in the presence of helper BaMV. CAT expression reached approximately 2μg per g of leaf tissue.

Natural or artificially-constructed dRNAs from the genera *Tombusvirus* (type member TBSV) and *Tobamovirus* (type member TMV) have been investigated as a means of foreign gene expression (Burgyan et al., 1994; Lewandowski et al., 1998). Though replication of the dRNA by the appropriate helper virus has been demonstrated, there is no report as yet demonstrating the general utility of these systems for gene expression in plants.

5. CONCLUSIONS

Despite their comparatively recent development, vectors based on RNA plant viruses have achieved fairly widespread use within the past 5 years. Though originally envisaged as systems for the high level expression of foreign peptides and polypeptides, they have also found applications which were quite unforeseen at the time the vectors were developed (e.g. through their ability to alter host gene expression through VIGS).

As described above, there have been numerous examples of the successful expression of both peptides and polypeptides using plant virus-based vectors. There are, however, still a number of challenges which will have to be met if virus-based systems are to achieve widespread use as expression systems. For example, it is clear that current virus-based systems have distinct limits on the size of insertion which they will tolerate and this will impose limits on the utility of such systems unless it is addressed. There are also problems associated with the possible genetic instability of the inserted sequence during multiple rounds of replication. Though the initial fears regarding the rates of genetic drift in RNA viruses proved to be exaggerated, there is still the problem that changes to the sequence of an virus-expressed protein may gradually accumulate. Another possible concern is that present virus-based expression systems generally utilize wild-type viruses which have the capacity to spread in the environment. To overcome this, there will be a need develop systems (e.g. the use of defective viruses) which will reduce or eliminate such spread.

REFERENCES

Ahlquist P and Janda M, 1984. cDNA cloning and *in vitro* transcription of the complete brome mosaic virus genome. Mol Cell Biol, 4:2876-2882

Ahlquist P, French R, Janda M and Loesch-Fries S, 1984. Multicomponent RNA plant virus infection derived from cloned viral cDNA. Proc Natl Acad Sci USA, 81:7066-7070

Baulcombe DC, 1999. Fast forward genetics based on virus-induced gene silencing. Curr Opin Plant Biol, 2:109-113

Baulcombe DC, Chapman S and SantaCruz SS, 1995. Jellyfish green fluorescent protein as a reporter for virus-infections. Plant J, 7:1045-1053

Bendahmane M, Koo M, Karrer E and Beachy RN, 1999. Display of epitopes on the surface of tobacco mosaic virus:Impact of charge and isoelectric point of the epitope on virus-host interactions. J Mol Biol, 290:9-20

Brennan FR, Jones TD, Gilleland LB, Bellaby T, Xu F, North PC, Thompson A, Staczek J, Lin T, Johnson JE, Hamilton WDO and Gilleland HE, 1999. *Pseudomonas aeroginosa* outer-membrane protein F epitopes are highly immunogenic when expressed on a plant virus. Microbiology, 145:211-220

Burgyan J, Salanki K, Dalmay T and Russo M, 1994. Expression of homologous and heterologous viral coat protein-encoding genes using recombinant DI RNA from cymbidium ringspot tombusvirus. Gene, 138:159-163

Chapman S, Kavanagh T and Baulcombe D, 1992. Potato virus-X as a vector for gene-expression in plants. Plant J, 2:549-557

Choi IR, Stenger DC, Morris TJ and French R, 2000. A plant virus vector for systemic expression of foreign genes in cereals. Plant J, 23:547-555

Dalsgaard K, Uttenthal Å, Jones TD, Xu F, Merryweather A, Hamilton WDO, Langeveld JPM, Boshuizen RS, Kamstrup S, Lomonossoff GP, Porta C, Vela C, Casal JI, Meloen RHand

Rodgers PB, 1997. Plant-derived vaccine protects target animals against a virus disease. Nat Biotechnol, 15:248-252

Dawson WO, Lewandowski DJ, Hilf ME, Bubrick P, Raffo AJ, Shaw JJ, Grantham GL and Desjardins PR, 1989. A tobacco mosaic virus-hybrid expresses and loses an added gene. Virology, 172:285-292

Dolja VV, McBride HJ and Carrington JC, 1992. Tagging of plant potyvirus replication and movement by insertion of beta-glucuronidase into the viral polyprotein. Proc Natl Acad Sci USA, 89:10208-10212

Donson J, Kearney CM, Hilf ME and Dawson WO, 1991. Systemic expression of a bacterial gene by a tobacco mosaic virus-based vector. Proc Natl Acad Sci USA, 88:7204-7208

Fernandez-Fernandez MR, Mourino M, Rivera J, Rodriguez F, Plana-Duran J and Garcia, JA, 20 01. Protection of rabbits against rabbit hemorrhagic disease virus by immunization with the VP60 ptotein expressed in plants with a potyvirus-based vector. Virology, 280:283-291

FitchenJ, BeachyRN, and Hein MB, 1995. Plant virus expressing hybrid coat protein with added murine epitope elicits autoantibody response. Vaccine, 13:1051-1057

FranconiA, RoggeroP, PirazziP, AriasFJ, DesiderioA, BittiO, PashkoulovD, MatteiB, BracciL, MasengaV, MilneRG and Benvenuto E 1999. Functional expression in bacteria and plants of an scFv antibody fragment against tospoviruses. Immunotechnology, 4:189-201

French R, Janda M and Ahlquist P,1986. Bacterial gene inserted in an engineered RNA virus - efficient expression in monocotyledonous plant-cells. Science, 231:1294-1297

Gopinath K, Wellink J, Porta C, Taylor KM, Lomonossoff GP and van Kammen A, 2000. Engineering cowpea mosaic virus RNA-2 into a vector to express heterologous proteins in plants. Virology, 267:159-173

Guo HS Lopez-Moya JJ and Garcia JA, 1998. Susceptibility to recombination rearrangements of a chimeric plum pox potyvirus genome after insertion of a foreign gene. Virus Res, 57:183-195

Hamamoto H, Sugiyama Y, Nakagawa N, Hashida E, Matsunaga Y, Takemoto S, Watanabe Y, and Okada Y, 1993. A new tobacco mosaic virus vector and its use for the systemic production of angiotensin-I-converting enzyme inhibitor in transgenic tobacco and tomato. Bio/Technology, 11:930-932

Haynes JR, Cunningham J, von Seefried A, Lennick M, Garvin RT and Shen S-H, 1986. Development of a genetically engineered, candidate polio vaccine employing the self-assembling properties of the tobacco mosaic virus coat protein. Bio/Technology, 4:637-641

Hendy S, Chen ZC, Barker H, SantaCruz S, Chapman S, Torrance L, Cockburn W and Whitelam, GC, 1999. Rapid production of single-chain Fv fragments in plants using a potato virus X episomal vector. J Immunol Methods, 231:137-146

Hull R, 1978. The possible use of plant viral DNAs in genetic manipulation in plants. Trends Biochem Sci, 3:254-256

Joelson T, Åkerblom L, Oxelfelt P, Strandberg B, Tomenius K and Morris TJ, 1997. Presentation of a foreign peptide on the surface of tomato bushy stunt virus. J Gen Virol, 78:1213-1217

Kearney CM, Donson J, Jones GE and Dawson WO, 1993. Low level of genetic drift in foreign sequences replicating in a RNA virus in plants. Virology, 192:11-17

Koo M, Bendahmane M, Lettieri GA, Paoletti AD, Lane TE, Hitchen JH, Buchmeier MJ and Beachy RN, 1999. Protective immunity against murine hepatitis virus (MHV) induced by intranasal or subcutaneous administration of hybrids of tobacco mosaic virus that carries an MHV epitope. Proc Natl Acad Sci USA, 96:7774-7779

Krebitz M, Wiedermann U, Essl D, Steinkellner H, Wagner B, Turpen TH, Ebner C, Scheiner O and Breiteneder H, 2000. Rapid production of the major birch pollen allergen Betv1 in *Nicotiana benthamiana* plants and its immunological *in vitro* and *in vivo* characterization. FASEB J, 14:(10) 1279-1288

Kumagai MH, Turpen TH, Weinzettl N, Della-Cioppa G, Turpen AM, Donson J, Hilf ME, Grantham GL, Dawson WO, Chow TP, Piatak M and Grill, LK, 1993. Rapid, high-level expression of biologically-active alpha-trichosanthin in transfected plants by an RNA viral vector. Proc Natl Acad Sci USA, 90:427-430

Kumagai MH, Donson J, Della-Cioppa G and Grill LK, 2000. Rapid, high-level expression of glycosylated rice alpha-amylase in transfected plants by an RNA viral vector. Gene, 245:169-174

Lewandowski DL and Dawson WO, 1998. Deletion of internal sequences results in tobacco mosaic virus defective RNAs that accumulate to high levels without interfering with replication of the helper virus. Virology, 251:427-437

Lin NS, Lee YS, Lin BY, Lee CW and Hsu, YH, 1996. The open reading frame of bamboo mosaic potexvirus satellite RNA is not essential for its replication and can be replaced with a bacterial gene. Proc Natl Acad Sci USA, 93:3138-3142

Lin T, Porta C, Lomonossoff GP and Johnson JE, 1996. Structure-based design of peptide presentation on a viral surface: The crystal structure of a plant/animal virus chimaera at 2.8Å resolution. Folding and Design, 1:179-187

Lomonossoff GP and Hamilton WDO, 1999. Cowpea mosaic virus-based vaccines. In *Current topics in microbiology and immunology*. Eds. Hammond J, McGarvey P and Yusibov V, Springer, Berlin, pp 177-189

Lomonossoff GP and Johnson JE, 1996. Use of macromolecular assemblies as expression systems for peptides and synthetic vaccines. Curr Opin Struct Biol, 6:176-182

MacFarlane SA and Popovich AH, 2000. Efficient expression of foreign proteins in roots from tobravirus vectors. Virology, 267:29-35

Masuta C, Yamana T, Tacahashi Y, Uyeda I, Sato M, Ueda S and Matsumura T, 2000. Development of clover yellow vein virus as an efficient, stable gene-expression system for legume species. Plant J, 23:539-546

McCormick AA, Kumagai MH, Hanley K, Turpen TH, Hakim I, Grill LK, Tuse D, Levy S and Levy R, 1999. Rapid production of specific vaccines for lymphoma by expression of the tumor-derived single-chain Fv epitopes in tobacco plants. Proc Natl Acad Sci USA, 96:703-708

McLain L, Porta C, Lomonossoff GP, Durrani Z and Dimmock NJ, 1995. Human immunodeficiency virus type 1 neutralizing antibodies raised to a gp41 peptide expressed on the surface of a plant virus. AIDS Res Hum Retroviruses, 11:327-334

Modelska A, Dietzschold B, Sleysh N, Fu ZF, Steplewski K, Hooper DC, Koprowski H and Yusibov V, 1998. Immunization against rabies with plant-derived antigen. Proc Natl Acad Sci USA, 95:2481-2485

Mori M, Zhang GH, Kaido M, Okuno T and Furusawa I, 1993. Efficient production of human gamma-interferon in tobacco protoplasts by genetically-engineered brome mosaic-virus RNAs. J Gen Virol, 74:1255-1260

O'Brien GJ, Bryant CJ, Voogd C, Greenberg HB, Gardner RC and Bellamy AR, 2000. Rotavirus VP6 expressed by PVX vectors in *Nicotiana benthamiana* coats PVX rods and also assembles into virus-like particles. Virology, 270:444-453

Porta C and Lomonossoff GP, 1996. Use of viral replicons for the expression of genes in plants. Mol Biotechnol, 5:209-211

Porta C and Lomonossoff GP, 1998. Scope for using plant viruses to present epitopes from animal pathogens. Reviews in Medical Virology, 8:25-41

Porta C and Lomonossoff GP, 2001. Viruses as vectors for the expression of foreign sequences in Plants. Biotechnol Genet Eng Rev, 19:(in press)

Porta C, Spall VE, Loveland J, Johnson JE, Barker PJ, and Lomonossoff GP, 1994. Development of cowpea mosaic virus as a high-yielding system for the presentation of foreign peptides. Virology, 202:949-955

Saitoh H, Kiba A, Nishihara M, Yamamura S, Suzuki K and Terauchi R, 2001. Production of antimicrobiol defensin in *Nicotiana benthamiana* with a potato virus X vector. Mol Plant-Microbe Interact, 14:111-115

SantaCruz S, Chapman S, Roberts AG, Roberts IM, Prior DAM and Oparka KJ, 1996. Assembly and movement of a plant virus carrying a green fluorescent protein overcoat. Proc Natl Acad Sci USA, 93:6286-6290

Scholthof HB, Morris TJ and Jackson AO, 1993. The capsid protein gene of tomato bushy stunt virus is dispensable for systemic movement and can be replaced for localized expression of foreign genes. Mol Plant-Microbe Interact, 6:309-322

Scholthof HB, 1999. Rapid delivery of foreign genes into plants by direct rub-inoculation with intact plasmid DNA of a tomato bushy stunt virus gene vector. J Virology, 73:7823-7829

Siegel A, 1985. Plant-virus-based vectors for gene transfer may be of considerable use despite a presumed high error frequency during RNA synthesis. Plant Mol Biol, 4:327-329

Sjölander S, Hansen J-ES, Lövgren-Bengtsson Åkerblom L and Morein B, 1996. Induction of homologous virus neutralizing antibodies in guinea pigs immunized with two human immunodeficiency virus type I glycoprotein gp120-iscom preparations. A comparison with other adjuvant systems. Vaccine, 14:344-352

Smolenska L, Roberts IM, Learmonth D, Porter AJ, Harris WJ, Wilson TMA and SantaCruz S, 1998. Production of a functional single chain antibody attached to the surface of a plant virus. FEBS Lett, 441:379-382

Sugiyama Y, Hamamoto H, Takemoto S, Watanabe Y and Okada Y, 1995. Systemic production of foreign peptides on the particle surface of tobacco mosaic virus. FEBS Lett, 359:247-250

Szeto WW, Hamer DH, Carlson PS and Thomas CA, 1977. Cloning of cauliflower mosaic virus (CaMV) DNA in *Escherichia coli*. Science, 196:210-212

Takamatsu N, Ishikawa M, Meshi T and Okada Y, 1987. Expression of bacterial chloramphenicol acetyltransferase gene in tobacco plants mediated by TMV-RNA. EMBO, 6:307-311

Takamatsu N, Watanabe Y, Yanagi H, Meshi T, Shiba T, Okada Y, 1990. Production of enkephalin in tobacco protoplasts using tobacco mosaic virus RNA vector. FEBS Lett, 269:73-76

Taylor KM, Lin T, Porta C, Mosser AG, Giesing HA, Lomonossoff GP and Johnson JE, 2000. Influence of three-dimensional structure on the immunogenicity of a peptide expressed on the surface of a plant virus. J Mol Recognit, 13:71-82

Taylor KM, Porta C, Lin T, Johnson JE, Barker PJ and Lomonossoff GP, 1999. Position-dependent processing of peptides presented on the surface of cowpea mosaic virus. Biol Chem, 380:387-392

Turpen TH, Reinl SJ, Charoenvit Y, Hoffman SL, Fallarme V and Grill LK, 1995. Malarial epitopes expressed on the surface of recombinant tobacco mosaic virus. Bio/Technology, 13:53-57

Usha R, Rohll JB, Spall VE, Shanks M, Maule AJ, Johnson JE, and Lomonossoff GP, 1993. Expression of an animal virus antigenic site on the surface of a plant virus particle. Virology, 197:366-374

van Vloten-Doting L, Bol JF and Cornelissen B, 1985. Plant virus-based vectors for gene transfer will be of limited use because of the high error frequency during viral RNA synthesis. Plant Mol Biol, 4:323-326

Verch T, Yusibov V and Koprowski H, 1998. Expression and assembly of a full-length monoclonal antibody in plants using a plant virus vector. J Immunol Methods, 220:69-75

Verver J, Wellink J, Van Lent J, Gopinath K and van Kammen A, 1998. Studies on the movement of cowpea mosaic virus using the jellyfish green fluorescent protein. Virology, 242:22-27

Yusibov V, Modelska A, Steplewski K, Agadjanyan M, Weiner D, Hooper DC and Koprowski H, 1997. Antigens produced in plants by infection with chimeric plant viruses immunize against rabies virus and HIV-1. Proc Natl Acad Sci, USA, 94:5784-5788

Zhang GC, Leung C, Murdin L, Rovinski B and White KA, 2000. *In planta* expression of HIV-1 p24 protein using an RNA plant virus-based expression vector. Molec Biotechnol, 14:99-107

Ziegler A, Cowan GH, Torrance L, Ross HA and Davies HV, 2000. Facile assessment of cDNA constructs for expression of functional antibodies in plants using the potato virus X vector. Mol Breed, 6:327-335

Chapter 4

CONTROL OF THE N-GLYCOSYLATION OF THERAPEUTIC GLYCOPROTEINS PRODUCED IN TRANSGENIC PLANTS: A NEW CHALLENGE FOR GLYCOBIOLOGISTS

P. Lerouge, M. Bardor, S. Pagny, V. Gomord, A-C . Fitchette and L. Faye
Laboratoire des Transports Intracelllulaires, CNRS-UMR 6037, IFRMP 23, Universite de Rouen, 76821 Mont Saint Aignan Cedex, France

Keywords: transgenic plants, therapeutic glycoproteins, N-glycosylation, immunogenicity, humanization.

Abstract The protein synthesis and folding machinery in plant and animal cells are so similar that many therapeutic proteins have already been successfully produced in transgenic plants. Most of these recombinant proteins are indistinguishable from their mammalian counterpart, as far as amino acid sequence, conformation and eventually biological activity are concerned. With regards to post-translational modifications such as glycosylation, mammalian glycoproteins are found to be glycosylated when they are produced in transgenic plants. However, plants, as other eukaryotic expression systems, are not ideal for production of pharmaceutical proteins because they produce molecules with N-glycans that differ from those on animal glycoproteins. This could represent a limitation to the use of plant-derived recombinant glycoproteins devoted to therapeutic applications since the presence of plant-specific glyco-epitopes on these molecules may elicit immune responses in humans. This has highlighted that controlling the N-glycosylation of glycoproteins is therefore a major scientific challenge on the way to obtain plant-derived recombinant glycoproteins compatible with therapeutic uses. Therefore, it seems timely to provide an update on current aspects on N-glycan processing in plants and on emerging glycobiology research on therapeutic proteins produced in transgenic plants. In this review, the first part will draw a broad overview on many aspects of the structure and the biosynthesis of plant N-glycans, as well as of the analytical tools that have been developed for the identification of these oligosaccharides. The second part will be focused on the N-glycosylation of therapeutic glycoproteins produced in transgenic plants and on the strategies that are currently developed to engineer the glycosylation in plants

L. Erickson et al. (eds.),
Molecular Farming of Plants and Animals for Human and Veterinary Medicine, 73–109.

to obtain recombinant glycoproteins with human-like N-glycans.

1. PLANT N-LINKED GLYCANS AND THEIR DIFFERENCES WITH THEIR MAMMALIAN COUNTERPARTS

1.1 Structures of Plant N-linked Glycans

In plants, as in other eukaryotes, N-glycans are covalently linked to specific Asn residues constitutive of tripeptide Asn-X-Ser/Thr N-glycosylation sites of the protein, where X can be any amino acid except proline and aspartic acid (Kornfeld and Kornfeld, 1985). All N-glycans share a common minimal structure $Man_3GlcNAc_2$ constituted of a N, N'-diacetyl chitobiose unit, a β-mannose residue linked to the chitobiose and two α-mannose residues linked to hydroxyl 3 and 6 of the β-mannose. According to the substitutions of this core, plant N-glycans are classified into the four following classes: high-mannose-, complex-, paucimannosidic- and hybrid-type N-glycans (Lerouge et al., 1998).

1.1.1 High-mannose-type N-glycans

High-mannose-type N-glycans from $Man_5GlcNAc_2$ to $Man_9GlcNAc_2$ (Figure 9a) are synthesized during the first steps of the N-glycan processing occurring in the endoplasmic reticulum (ER) and in the early compartments of the Golgi apparatus. As these steps are common to plants and mammals (see section 1.2.1), high-mannose-type N-glycans are common to both organisms. High-mannose-type N-glycans were first identified in plants in soybean agglutinin (Lis and Sharon, 1998). They were then found N-linked to various extracellular and vacuolar glycoproteins. High-mannose-type N-glycans have also recently been identified as the unique N-linked oligosaccharides of spinach (Navazio et al., 1996) or maize calreticulin (Pagny et al., 2000), a glycoprotein specific chaperone that resides in the plant ER.

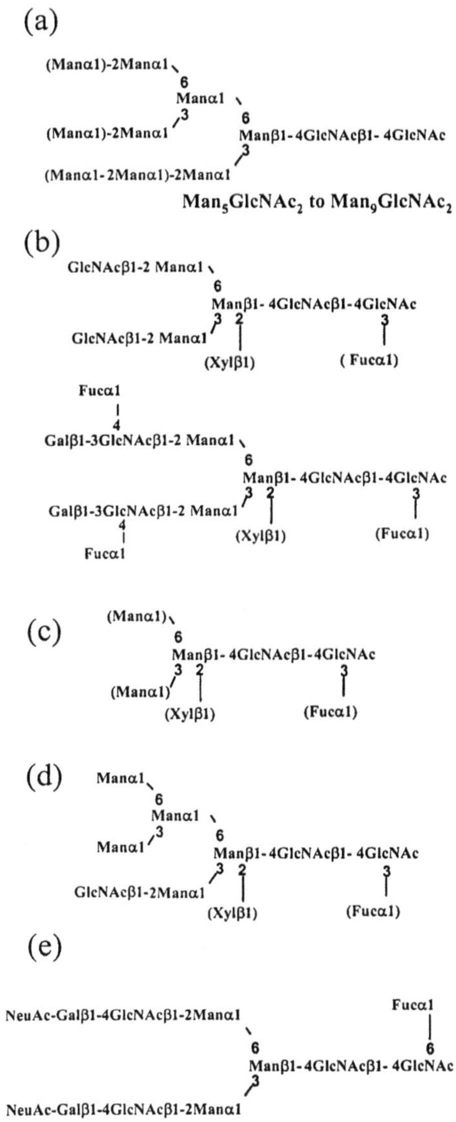

Figure 9. Structures of a) high-mannose-type N-glycans that are common to plants and animals, b) plant complex-type N-glycans, c) plant paucimannosidic-type N-glycans, d) plant hybrid-type N-glycans and e) mammalian sialylated bi-antennary N-glycan.

1.1.2 Complex-type N-glycans

As in other eukaryotic cells, plant complex-type N-glycans result from an extended processing of high-mannose-type N-glycans by the action of specific glycosidases and glycosyltransferases in the Golgi apparatus. Complex-type plant N-glycans are characterized by the presence of α(1, 3)-fucose and/or a β(1, 2)-xylose residues respectively linked to the proximal N-acetyl glucosamine and to the β-mannose residues of the core (also named core α(1, 3)-fucose and core β(1, 2)-xylose). These two glycan sequences are absent in mammals. However, it should be noted that these two epitopes are not specific to plants since the α(1, 3)-fucose epitope is also found in insect and nematode glycoprotein N-glycans (Altmann, 1997; van Die et al., 1999) and β(1, 2)-xylose residues have been characterized in snail and nematode glycoproteins (Kamerlink, 1991; Khoo et al., 1997; van Die et al., 1999). In addition to the core α(1, 3)-fucose and the core β(1, 2)-xylose residues, plant complex-type N-glycans contain β(1, 2)-N-acetyl glucosamine residues linked to the α-mannose units, as well as additional α(1, 4)-fucose and β(1, 3)-galactose residues linked to these terminal N-acetyl glucosamine units (Fitchette-Lainé et al., 1997; Melo et al., 1997). These later decorations yield Galb1-3(Fuca1-4)GlcNAc sequences known as Lewis a (Lewisa) antigens and usually found on cell-surface glycoconjugates in mammals (Figure 9b).

1.1.3 Paucimannosidic-type N-glycans

This nomenclature, previously proposed for truncated insect N-linked glycans deprived of terminal N-acetyl glucosamine residues linked to the α-mannose residues of the core (Altmann, 1997), was extended to modified plant oligosaccharides having only an α(1, 3)-fucose and/or a β(1, 2)-xylose residue linked to the core $Man_3GlcNAc_2$, or to the restricted core $ManGlcNAc_2$ and $Man_2GlcNAc_2$ (Figure 1c) (Lerouge et al., 1998). As paucimannosidic-type N-glycans result from the elimination in, or on the way to, the plant vacuole of terminal residues from complex-type N-glycans, they can be considered as typical vacuole-type N-glycans as developed later in this review (section 1.2.3).

1.1.4 Hybrid-type N-glycans

Hybrid-type N-glycans derive from the processing of only the α(1,

3)-mannose branch of the intermediate $Man_5GlcNAc_2$ leading to oligosaccharides having α(1, 3)-fucose and/or a β(1, 2)-xylose residues linked to GlcNAcMan5$GlcNAc_2$ (Figure 1d) (Oxley et al., 1996). Structures having $Man_{4-5}GlcNAc_2$ cores substituted by an α(1, 3)-fucose and/or a β(1, 2)-xylose residues were also identified in various plant glycoproteins (Kimura et al., 1988; Altmann et al., 1998; Shimazaki et al., 1999; van Ree et al., 2000). These N-linked glycans probably result from the post-Golgi removal of terminal GlcNAc residues from hybrid N-glycans as observed for paucimannosidic-type N-glycans.

1.2 Processing of N-linked Glycans in Plants

1.2.1 First Steps of the N-glycan Processing Are Common to Plants and Mammals

The first steps of the N-glycan processing, occurring in the ER and in the early Golgi apparatus, are common to plants and mammals. The N-glycosylation of plant proteins starts in the ER with the transfer by the oligosaccharyl transferase of the oligosaccharide precursor $Glc_3Man_9GlcNAc_2$ from a dolichol lipid carrier to specific Asn residues on the nascent polypeptide chain. The precursor is then modified by glycosidases and glycosyltransferases during the transport of the glycoprotein downstream the secretory pathway to its final localization. As illustrated on Figure 10, the oligosaccharide precursor $Glc_3Man_9GlcNAc_2$ first undergoes an early trimming of the three terminal glucose units catalyzed by the glucosidases I and II in the ER (Szumilo et al., 1986a; Kaushal et al., 1990a). A transient reglucosylation by an ER UDP-glucose: glycoprotein glucosyltransferase may occur after the removal of these three glucose residues (Parodi et al., 1984; Trombetta et al., 1989). This reglucosylation has been shown to act on unfolded proteins and to be involved in the quality control of glycoproteins in the ER (Hammond et al., 1994). In mammals, an ER-mannosidase specifically removes a single mannose residue to yield $Man_8GlcNAc_2$. Such an ER-mannosidase has not been detected in plants so far. However, the structures of the major N-linked glycans of an ER resident chaperone, calreticulin, purified either from the spinach (Navazio et al., 1996) or from maize (Pagny et al., 2000), were identified as $Man_8GlcNAc_2$ and $Man_{8-9}GlcNAc_2$, respectively, which indicates that a specific mannosidase could also be involved in the processing of plant N-linked glycans within the ER.

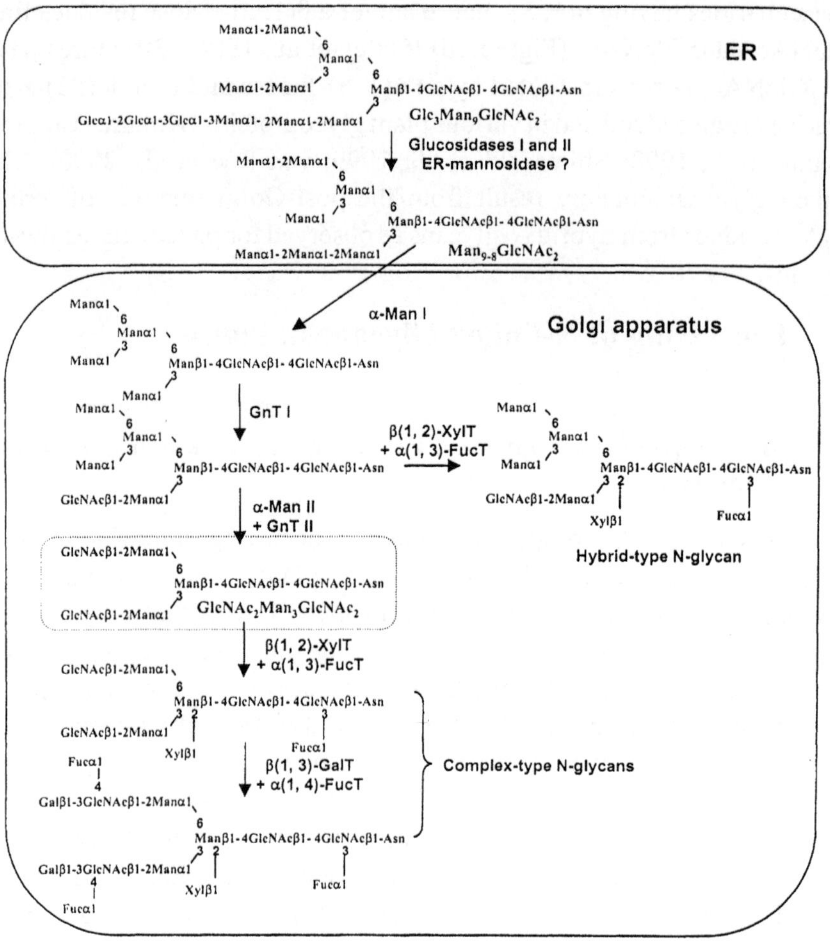

Figure 10. Processing of plant N-linked glycans. α-Man-I and II: α -mannosidases I and II; GnT I and II: N-acetylglucosaminyltransferases I and II; β(1, 2)-XylT: β(1, 2)- xylosyltransferase; α(1, 3)-FucT: α(1, 3)-fucosyltransferase;β(1, 3)-GalT:β(1, 3)-galactosyltransferase;α(1, 4)-FucT: α(1, 4)-fucosyltransferase.

As illustrated in Figure 10, mannose residues constitutive of high-mannose-type N-glycans can be further trimmed in the Golgi apparatus by the action of α-mannosidase I (α-Man I) which removes one to four α(1, 2)-mannose residues and converts $Man_{8\text{-}9}GlcNAc_2$ to $Man_5GlcNAc_2$ (Sturm et al., 1987a; Szumilo et al., 1986b). Then, the biosynthesis of complex-type N-glycans starts with the addition of new residues. A first N-acetylglucosamine residue is initially added to the α(1, 3)-mannose branch of the $Man_5GlcNAc_2$ by the N-acetylglucosaminyltransferase I (GnT I) (Johnson and Chrispeels, 1987; Tezuka et al., 1992). Two additional mannose residues are then removed by α-mannosidase II (α-Man II) (Kaushal et al., 1990b) and another outer N-acetylglucosamine residue is transferred by N-acetylglucosaminyltransferase II (GnT II) to the α(1, 6)-mannose branch yielding $GlcNAc_2Man_3GlcNAc_2$ (Johnson and Chrispeels, 1987; Tezuka et al., 1992).

1.2.2 Late Golgi Modifications in Plants Yield Complex N-linked Glycans Differing from Those Found in Mammals

$GlcNAc_2Man_3GlcNAc_2$ oligosaccharide (encircled in Figure 10) is the last common N-glycan on the way to complex-type N-glycans in plants and mammals. In mammals, this glycan is further modified by the action of N-acetylglucosaminyltransferases III to V, α(1, 6)-fucosyltransferase and β(1, 4)-galactosyltransferase giving rise to di-, tri- and tetra-antennary complex N-glycans that are specific to mammalian cells. These polyantennary complex N-glycans are then usually further elongated by transfer of sialic acids onto terminal galactose residues (see Figure 9e for the structure of a typical bi-antennary sialylated mammalian N-glycan). In plants, $GlcNAc_2Man_3GlcNAc_2$ N-glycan is converted into plant complex N-glycans by the action of an α(1, 3)-fucosyltransferase (α(1, 3)-FucT) and a β(1, 2)-xylosyltransferase (β(1, 2)-XylT) (Figure 10). Alternatively, α(1, 3)-FucT and β(1, 2)-XylT could also act on $GlcNAcMan_5GlcNAc_2$ substrate leading to plant hybrid-type N-glycans. The study of the substrate specificity of α(1, 3)-FucT and β(1, 2)-XylT has shown that the presence of at least one terminal N-acetylglucosamine residue is a prerequisite for the transfer of α(1, 3)-fucose and β(1, 2)-xylose (Johnson and Chrispeels, 1987; Tezuka et al., 1992; Staudacher et al., 1995; Zeng et al., 1997). The sequence of the xylosylation and the fucosylation events is not completely understood. Plant N-linked glycans having only a β(1, 2)-xylose or only an α(1, 3)-fucose residue have been identified in plant glycoproteins (Lerouge et al., 1998). Moreover, the substrate specificities of the α(1, 3)-FucT and β(1, 2)-XylT are not affected, respectively, by the absence of the β(1, 2)-xylose or α(1, 3)-fucose residues linked to the core (Johnson and Chrispeels, 1987; Tezuka et al., 1992; Staudacher et al., 1995; Zeng et al., 1997). Consequently, the α(1, 3)-fucosylation and β(1, 2)-xylosylation appear to be independent events. Using an immunocytochemical approach to study the subcompartmentation of the β(1, 2)-xylosylation and α(1, 3)-fucosylation events, we have demonstrated that these two steps occur mostly in the medial and in the trans Golgi cisternae, respectively. This indicates that during the transport of the glycoprotein through the Golgi apparatus, the β(1, 2)-xylose transfer starts before the addition of the α(1, 3)-fucose residue to the core (Fitchette-Lainé et al., 1994).

After the transfers of xylose and fucose on the core, complex-type N-glycans can be further processed by addition of terminal fucose and galactose residues to yield one or two antennae constituted of Gal β1-3(Fuc α1-4)GlcNAc sequences as represented on Figure 9b (Fitchette-Lainé et al., 1997; Melo et al., 1997; Fitchette et al., 1999). This sequence, known as Lewisa antigen, is usually found on cell-surface glycoconjugates in mammals and is involved in cell-cell recognition and cell adhesion processes. Plant Lewisa antigens result from the addition of fucose and galactose residues by a β(1, 3)-galactosyltransferase (β(1, 3)-GalT) and an α(1, 4)-fucosyltransferase (α(1, 4)-FucT) on terminal N-acetylglucosamine residues of complex-type N-glycans. The study of the substrate specificity of the α(1, 4)-FucT has shown that this enzyme specifically transfers fucose from GDP-fucose to Gal β1-3GlcNAc (Crawley et al., 1989; Fitchette-Lainé et al., 1997; Melo et al., 1997).

Among all glycosidases and glycosyltransferases involved in the processing of plant N-glycans (Figure 10), four genes have been cloned to date: α-Man I from soybean (Nebenführ et al., 1999), GnT I from tobacco (Strasser et al., 1999), *A. thaliana* (Bakker et al., 1999) and potato (Wenderoth and von Schaewen, 2000), α(1, 3)-FucT from mung bean (Leiter et al., 1999) and β(1, 2)-XylT from *A. thaliana* (AF272852; AJ272121; AJ277603). All these enzymes are type II membrane proteins as observed for Golgi glycosyltransferases from yeast and mammals. A Golgi localization was shown for soybean α-Man I (Nebenführ et al., 1999) and tobacco GnT I (Essl et al., 1999). GnT I (Strasser et al., 1999) and α(1, 3)-FucT (Leiter et al., 1999), expressed in insect cells using a baculovirus vector, show the expected specificity for N-glycan substrates. Furthermore, GnT I cDNA was found to be able to complement the N-acetylglucosaminyl-transferase I deficiencies in CHO Lec1 cells (Bakker et al., 1999) or in the *A. thaliana* cgl mutant (Wenderoth and von Schaewen, 2000).

1.2.3 Post-Golgi Modifications of Plant N-glycans

After maturation in the ER and the Golgi apparatus, complex-type N-glycans can be further modified during the glycoprotein transport to, or in, the compartment of its final destination, leading to N-glycans that are representative of the final glycoprotein localization within the plant cell (Figure 11). For instance, most vacuolar glycoproteins and seed storage glycoproteins, described so far, were found to be N-glycosylated with truncated N-glycans containing fucose and/or xylose residues but devoid of terminal glucosamine residues. As mentioned earlier, we propose to name paucimannosidic-type N-glycans these typical vacuole-type oligosaccharide side chains (Figure 9c). Given the fact that the presence of terminal glucosamine residues on N-glycans is a prerequisite for the transfer of the α(1, 3)-fucose and the β(1, 2)-xylose residues, paucimannosidic-type N-glycans can only result from post Golgi modifications occurring on complex-type N-glycans. The analysis of the maturation of vacuolar lectins, i.e. phytohemagglutinin and phaseolin, has shown that the terminal glucosamine residues attached to the complex-type N-glycan of these glycoproteins are removed during the transport to the vacuole, or within this compartment, by the action of a N-acetylglucosaminidase giving rise to truncated N-glycans as shown in Figure 11 (Vitale and Chrispeels, 1984; Sturm et al., 1987b). Paucimannosidic-type N-glycans linked to vacuolar glycoproteins could also result from the degradation of larger Lewisa-containing N-glycans by successive action of exoglycosidases either in a prevacuolar compartment or in the vacuole itself (Fitchette et al., 1999). Moreover, in some vacuolar glycoproteins, additional mannose residues can be hydrolyzed by vacuolar α-mannosidases yielding truncated N-glycans containing two mannose residues or even only the β-mannose of the core (Kimura et al., 1999). By a similar process, hybrid $Man_4Xyl(Fuc)GlcNAc_2$ and $Man_5Xyl(Fuc)GlcNAc_2$ (Kimura et al., 1988; Altmann et al., 1998; Shimazaki et al., 1999; van Ree et al., 2000) could also result from the degradation in the vacuole of the intermediate hybrid-type N-glycan described in Figure 10. In contrast to vacuolar glycoproteins, extracellular glycoproteins were found to be N-glycosylated mostly by complex-type N-glycans including Lewisa-containing oligosaccharides as illustrated in Figure 11 (Fitchette et al., 1999). The complex oligosaccharides may also be partially degraded by exoglycosidases in the extracellular compartment (Tezuka et al., 1992) but the rate of degradation appears to be lower than observed in the vacuole.

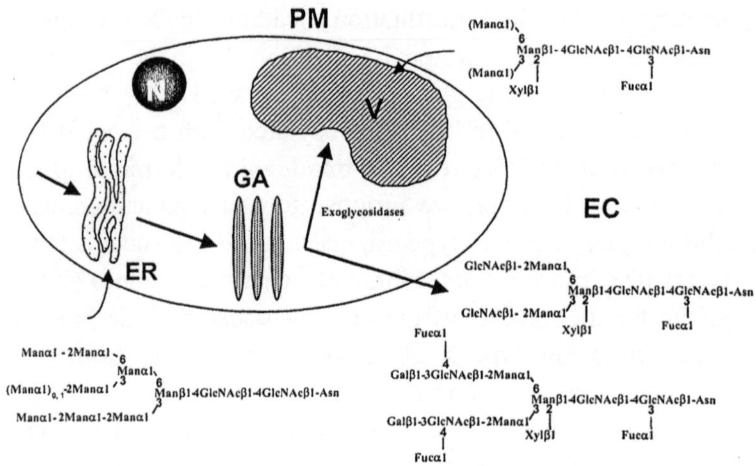

Figure 11. Distribution of N-glycan structures in the plant cell. EC: extracellular compartment; ER: Endoplasmic reticulum; GA: Golgi apparatus; N: nucleus; PM: plasma membrane; V: vacuole.

1.3 N-glycosylation and Folding of Proteins

The properties of a protein, such as stability, solubility, folding and biological activity, could be affected by glycosylation. As developed in the previous section, the N-glycosylation pattern of a mature plant glycoprotein results from both co-translational transfer, in the ER, of oligosaccharide precursors onto specific Asn residues and from processing of these precursors in the ER and in the Golgi apparatus. Different factors, such as the polypeptide folding or position in the polypeptide sequence, could influence the use of potential N-glycosylation sites in the ER. As a consequence, when multiple N-glycosylation sites are present in a protein sequence, some sites could be glycosylated inefficiently, giving rise to various glycoforms, while some others are not used at all. For instance, one of the two glycosylation sites of the bean storage protein phaseolin is partially used during the biosynthesis. As a consequence, mature phaseolin bears either one or two oligosaccharide side chains (Bollini et al., 1983).

After co-translational transfer of the oligosaccharide precursor, the folding of the protein continues with the assistance of ER-resident molecular chaperones such as calreticulin and calnexin, which interact specifically with partially trimmed monoglucosylated N-linked oligosaccharide side chains (Crofts et al., 1998). When a glycoprotein leaves the ER, its final folding results from the use of its potential N-glycosylation sites, not only because oligosaccharide chains N-linked to the nascent polypeptide have an energetic contribution in directing the folding of the protein, but also because glycans are involved in the interaction of not-yet-folded glycoproteins with specific ER chaperones.

In the Golgi apparatus, the processing of the glycan side-chains is related to their accessibility to the Golgi enzymes. Glycans located on the protein surface will be matured into complex-type N-glycans, while other oligosaccharide side chains buried in the protein will stay unmodified as high-mannose-type N-glycans (Faye et al., 1986; Kimura et al., 1999). As a consequence, both the distribution of oligosaccharide side chains on the protein backbone and the maturation level of glycans on each glycosylation site are closely related to the folding of the mature glycoprotein.

1.4 Complex N-glycans Are Ubiquitous among the Plant Kingdom

To investigate whether the structures of N-linked glycans are conserved among the plant kingdom, we have recently analyzed the N-glycan patterns of glycoproteins isolated from various mono- and dicot plants. Western blot analysis of different plant extracts has shown that α(1, 3)-fucose and the β(1, 2)-xylose (Faye and Chrispeels, 1998) glyco-epitopes as well as Lewisa antigen (Fitchette et al., 1999) are highly conserved glycan sequences within the plant kingdom. Regarding the latter glycan sequence, the only exceptions are members of the cruciferae and particularly *A. thaliana* which does not show any response to anti-Lewisa antibodies (Rayon et al., 1999; Fitchette et al., 1999). Structural analysis of matured N-glycans prepared from the glycoproteins of *A. thaliana* has confirmed that the absence of Lewisa sequences results in the biosynthesis of a limited N-glycan pattern (Rayon et al., 1999) (Figure 12). In contrast, as illustrated in Figure 12, structural analysis of total protein extracts isolated from tobacco (Fitchette et al., 1999; Bakker et al., 2000), as well as from maize, onion and pea (unpublished results) show that the glycoproteins from these plants harbour a large variety of complex N-glycans ranging from truncated paucimannosidic N-glycans to highly matured oligosaccharides bearing two Lewisa epitopes. However, despite such a large variety of oligosaccharide chains, no significant structural differences were observed between the N-glycan patterns of plants analyzed to date.

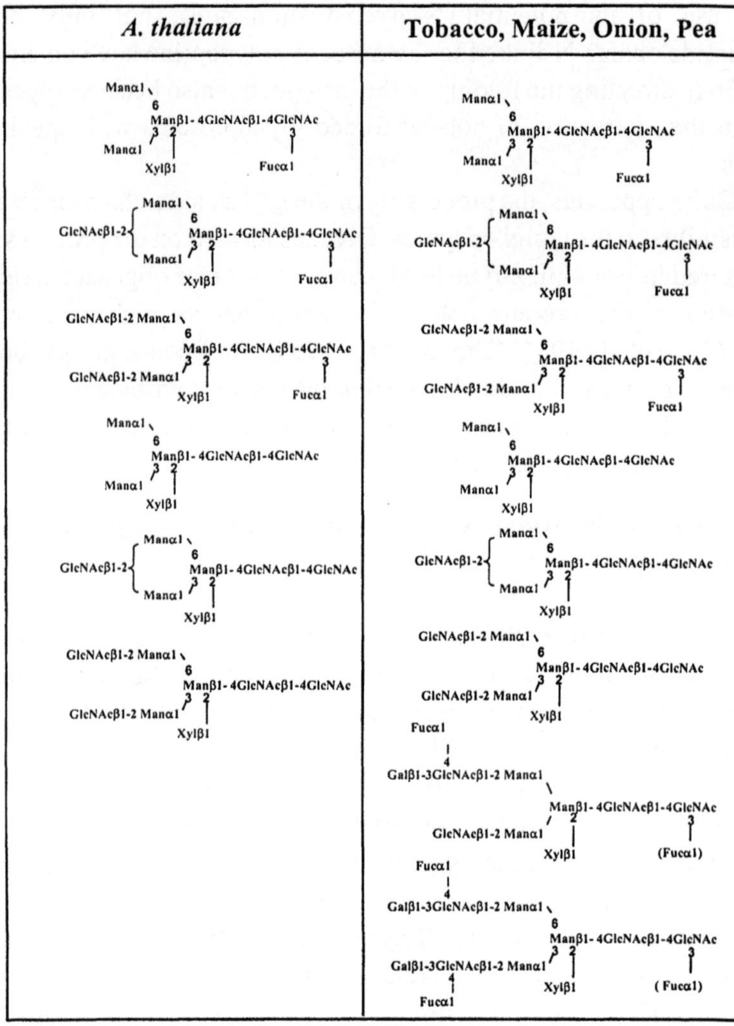

Figure 12. Structure of the most abundant matured – linked glycans isolated from plant glycoproteins.

1.5 Structural Analysis of Plant N-glycans

As described in section 1.3, both the distribution of oligosaccharide side chains on the protein backbone and the maturation level of glycans on each glycosylation site are closely related to the folding of the mature glycoprotein. Thus, the identification of the N-glycosylation pattern of a plant glycoprotein not only necessitates the elucidation of the oligosaccharide structures attached to this protein, but also includes the determination of their distribution on potential N-glycosylation sites of the polypeptide sequence. Several on-blot, chromatographic and spectroscopic techniques have been developed for fast and accurate characterization of plant N-linked glycans isolated from low amounts of material. Analytical tools used for the identification of plant N-glycan patterns are summarized in the following sections.

1.5.1 Affino- and Immunoblotting of N-linked Glycoproteins

Fast preliminary analysis of the N-glycosylation of plant glycoproteins can be achieved by affino- and immunodetection on blots using glycan-specific probes. Evidence for the presence of high-mannose-type N-glycans can be deduced by affinodetection with concanavalin A (ConA), a lectin which specifically binds to mannose sequences typical of high-mannose-type N-glycans (Faye and Chrispeels, 1985). On the other hand, immunodetection of modified glycans N-linked to a protein can be carried out with antibodies specific for plant-specific glycan epitopes (Figure 13), i.e. antibodies specific for core α(1, 3)-fucose or for core β(1, 2)-xylose residues (Faye et al., 1993) and antibodies specific for the Lewisa antigen (Fitchette-Lainé et al., 1997; Fitchette et al., 1999). These antibodies, that are specific for plant carbohydrate decorations, have been raised in rabbits. In addition to the affino- and immunodetection on blots, sensitivities of the glycoprotein to specific deglycosylating enzymes could afford additional data about N-glycan patterns. Endoglycosidase H (Endo H) is only able to release high-mannose-type N-glycans from plant glycoproteins by hydrolyzing the glycosidic bond between the two GlcNAc residues of the chitobiose unit. Peptide N-Glycosidases (PNGase) hydrolyze the Asn-GlcNAc bond of both high-mannose and complex N-glycans. However, PNGase F, which is widely used for analysis of mammalian glycoproteins, is active on complex plant N-glycans except those having an α(1, 3)-fucose residue linked to the proximal GlcNAc. As a consequence, information about the structure of glycans N-linked to glycoproteins (i.e. the presence of high-mannose-type or of non-fucosylated N-glycans) can be deduced from the sensitivity of a glycoprotein to these enzymes. This sensitivity is easily determined after enzymatic treatment either from an increased electrophoretic mobility or from a loss of reactivity of the glycoprotein on blots with glycan-specific probes.

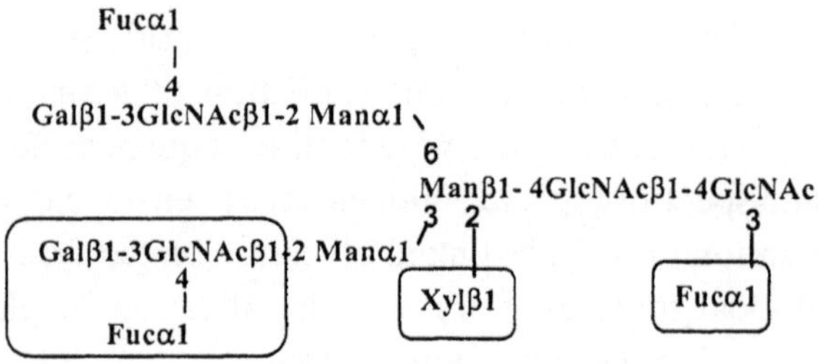

Figure 13. Encircled glycan sequences correspond to epitopes recognized by antibodies used for immunoblotting analysis of plant glycoproteins.

1.5.2 Chromatography and Electrophoresis of Plant N-linked Glycans

The determination of the N-glycosylation pattern of a plant glycoprotein, as well as of a glycopeptide, can be carried out by chromatographic analysis of N-linked oligosaccharides released from the peptide backbone by Endo H or PNGase (A or F) treatments (reviewed in Bardor et al., 1999a). If the plant glycoprotein contains both high-mannose-type and complex-type N-glycans, PNGase A treatment should be recommended since this enzyme is able to release all plant N-linked oligosaccharides. The resulting N-linked oligosaccharides can be directly separated by High-pH Anion Exchange Chromatography with a Pulsed Amperometric Detection (HPAEC-PAD) which allows the separation of closely related N-glycans and the visualization by amperometric detection of the reducing glycans without any prerequisite derivatization of the oligosaccharide mixture (Figure 14b). For detection by fluorescence, the reducing oligosaccharides must be labelled by reductive amination with 2-aminopyridine or 2-aminobenzamide before separation either by gel permeation or chromatography on amide or reverse phase columns. The structural identification of N-linked glycans by chromatography can be carried out by comparison of the retention times with standard glycans of known structures. Moreover, N-glycans can be isolated by chromatography and then characterized by exoglycosidase sequencing or Nuclear Magnetic Resonance.

Recently, an electrophoretic technique, the Fluorophore-Assisted Carbohydrate Electrophoresis (FACE), has been described for oligosaccharide analysis, in which reducing oligosaccharides released from the glycoprotein are labelled by reductive amination with the negatively charged fluorophore 8-amino-naphtalene-1, 3, 6-trisulfonic acid (ANTS). The resulting fluorescent derivatives are separated according to their hydrodynamic size with high resolution by polyacrylamide gel electrophoresis (Jackson, 1994; Starr et al., 1996). The electrophoretic mobilities of ANTS-labelled N-glycans are estimated and expressed in terms of glucose equivalents relative to a standard ladder of glucose oligomers. One major advantage of FACE is that experiments can be carried out using classical laboratory equipment such as electrophoresis supplies and a UV light box available in all laboratories equipped for protein and nucleic acid electrophoretic separations. We have recently adapted this electrophoretic approach to the analysis of plant N-linked glycans and shown that FACE allows plant N-glycosylation analysis or plant glycoprotein screening to be routinely performed in any non glyco-specialised laboratory (Bardor et al., 2000). As reported for recombinant glycoproteins produced in mammalian cells (Friedman and Higgins, 1995; Masada et al., 1996), FACE analysis will constitute a well-adapted method for rapid batch-to-batch quality control of the glycosylation of the recombinant proteins produced in plants.

For illustration of both chromatographic and electrophoretic analysis of plant N-linked glycans, the N-glycosylation pattern of the isolectin L of the bean phytohemagglutinin (PHA-L) was investigated by HPAEC-PAD and FACE (Figure 14). PHA-L was previously reported to be N-glycosylated by two N-glycan side chains: one paucimannosidic-type N-glycan located on Asn-60 and a mixture of high-mannose-type N-glycans from $Man_6GlcNAc_2$ to $Man_9GlcNAc_2$ located on Asn-12 (Sturm and Chrispeels, 1986; Sturm et al., 1992; Rayon et al., 1996). The profiles obtained by HPAEC-PAD and FACE for the N-glycans isolated from PHA-L are shown in Figures 14b and c. The retention times on HPAEC, as well as the electrophoretic mobilities on FACE, of oligosaccharides were assigned by comparison with standards of high-mannose-type N-glycans and of $Man_3XylFucGlcNAc_2$. Furthermore, in both the chromatography and the electrophoresis, the relative abundance of the different oligosaccharides can be estimated by quantification of the peak and band intensities. Taking advantage of the efficiency in the HPAEC-PAD separation of N-linked oligosaccharides from PHA-L, we have recently compared the N-glycosylation profiles of PHA-L expressed in different plant expression systems and in different organs of transgenic tobacco plants (Rayon et al., 1996, 1998)

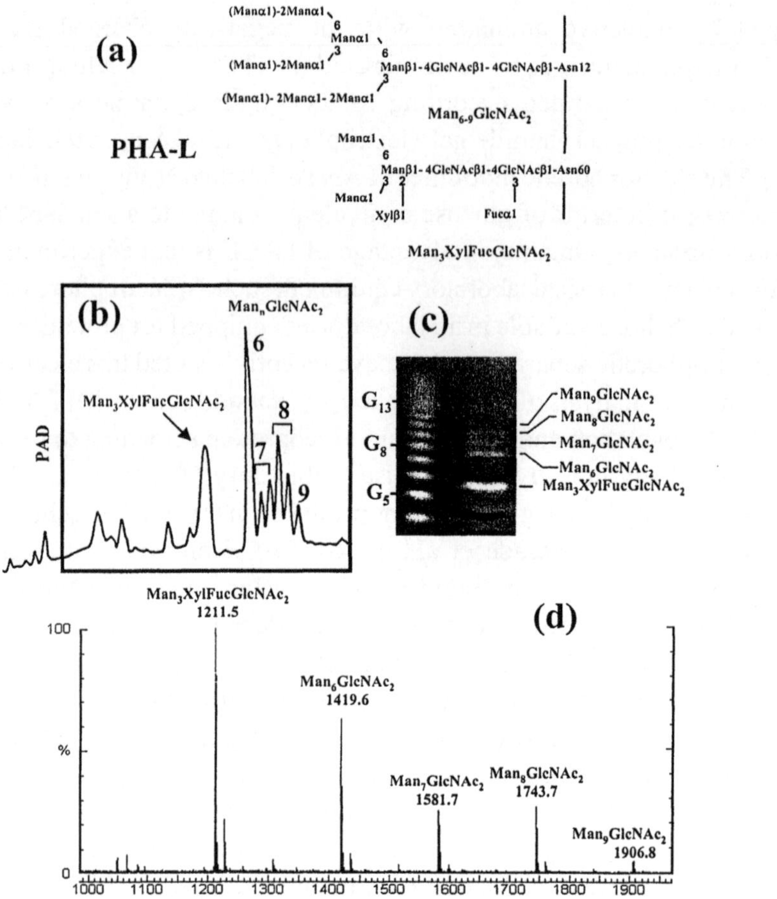

Figure 14. Structure (a), HPAEC-PAD profile (b), FACE analysis (c) and MALDI-TOF mass spectrum (d) of glycans N-linked to PHA-L.

1.5.3 Mass Spectrometry Analysis

Soft ionization mass spectrometry techniques, such as electrospray ionization (ESI) (Annan et al., 1997) and Matrix-Assisted Laser Desorption Ionization-Time of Flight (MALDI-TOF) (Kaufmann, 1995; Harvey, 1999), are major tools for characterization of the N-glycosylation patterns of glycoproteins from various origins. These techniques allow the determination of the native molecular mass, as well as the elucidation of the oligosaccharide structure and their localization on the protein sequence. Taking advantage of these techniques, the characterization of the N-glycan patterns of plant glycoproteins have been reported (reviewed in Bardor et al., 1999a). As an illustration, Figure 14d shows the MALDI-TOF mass spectrum of the mixture of N-glycans released from PHA-L. Paucimannosidic-type N-glycan $Man_3XylFucGlcNAc_2$ and high-mannose-type N-glycans ranging from $Man_6GlcNAc_2$ to $Man_9GlcNAc_2$ can be easily identified in the picomolar range without any chemical derivatization (Bardor et al., 1999b). Such spectra are often sufficient for elucidation of the oligosaccharide structures N-linked to plant glycoproteins. However, when the measured molecular ions correspond to different possible structural isomers, further information on the N-glycan structure can be obtained by tandem mass spectrometry after chemical derivatization of the oligosaccharide mixture (Melo et al., 1997; Rayon et al., 1999). For instance, tandem mass spectrometry of permethylated plant complex-type N-glycans affords non ambiguous indications on the localization of fucose residues on the oligosaccharide sequence, i.e. the α(1, 3)-fucose residue located on the proximal glucosamine residue and the α(1, 4)-fucose residue constitutive of terminal Lewisa (Figure 9b) (Melo et al., 1997).

The determination of the site-specific distribution of N-linked glycans on the polypeptide sequence can be directly deduced by mass spectrometry analysis of glycopeptides generated by endoprotease digestion of the glycoprotein. The N-glycosylation pattern can be established by directly analyzing the digest by liquid chromatography coupled to ESI mass spectrometry (Melo et al., 1997). The oligosaccharide distribution can be also determined after isolation by HPLC of the different endoprotease-generated glycopeptides and analysis by mass spectrometry of N-glycans released from these glycopeptides (Oxley and Bacic, 1995; Gray et al., 1996; Bardor et al., 1999b).

2. N-GLYCOSYLATION OF THERAPEUTIC GLYCOPROTEINS PRODUCED IN TRANSGENIC PLANTS: CURRENT ASPECTS AND FUTURE TRENDS

2.1 N-glycan Patterns of Therapeutic Glycoproteins

Produced in Transgenic Plants

Most mammalian proteins of therapeutic interest are N-glycosylated. As a consequence, when these proteins are expressed in plants, it becomes crucial to examine the N-glycosylation pattern introduced by the plant on recombinant glycoproteins. Preliminary analysis of the N-glycosylation of plant recombinant glycoproteins can be achieved on a few micrograms of purified material by affino- and immunodetection on blots using glycan-specific probes as mentioned in section 1.5.1. This approach was successfully applied to the preliminary characterization of the N-glycan pattern of plant-derived pharmaceutical glycoproteins (Zeitlin et al., 1998; Cabanes-Macheteau et al., 1999). As an illustration, Figure 15 shows the on-blot analysis of glycan N-linked to the dog gastric lipase produced in transgenic tobacco. In this study, we have analyzed the N-glycosylation of the recombinant gastric lipase fused to targeting signals and expressed in tobacco cells. According to the specificity of targeting signals, the recombinant protein was addressed and stored in different subcellular compartments, i.e. the extracellular compartment or the vacuole. As shown in Figure 15, the recombinant lipase was found to be N-glycosylated by both high-mannose-type and matured plant N-glycans. However, as discussed in section 1.2.3, the gastric lipase bears complex glycans having Lewisa epitopes only when the protein is secreted but not when it is accumulated in the plant vacuole. To investigate the detailed N-glycan pattern of therapeutic glycoproteins produced in transgenic plants, we have recently compared the structure of glycans N-linked to monoclonal antibodies produced in hybridoma cells and in transgenic tobacco plants (plantibodies). The first study was carried out on the Guy's 13 monoclonal antibody in collaboration with J. Ma (Guy's Hospital, London, UK) (Cabanes-Macheteau et al., 1999). As expected from the literature data (Raju et al., 2000), we have identified, attached to the mouse antibody, partially sialylated bi-antennary core $\alpha(1,6)$-fucosylated N-glycans. From the plantibody, we have then characterized an heterogeneous array of N-linked oligosaccharides including high-mannose-type N-glycans and complex N-glycans having core $\beta(1,2)$-xylose and core $\alpha(1,3)$-fucose residues.

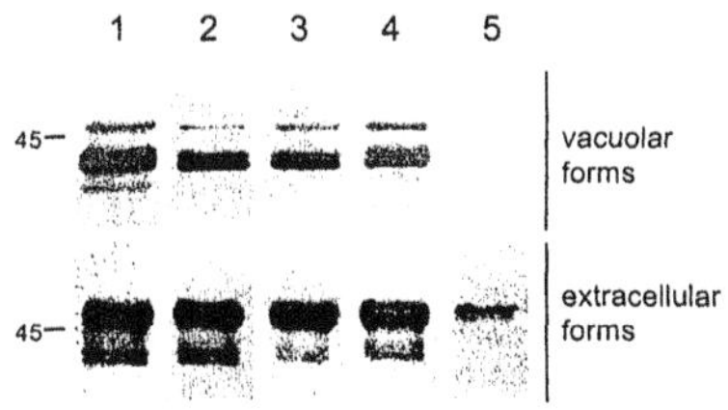

Figure 15. Immuno- and affinodetection of the gastric lipase expressed in transgenic tobacco plants. 1) Silver staining, 2) Immunodetection with antibodies specific for the gastric lipase, 3) Affinodetection of high-mannose-type N-glycans with ConA, 4)Immunodetection with antibodies specific for α(1, 3)-fucose and β(1, 2)-xylose epitopes and 5) Immunodetection with antibodies specific for the Lewis (a) sequence.

A second study carried out in collaboration with H. Bakker, D. Bosch and W. Jordi (Plant Research International, Wageningen, Holland), on the Mgr48 monoclonal IgG1, led to the same conclusions, confirming that expression of immunoglobulins in plants results in the production of plantibodies harbouring plant-specific N-linked glycans (Bakker et al., 2000). In both samples, the minor population of high-mannose-type glycans N-linked to the plantibodies was assumed to arise from the population of immature plantibodies not yet secreted by the tobacco plant cells.

2.2 Immunogenicity and Allergenicity of Plant N-glycans

Based on the data on the N-glycan processing in plants reported in section section 1.2, the presence of β(1, 2)-xylose and α(1, 3)-fucose residues on the N-glycans of the plantibodies was not surprising (Figure 10). These structural differences in the glycosylation do not affect the biological activity of plantibodies. However, glyco-epitopes including the β(1, 2)-xylose and α(1, 3)-fucose residues, that are absent in mammals, are known to be highly immunogenic in animals (Faye and Chrispeels, 1985; Faye et al., 1993).

Structural analysis of plant allergens have shown that complex-type plant N-glycans are frequently observed on pollen grain and food allergens of plant origin. Furthermore, these oligosaccharides have been reported to be included in the IgE-determinants of these plant allergens (Aalberse et al., 1981; Kurosaka et al., 1991; Garcia-Casado et al., 1996; Wilson et al., 1998; Fötisch et al., 1999; van Ree et al., 2000). In most allergic reactions, patients have IgE antibodies directed to ubiquitous plant glycans, i.e. the α(1, 3)-fucose and/or the β(1, 2)-xylose epitopes (Wilson et al., 1998; van Ree et al., 2000) leading to cross-reactivity between several foods and pollen allergens (Aalberse et al., 1981; Faye and Chrispeels, 1998; van Ree and Aalberse, 1993). Furthermore, the α(1, 3)-fucose residue also occurs in glycan N-linked to insect glycoproteins. The widespread occurrence of such glycoepitopes contributes to immunological cross-reactions among non-related glycoproteins and have been shown to be the major cause of cross-reactivities between environmental allergens (Aalberse et al., 1981; Faye and Chrispeels, 1985).

The complex glycans N-linked to plantibodies are not novel for humans who are exposed daily to such antigens in edible plant material. In this respect, for some therapeutic applications, the presence of plant complex-type glycans N-linked to recombinant glycoproteins will not introduce any drawback. For instance, the presence of plant-specific oligosaccharides on the plant Guy's 13 antibody, which is specific of a cell-surface protein of *Streptococcus mutans*, is not a limitation in the use of this antibody for oral applications to protect humans against dental caries. This was recently demonstrated by a completed clinical trial with the plant-produced Guy's 13 antibody for the prevention of oral colonization by *S. mutans* (Ma et al., 1998). Moreover, a monoclonal antibody produced in plants was found to be highly efficient in vivo for prevention of vaginal anti-herpes simplex virus 2 infection in mouse, without inducing safety problems or without eliciting an adverse immunological response (Zeitlin et al., 1998). In vivo assays using an antibody produced in alfalfa have shown that the plantibody is as stable as the corresponding hybridoma antibody in the blood stream of mice (Khoudi et al., 1999). All these preliminary data indicate that plant-derived antibodies are suitable and inexpensive products for therapeutic applications in humans. However, if people have prolonged exposure to large quantities of highly immunogenic plant N-glycans that are attached to these plantibodies, as may be required for certain in vivo therapies, elicitation of immune responses in humans by specific plant glyco-antigens may occur.

2.3 Metabolic Engineering of the Glycosylation in Plants to Obtain a Perfect Copy of Mammalian N-glycans

The presence of immunogenic glycoepitopes on glycoproteins produced in transgenic plants is considered as a major limitation for their use in human therapy. Recent results described in section 1.4 demonstrate that immunogenic plant glycoepitopes are ubiquitous in plants and thus changing a plant expression system for another will not resolve the problem. In contrast, the metabolic engineering of the plant N-glycosylation appears as the most promising way to erase the structural differences that exist between plant and mammal N-glycans. One of the advantages of plants as production systems for recombinant mammalian glycoproteins is our good knowledge of the protein N-glycosylation machinery in plants. This allows us to define strategies to produce humanized recombinant proteins with more mammalian-like N-glycans. To date, different strategies emerge either to prevent the formation of highly immunogenic plant N-glycans on recombinant proteins, or to produce plant glycoproteins harbouring mammalian-like N-glycans. These strategies are summarized in Figure 16 and described in the following sections.

2.3.1 Retention in the Endoplasmic Reticulum

The first developed strategy for obtaining recombinant glycoproteins with non-immunogenic N-glycans is to store the protein in the ER. It was shown previously that the addition of H/KDEL sequences at the C-terminal end of a recombinant protein is sufficient to get it efficiently retained in the plant ER (Gomord et al., 1997). In addition, it was observed for many recombinant proteins that the ER is the compartment where they have the higher stability in the plant secretory pathway. Preliminary information on the ER glycosylation in plants was recently obtained from the analysis of natural proteins that reside in the ER (reticuloplasmins) such as calreticulins. In maize (Pagny et al., 2000) and spinach (Navazio et al., 1996), this ER chaperone has high-mannose-type N-glycans containing 8-9 mannose residues (Figure 9), structures which exclusively result from an ER processing as illustrated in Figure 11. These structures are common to plants and mammals and probably non immunogenic. However, we have recently shown that, in contrast with natural reticuloplasmins, ER retention of recombinant proteins fused with HDEL exclusively relies upon the efficiency of the recycling machinery between the ER and the Golgi apparatus (Pagny et al., 2000). Consistent with their recycling from the Golgi apparatus back to the ER, we have observed that recombinant glycoproteins fused with HDEL exhibit N-glycan modifications, such as α(1, 3)-fucosylation and β(1, 2)-xylosation, that are known to occur in the Golgi apparatus (Pagny et al, 2000). Hence, expression in plants of HDEL-fused recombinant glycoproteins can no longer be considered as a reliable way to produce in transgenic plants mammalian glycoproteins bearing non-immunogenic oligosaccharides.

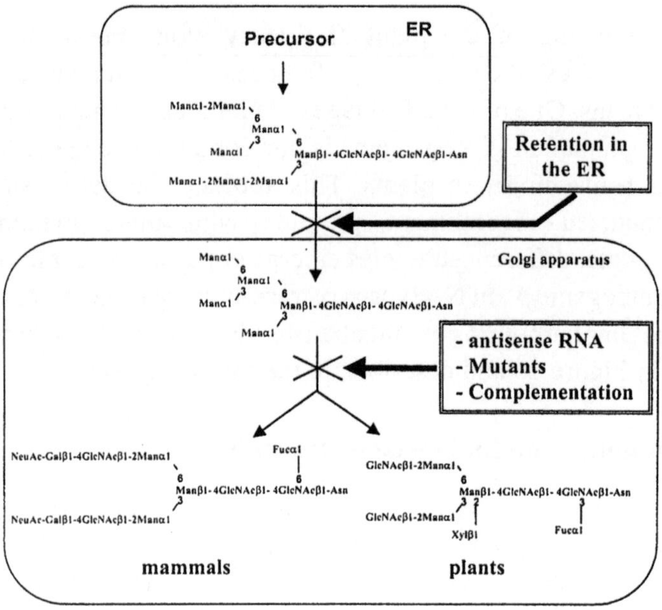

Figure 16. Strategies to metabolically engineer the N-glycosylation in plants.

2.3.2 Expression in Plant Mutants

A second strategy to produce non-immunogenic recombinant glycoproteins in plant cells is to use plant mutants that are affected in the N-glycan biosynthesis pathway. Although a rapidly increasing number of mutants are available in several plant species, and particularly in *A. thaliana*, only two of them show a clearly identified mutation affecting the biosynthesis of N-linked glycans. The first one, the *mur1* mutant is affected in the gene encoding for a GDP-D-mannose-4, 6-dehydratase, an enzyme involved in the biosynthesis of L-fucose (Bonin et al., 1997), and does not synthesize L-fucose (Reiter et al., 1993). This mutant was identified by screening chemically mutagenized *A. thaliana* plants for changes in the monosaccharide composition of cell wall material. We have analyzed the structures of N-linked oligosaccharides in the *mur1* mutant and found that L-fucose is partially replaced by L-galactose (Rayon et al., 1999). This demonstrates that, in absence of L-fucose, the α(1, 3)-fucosyltransferase is able to partially transfer L-galactose from GDP-L-galactose, instead of fucose from GDP-L-fucose, to the proximal N-acetyl glucosamine residue of the core. However, about 95% of complex glycans were found to be non fucosylated as represented in Figure 17. Another *A. thaliana* mutant, the cgl mutant which lacks the N-acetylglucosaminyltransferase I (GnT I) activity, is unable to synthesize complex-type N-glycans and accumulates $Man_5GlcNAc_2$, a N-glycan structure common to plants and mammals (Figure 17) (von Schaewen et al., 1993). This mutant was identified by screening chemically mutagenized *A. thaliana* plants for absence of detection with antibodies specific for α(1, 3)-fucose/β(1, 2)-xylose glyco-epitopes. The *cgl* mutant plants are able to normally complete their development and show no phenotype which indicates that complex-type N-glycans are not essential for plants. In the future, efforts have to be made to screen other mutagenized plant species to isolate N-glycan mutants that are more adapted to high scale production of therapeutic glycoproteins.

2.3.3 Antisense RNA Expression of Glycosidases or Glycosyltransferases

Knocking out specific Golgi enzymes by an antisense strategy appears as an alternative way to obtain plants with a modified enzymatic machinery for N-glycan maturation. As an illustration, GnT I is a key glycosyltransferase in the maturation of precursor oligosaccharides into complex-type N-glycans (Figure 10). This enzyme appears as a major target for knocking out strategies. GnT I cDNA was recently cloned from tobacco (Strasser et al., 1999), *A. thaliana* (Bakker et al., 1999) and potato (Wenderoth and von Schaewen, 2000) and first attempts of a GnT I knocking out approach were carried out by two groups. First results reported by Steinkellner (1999) have shown that a reduction by up to 70% of GnT I activity has little effect on the level of N-glycan xyloxylation or fucosylation in transgenic tobacco plants. More recently, Wenderoth and von Schaewen (2000) have confirmed that expression of GnT I antisense contructs in transgenic tobacco plants have a detectable but limited effect on the level of N-glycan maturation. In addition to GnT I, cDNA of three other N-glycan processing enzymes, α-Man I (Nebenführ et al., 1999), α(1, 3)-FucT from mung bean (Leiter et al., 1999) and β(1, 2)-XylT from *A. thaliana* (AF272852; AJ272121; AJ277603), were recently cloned. These enzymes also appear as good candidates for antisense strategies to block the N-glycan processing and prevent the formation of highly immunogenic plant N-glycans on recombinant proteins (Figure 10).

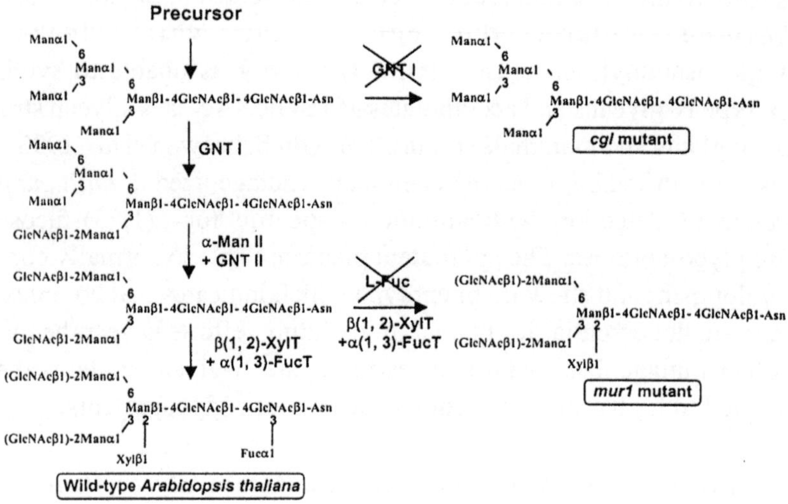

Figure 17. Processing of the N-linked glycans in wild-type *A. thaliana*, and in the cgl and the mur1 mutant.

2.3.4 Reducing Differences Between Plant and Mammalian N-glycans: Complementation with Human Glycosyltransferases.

As discussed in previous sections, the carbohydrate side chains play crucial roles in protein folding and antigenicity of glycoproteins. In mammals, in addition to these two roles, N-linked glycans are also known to be involved in numerous other functions such as secretion and cell-cell interaction. For that, specific N-glycan sequences are required for the functionality of mammalian glycoproteins. For instance, the presence of terminal β(1, 4)-galactose residues on glycans N-linked to immunoglobulins is crucial as these residues play a role in the effector function of these molecules (Boyd et al., 1995; Wright and Morrisson, 1997). Furthermore, the presence of terminal sialic acids on N-glycans is important for the in vivo half life of most mammalian circulatory glycoproteins. As a consequence, for certain applications, metabolic engineering of the N-glycosylation of plants producing therapeutic glycoproteins must be carried out not only to suppress the immunogenicity of the N-glycans but also to achieve the right N-glycosylation. To this end, in contrast to strategies developed in section 2.3.2 and 3, the plant N-glycosylation must not be restricted by blocking one of the key steps of the N-glycan processing, but must be extended to acquire the appropriate glycan sequence.

Getting extended N-glycan patterns can be achieved by introducing in plants mammalian glycosyltransferases in order to switch the plant endogenous N-glycan processing to the synthesis of mammalian-like structures. Some mammalian glycosyltransferases have been successfully expressed in plants, such as GnT I (Gomez and Chrispeels, 1994) or α(2, 6)-sialyltransferase (Wee et al., 1998). These enzymes have been shown to be correctly targeted to the Golgi apparatus (Wee et al., 1998) and to be active when produced in the secretory pathway of plant cells (Gomez and Chrispeels, 1994; Wee et al., 1998). These results have illustrated that the targeting signals of Golgi enzymes are probably relatively well-conserved between plants and mammals and that transformation of plants with mammalian glycosyltransferases may result in the production of glycan-humanized glycoproteins, assuming that nucleotide-sugars and glycan substrates are available in the plant cell. Preliminary attempts to engineer plant N-glycans were recently reported by Palacpac et al. (1999) who have expressed the human β(1, 4)-galactosyltransferase (β(1, 4)-GalT) in tobacco BY2 cells. This glycosyltransferase is able to transfer galactose residues onto terminal N-acetylglucosamine residues of intermediate N-linked glycans (Figure 18). Compared to N-glycans isolated from wild-type tobacco cells, about 40% of oligosaccharides were found to be modified after transformation with the human β(1, 4)-GalT into partially humanized hybrid structure (Figure 18). In collaboration with our colleagues from the International Plant Research Center in Wageningen, we have recently expressed the human β(1, 4)-GalT in tobacco plants and compared the structure of N-glycans from the transformed and non-transformed plants. We found that about 15% of complex N-glycans in the transformed plants were galactosylated. Despite the fact that through the modified glycosylation machinery, numerous proteins have acquired unusual N-glycans with terminal β(1, 4)-galactose residues, no obvious changes were observed in the physiology of these transgenic plants. As illustrated in Figure 18, modified oligosaccharides isolated from transformed tobacco plants expressing human β(1, 4)-GalT were found to be mono- and di-galactosylated and to bear β(1, 2)-xylose and α(1, 3)-fucose residues.

β(1, 4)-GalT is a N-glycan processing enzyme which is located in the median and trans Golgi apparatus of mammalian cells (Evans et al., 1995). These two Golgi compartments were found to also contain the β(1, 2)-xylosyl- and α(1, 3)-fucosyltransferases in the plant cell (Fitchette-Lainé et al., 1994). Modified N-glycans, isolated from tobacco plants transformed with β(1,

4)-GalT, have both β(1, 4)-galactose, β(1, 2)-xylose and α(1, 3)-fucose residues which indicates that, in tobacco plants, the β(1, 4)-GalT is probably properly located in the plant Golgi apparatus and acts in competition with endogenous enzymes on a common $GlcNAc_2Man_3GlcNAc_2$ substrate (Figure 18). In contrast, the major β(1, 4)-galactose-containing N-glycan formed in BY2 tobacco cell expressing human β(1, 4)-GalT is an hybrid structure resulting from the action of human β(1, 4)-GalT onto $GlcNAcMan_5GlcNAc_2$ (Palacpac et al., 1999). This indicates that β(1, 4)-GalT acts earlier in the Golgi apparatus in BY2 cells than observed in tobacco plants. As the presence of at least one terminal N-acetylglucosamine residue is a prerequisite for the action of α-Man II and for the transfer of α(1, 3)-fucose and β(1, 2)-xylose onto plant N-glycans (Johnson and Chrispeels, 1987; Tezuka et al., 1992; Staudacher et al., 1995; Zeng et al., 1997), the early transfer of a terminal galactose residue onto $GlcNAcMan_5GlcNAc_2$ prevents the processing of the α(1, 6)-mannose arm and inhibits further transfer of xylose and fucose on the core. Compared to the results that we obtained (Bakker et al., 2000), the different effect of human β(1, 4)-GalT expression in BY2 cells can be related either to differences in the glycosyltranferase sub-Golgi targeting between suspension-cultured BY2 cells and tobacco plant cells or to a very high expression level of β(1, 4)-GalT in BY2 cells which was not reached in the tobacco plants.

The early action of human β(1, 4)-GalT in the Golgi apparatus of BY2 cells results in the inhibition of the transfer of xylose and fucose residues onto complex N-glycans. This demonstrates that both humanization of plant N-glycans and inhibition of the plant immunogenic glyco-epitope biosynthesis can be obtained in a single step. However, hybrid structures described by Palacpac et al. (1999) are not consistent with most glycans N-linked to therapeutic glycoproteins. In contrast, the transformed tobacco plants have a more promising galactosylation profile in a humanization context, as bi-antennary bi-galactosylated structures are synthesized. Except for the presence of α(1, 3)-fucose and β(1, 2)-xylose residues, the structures obtained in tobacco plants transformed with the human β(1, 4)-GalT are similar to human complex N-glycans and are the appropriate support for further modifications by transfer of sialic acids (see Figure 9e).Antibodies contain one conserved site of N-glycosylation located in the Fc domain of both heavy chains where are attached bi-antennary N-glycans having terminal β(1, 4)-galactose residues but low contents of sialic acid (Raju et al., 2000). The presence of the β(1, 4)-galactose on these glycans is crucial as these residues play a role in the functionality of antibodies (Boyd et al., 1995; Wright and Morrisson, 1997).

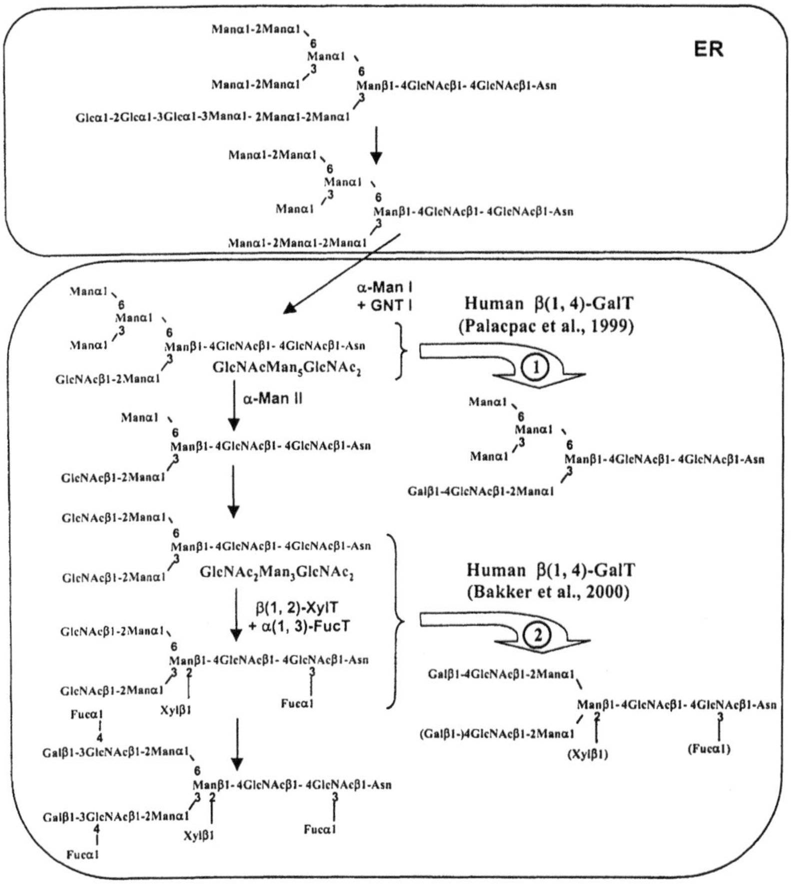

Figure 18. Humanization of plant N-glycans by expression of the humanβ(1, 4)-galactosyltransferase in tobacco BY2 cells (1) and tobacco plants (2).

We have expressed the murine Mgr48 antibody in tobacco plants transformed with human β(1, 4)-GalT cDNA (see section 2.1) and we have analyzed the structure of glycans N-linked to the plantibody produced in these conditions (Bakker et al., 2000). The analysis of the plantibody expressed in β(1, 4)-GalT transformed tobacco plants has shown that 30% of the N-linked glycans are galactosylated. With respect to the galactosylation, the profile of N-glycans in immunoglobulins produced either in hybridoma cells or in tobacco plants expressing human β(1, 4)-GalT is comparable (Figure 19). This result demonstrates the feasibility of such a complementation strategy for the production of immunoglobulins, or any other recombinant therapeutic glycoproteins, harbouring humanized glycan.

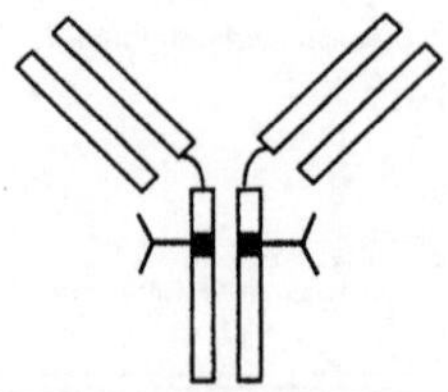

hybridoma cells	Fucα1 (Galβ1-4)GlcNAcβ1-2Manα1 \ │ 6 6 Manβ1-4GlcNAcβ1-4GlcNAc 3 GlcNAcβ1- 2Manα1 /
tobacco plants	GlcNAcβ1-2 Manα1 \ Manβ1- 4GlcNAcβ1-4GlcNAc / 2 3 GlcNAcβ1-2 Manα1 │ │ (Xylβ1) (Fucα1)
tobacco plants expressing human β(1 ,4)-GalT	(Galβ1)-4GlcNAcβ1-2Manα1 \ Manβ1- 4GlcNAcβ1-4GlcNAc / 2 3 (Galβ1)-4GlcNAcβ1-2Manα1 │ │ (Xylβ1) (Fucα1)

Figure 19. Structure of glycans N-linked to Mgr48 expressed in hybridoma cells, tobacco plants and tobacco plants expressing the human β(1, 4)-galactosyltransferase.

3. FUTURE PERSPECTIVES

Numerous efforts have to be made to reduce the differences between human and plant N-glycans and to produce plant recombinant glycoproteins having oligosaccharides that are consistent with a therapeutic use in humans. Regarding the expression in plants of human glycosyltransferases, such as β(1, 4)-GalT, higher expression levels of the human enzymes should be obtained to get fully modified oligosaccharides and to decrease the high diversity of glycan structures resulting from the incomplete action of the recombinant glycosyltransferase. Attempts to engineer in planta the plant N-glycosylation, by expression of human β(1, 4)-GalT in tobacco BY2 cells or tobacco plants, have demonstrated the feasibility of such a strategy. However, further experiments will have to be carried out to investigate whether the introduction of human glycan processing enzymes in plant Golgi apparatus is sufficient to obtain perfect copies of human N-glycans or whether both introduction of human glycosyltransferases and knock out of endogenous Golgi enzymes, that are responsible for transfers of non-human epitopes, have to be combined for a complete humanization of plant N-linked glycans.

Except for the presence of α(1, 3)-fucose and β(1, 2)-xylose residues, N-glycans identified in tobacco plants expressing the human β(1, 4)-GalT are the appropriate support for obtaining a perfect copy of mammalian glycans after addition of terminal sialic acids. Such sialic acids are missing in plants but are important for the in vivo half life of most mammalian circulatory glycoproteins. For instance, expression of erythropoietin in BY2 tobacco cells yielded a recombinant glycoprotein exhibiting the expected biological activity in vitro. However, no in vivo activity was detected, probably because the plant-derived erythropoietin is rapidly eliminated from the circulation through binding to asialoglycoprotein receptors located on the hepatocyte surface (Matsumoto et al., 1995). Based on our current knowledge, obtaining sialylated plant N-glycans by adapting the plant cell glycan maturation machinery would require the transfer of at least five different heterologous genes encoding for enzymes involved the biosynthesis of sialic acid in the cytosol and its transport into the Golgi apparatus. All these "missing" enzymes of this metabolic pathway need to be steadily expressed, correctly targeted and active in the plant cell. In this respect, obtaining recombinant glycoproteins sialylated in planta is highly challenging and reaches the upper limits of what is technically feasible today.

In vitro remodelling of plant N-glycans also appears as an alternative way to obtain plant-derived therapeutic glycoproteins having human-like N-glycans. This strategy consists in the removal of glycan epitopes by treatment with exoglycosidases and/or elongation of N-glycans by treatment with glycosyltransferases and appropriate nucleotide sugars. With regard to glycoproteins produced in transgenic plants, no attempt at enzymatic remodelling of N-glycans has been reported to date. The removal of the α(1, 3)-fucose and β(1, 2)-xylose residues attached to matured plant N-glycans can be achieved by treatment with fucosidase and xylosidase (Costa et al., 1997). However, such exoglycosidase degradations are known to be efficient on isolated oligosaccharides but much more limited on glycans N-linked to glycoproteins. On the other hand, it appears feasible to remodel plant N-glycans of therapeutic glycoproteins produced in transgenic plants by in vitro treatments with glycosyltransferases. For instance, sialylated biantennary N-glycans could be obtained by action of commercially available β(1, 4)-galactosyltransferase and sialyltransferase on recombinant glycoproteins from plant origin or by action of sialyltransferases on therapeutic glycoproteins produced in plant transformed with human β(1, 4)-GalT.

So far, most efforts have been focussed on the N-glycosylation of recombinant proteins of mammalian origin produced in transgenic plants. However, for some mammalian proteins, expression of a functional recombinant protein could also be dependent on the capacity of the plant to introduce other post-translational modifications, such as O-glycosylation, lipid anchoring, C-terminal amidation, sulphatation. Data on the extent of conservation, between plants and mammals, of these protein post-translational processes are still limited. For instance, glycans O-linked to plant glycoproteins have been characterized. However, their structures completely differ from mucin-type O-glycans found on mammalian glycoproteins such as interferons or interleukins. In addition to glycosylation, some mammalian membrane proteins are anchored in the lipid membrane via a glycosylphosphatidylinositol anchor (GPI anchor) or via acylation of specific amino acids. Such membrane protein modifications have been recently characterized in plants. However, animal membrane-anchored proteins have to be expressed in plants to investigate if molecular mechanisms controlling these protein maturations are conserved enough between plants and mammals, to obtain an efficient production of correctly matured lipid-modified recombinant proteins.

ACKNOWLEDGEMENTS

Part of this research was supported by the CNRS, University of Rouen and the European Community (BMH4-CT-97-2345 and FAIR-CT-97-3110). We also thank Dany Marfaux and Hatim Jawhari from Meristem Therapeutics (Clermont Ferrand, France) for the fruitful collaboration on gastric lipase and maize N-glycan characterization.

REFERENCES

Aalberse RC, Koshte V and Clemens JG, 1981. Cross-reactions between vegetable foods, pollen and bee venom due to IgE antibodies to a ubiquitous carbohydrate determinant. Int Arch Allergy Appl Immunol, 66: 259-260

Altmann F, 1997. More than silk and honey-or, can insect cells serve in the production of therapeutic glycoproteins? Glycoconjugate J, 14:643-6

Altmann F, Paschinger K, Dalik T and Vorauer K, 1998. Characterisation of peptide-N4-(N-acetyl-β-glucosaminyl)asparagine amidase A and its N-glycans. Eur J Biochem, 252: 118-123

Annan RS and Carr SA, 1997. The essential role of mass spectrometry in characterizing protein structure: mapping posttranslational modifications. J Protein Chem, 16:391-402

Bakker H, Lommen A, Jordi W, Stiekema,W and Bosch D, 1999. An *Arabidopsis thaliana* cDNA complements the N-acetylglucosaminyltransferase I deficiency of CHO Lec1 cells. Biochem Biophys Res Commm, 261:829-832

Bakker H, Bardor M, Molthoff JW, Gomord V V, Elbers I I, Stevens LH, Jordi W, Lommen A, Faye L, Lerouge P and Bosch D, 2001. Galactose-extended glycans of antibodies produced by transgenic plants. Proc Natl Acad Sci U S A, 98:2899-2904

Bardor M, Faye L and Lerouge P, 1999a. Analysis of the N-glycosylation of recombinant glycoproteins produced in transgenic plants. Trends in Plant Science 9:376-380

Bardor M, Loutelier-Bourhis C, Marvin L, Cabanes-Macheteau M, Lange C, Lerouge P and Faye L, 1999b. Analysis of plant glycoproteins by Matrix-Assisted Laser Desorption Ionization Mass Spectrometry: Application to the N-glycosylation analysis of bean phytohemagglutinin, Plant Physiol Biochem, 37:319-325

Bardor M, Cabanes-Macheteau M, Faye L and Lerouge P, 2000. Monitoring the N-glycosylation by fluorophore-assisted carbohydrate electrophoresis. Electrophoresis, 12:2550-2556

Bollini R, Vitale A and Chrispeels MJ, 1983. In vivo and in vitro processing of seed reserve protein in the endoplasmic reticulum: Evidence for two glycosylation steps. J Cell Biol 96:999-1007

Bonin CP, Potter I, Vanzin GF and Reiter WD, 1997. The mur 1 gene of *Arabidopsis thaliana* encodes an isoform of GDP-D-mannose-4, 6-dehydratase, catalyzing the first step in the de novo synthesis of GDP-L-fucose. Proc Natl Acad Sci USA, 94:2085-2090

Boyd PN, Lines AC and Patel AK, 1995. The effect of the removal of sialic acid, galactose and total carbohydrate on the functional activity of Campath-1H. Mol Immunol, 32:311-318

Cabanes-Macheteau M, Fitchette-Lainé AC, Loutelier-Bourhis C, Lange C, Vine N, Ma J, Lerouge P and Faye L, 1999. N-glycosylation of a mouse IgG expressed in transgenic tobacco plants. Glycobiology, 9:365-372

Costa J, Ashford DA, Nimtz M, Bento I, Frazao C, Esteves CL, Faro CJ, Kervinen J, Pires E,

Verissimo P, Wlodawer A and Carrondo MA, 1997. The glycosylation of the aspartic proteinases from barley (*Hordeum vulgare* L.) and cardoon (*Cynara cardunculus* L.). Eur J Biochem, 243:695-700

Crawley SC, Hindsgaul O, Ratcliffe RM, Lamontagne LR and Palcic MM, 1989. A plant fucosyltransferase with human lewis blood-group specificity. Carbohyd Res, 193:249-256

Crofts AJ, Leborgne-Castel N, Pesca M, Vitale A and Denecke J, 1998. Bip and calreticulin form an abundant complex that is independent of endoplasmic reticulum stress. Plant Cell, 10:813-823

Essl D, Dirnberger D, Gomord V, Strasser R, Faye L, Glossl J and Steinkellner H, 1999. The N-terminal 77 amino-acids from tobacco N-acetylglucosaminyltransferase I are sufficient to target a reporter protein to the Golgi apparatus of *Nicotiana benthamiana*. FEBS Lett, 453:169-173

Evans SC, Youakim A and Shur BD, 1995. Biological consequences of targeting beta 1,4-galactosyltransferase to two different subcellular compartments. Bioessays, 17:261-268

Faye L and Chrispeels MJ, 1985. Characterization of N-linked oligosaccharides by affinoblotting with concanavalin A-peroxidase and treatment of the blots with glycosidases. Anal Biochem, 149:218-224

Faye L and Chrispeels MJ, 1998. Common antigenic determinants in the glycoproteins of plants, molluscs and insects. Glycoconjugate J, 5:245-256

Faye L, Sturm A, Bollini R, Vitale A and Chrispeels MJ, 1986. The position of the oligosaccharide side-chains of phytohemagglutinin and their accessibility to glycosidases determines their subsequent processing in the Golgi. Eur J Biochem, 158:655-661

Faye L, Gomord V, Fitchette-Lainé AC and Chrispeels MJ, 1993. Affinity purification of antibodies specific for Asn-linked glycans containing α1->3 fucose or β1->2 xylose. Anal Biochem, 209:104-108

Fitchette-Lainé AC, Gomord V, Chekkafi A and Faye L, 1994. Distribution of xylosylation and fucosylation in the plant Golgi apparatus. Plant J, 5:673-682

Fitchette-Lainé AC, Gomord V, Cabanes M, Michalski JC, Saint-Macary M, Foucher B, Cavelier B, Hawes C, Lerouge P and Faye L, 1997. N-glycans harboring the lewis a epitope are expressed at the surface of plant cells. Plant J, 12:1411-1417

Fitchette AC, Cabanes-Macheteau M, Marvin L, Martin B, Satiat-Jeunemaitre B, Gomord V, Crooks K, Lerouge P, Faye L and Hawes C, 1999. Biosynthesis and immunolocalization of Lewis A-containing N-glycans in the plant cell. Plant Physiol, 121:333-343

Fötisch K, Altmann F, Haustein D and Vieths S, 1999. Involvement of carbohydrate epitopes in the IgE response of celery-allergic patients. Int Arch Allergy Immunol, 120:30-42

Friedman Y and Higgins EA, 1995. A method for monitoring the glycosylation of recombinant glycoproteins from conditioned medium, using fluorophore-assisted carbohydrate electrophoresis. Anal Biochem, 228:221-225

Garcia-Casado G, Sanchez-Monge R, Chrispeels MJ, Armentia A, Salcedo G and Gomez L, 1996. Role of complex asparagine-linked glycans in the allergenicity of plant glycoproteins.

Glycobiology, 6:471-477

Gomez L and Chrispeels MJ, 1994. Complementation of an *Arabidopsis thaliana* mutant that lacks complex asparagine-linked glycans with the human cDNA encoding N-acetylglucosaminyltransferase I. Proc Natl Sci USA, 91:1829-1833

Gomord V, Denmat LA and Fitchette-Lainé, AC, Satiat-Jeunemaitre, B, Hawes C and Faye L, 1997. The C-terminal HDEL sequence is sufficient for retention of secretory proteins in the endoplasmic reticulum (ER) but promotes vacuolar targeting of proteins that escape the ER. Plant J, 11:101-103

Gray JSS, Yang BY, Hull SR, Venzke DP and Montgomery R, 1996. The glycans of soybean peroxidase.Glycobiology, 6:23-32

Hammond C, Braakman I and Helenius A, 1994. Role of N-linked oligosaccharide recognition, glucose trimming, and calnexin in glycoprotein folding and quality control. Proc Natl Acad Sci USA, 91:913-917

Harvey JH, 1999. Matrix-assisted laser desorption/ionization mass spectrometry of carbohydrates. Mass Spectrometry Reviews, 18:349-451

Hein MB, Tang Y, McLeod DA, Janda KD and Hiatt A, 1991. Evaluation of immunoglobulins from plant cells. Biotechnol Prog,, 7:455-461

Jackson P, 1994. High-resolution polyacrylamide gel electrophoresis of fluorophore-labeled reducing saccharides, Methods in Enzymol, 230:250-265

Johnson KD and Chrispeels MJ, 1987. Substrate specificities of N-acetylglucosaminyl-, fucosyl-, and xylosyltransferases that modify glycoproteins in the Golgi apparatus of bean cotyledons. Plant Physiol, 84:1301-1308

Kamerlink JP,1991. Xylose-containing carbohydrate chains derived from glycoproteins. Pure Appl Chem, 63:465-472

Kaufmann R,1995. Matrix-assisted laser desorption ionisation (MALDI) mass spectrometry: a novel analytical tool in molecular biology and biotechnology. J Biotechnol, 41:155-175

Kaushal G, Pastuszak I, Hatanaka KI, and Elbein AD, 1990a. Purification to homogeneity and properties of glucosidase II from mung bean seedlings and suspension-cultured soybean cells. J Biol Chem, 265:16271-16279

Kaushal G, Szumilo T, Pastuszak I and Elbein AD, 1990b. Purification to homogeneity and properties of mannosidase II from mung bean seedlings. Biochemistry, 29:2168-2176

Khoo KH, Chatterjee D, Caufield JP, Morris HR and Dell A, 1997. Structural mapping of the glycans from the egg glycoproteins of *Schistosoma mansoni* and *Schistosoma japonicum*: identification of novel core structures and terminal sequences. Glycobiology, 7:663-667

Khoudi H, Laberge S, Ferullo J-M, Bazin R, Darveau A, Castonguay Y, Allard G, Lemieux R and Vézina L-P, 1999. Production of a diagnostic monoclonal antibody in perennial alfalfa plants. Biotech Bioengin, 64:135-143

Kimura Y, Hase S, Kobayashi Y, Kyogoku Y, Ikenaka T and Funatsu G, 1988. Structures of sugar of ricin D. J Biochem, 103:944-949

Kimura Y, Hess D and Sturm A, 1999. The N-glycans of jack bean α-mannosidase. Eur J Biochem, 264:168-175

Kornfeld R and Kornfeld S, 1985:Assembly of asparagine-linked oligosaccharides. Ann Rev Biochem, 54:631-664

Kurosaka A, Yano A, Itoh N, Kuroda Y, Nakagawa T and Kawasaki T, 1991. The structure of a neural specific carbohydrate epitope of horseradish peroxidase recognized by anti-horseradish peroxidase antiserum. J Biol Chem, 266:4168-4172

Leiter H, Mucha J, Staudacher E, Grimm R, Glössl J and Altmann FJ, Purification, 1999. cDNA cloning, and expression of GDP-L-Fuc:Asn-linked GlcNAc alpha1,3-fucosyltransferase from mung beans. J Biol Chem, 274:21830-21839

Lerouge P, Cabanes-Macheteau M, Rayon C, Fitchette-Lainé, A-C, Gomord V and Faye L, 1998. N-glycoprotein biosynthesis in plants: recent developments and future trends. Plant Mol Biol, 38:31-48

Lis H and Sharon N, 1998. Soybean agglutinin - a plant glycoprotein. J Biol Chem, 253:3468-3476

Ma J, Hikmat BY, Vine ND, Chargelegue D, Yu L, Hein MB and Lehner T, 1998. Characterization of a recombinant plant monoclonal secretory antibody and preventive immunotherapy in humans. Nat Med, 4:601-606

Masada RI, Skop E and Starr CM, 1996. Fluorophore-assisted carbohydrate electrophoresis (FACE registered) for quality control of recombinant-protein glycosylation. Biotechnol Appl. Biochem, 24:195-205

Matsumoto S, Ikura K, Ueda M. and Sasaki R, 1995. Characterization of a human glycoprotein (erythropoietin) produced in cultured tobacco cells. Plant Mol Biol, 27:1163-1172

Melo NS, Nimtz M, Conradt HS, Fevereiro PS and Costa J, 1997. Identification of the human Lewisa carbohydrate motif in a secretory peroxidase from a plant cell suspension culture (*Vaccinium myrtillus* L.). FEBS Lett, 415:186-191

Navazio L, Baldan B, Mariani P, Gerwig GJ and Vliegenthart JFG, 1996. Primary structure of the N-linked carbohydrate chains of calreticulin from spinach leaves. Glycoconjugate J, 13:977-983

Nebenführ A, Gallagher, LA, Dunahay TG, Frohlick JA, Mazurkiewicz AM, Meehl JB and Staehelin LA, 1999. Stop-and-go movements of plant Golgi stacks are mediated by the acto-myosin system. Plant Physiol, 121:1127-1141

Oxley D and Bacic A, 1995. Microheterogeneity of N-glycosylation on a stylar self-incompatibility glycoprotein of *Nicotiana alata*. Glycobiology, 5:517-523

Oxley D, Munro SLA, Craik DJ and Bacic A, 1996. Structure of N-glycans on the S3- and S6-allele stylar self-incompatibility ribonucleases of *Nicotiana alata*. Glycobiology, 6:611-618

Pagny S, Cabanes-Macheteau M, Gillikin JW, Leborgne-Castel N, Lerouge P, Boston RS, Faye L and Gomord V, 2000. Protein recycling from the Golgi to the endoplasmic reticulum is very active in plants but has a minor contribution to calreticulin retention. Plant Cell,

12:739-755

Palacpac NQ, Yoshida S, Sakai H, Kimura Y, Kazuhito K, Yoshida T and Seki T, 1999. Stable expression of human beta 1,4- galactosyltransferase in plant cells modifies N-linked glycosylation patterns. Proc Natl Acad Sci USA, 8:4692-4697

Parodi AJ, Mendelzon DH, Lederkremer GH and Martin-Barrientos J, 1984. Evidence that transient glucosylation of protein-linked $Man_9GlcNAc_2$, $Man_8GlcNAc_2$ and $Man_7GlcNAc_2$ occurs in rat liver and *Phaseolus vulgaris* cells. J Biol Chem, 259:6351-6357

Raju TS, Briggs JB, Borge SM and Jones AJS, 2000. Species-specific variation in glycosylation of IgG: evidence for the species-specific sialylation and branch-specific galactosylation and importance for engineering recombinant glycoprotein therapeutics. Glycobiology, 10:477-486

Rayon R, Gomord V, Faye L and Lerouge P, 1996. N-glycosylation of phytohemagglutinin expressed in bean cotyledons or in transgenic tobacco plants. Plant Physiol Biochem, 34:273-281

Rayon C, Lerouge P and Faye L, 1998. The protein N-glycosylation in plants. J Exp Bot, 49:1463-1472

Rayon C, Cabanes-Macheteau M, Loutelier-Bourhis C, Saliot-Maire I, Lemoine J, Reiter WD, Lerouge P and Faye L, 1999. Characterization of N-glycans from *Arabidopsis thaliana*. Application to a fucose-deficient mutant. Plant Physiol, 119:725-733

Reiter WD, Chapple CCS and Somerville CR, 1993. Altered growth and cell walls in a fucose-deficient mutant of Arabidopsis. Science, 261:1032-1035

Shimazaki A, Makino Y, Omichi K, Odani S and Hase S, 1999. A new sugar chain of the proteinase inhibitor from latex of *Carica papaya*. J Biochem, 125:560-565

Starr C, Masada IR, Hague C, Skop E and Klock JC, 1996. Fluorophore-assisted carbohydrate electrophoresis in the separation, analysis, and sequencing of carbohydrates, J Chromatogr, 720:295-321

Staudacher E, Dalik T, Wawra P, Altmann F and März L, 1995. Functional purification and characterization of a GDP-fucose : β-N-acetylglucosamine (Fuc to Asn linked GlcNAc) α-1,3-fucosyltransferase from mung beans. Glycoconjugate J, 12:780-786

Steinkellner H, 1999. Reduction of GnTI activity in transgenic plants: effects on glycosylation patterns of recombinant proteins. International Molecular Farming Conference, London, Ontario, Canada

Strasser R, Mucha J, Schwihla H, Altmann F, Glössl, J and Steinkellner H, 1999. Molecular cloning and characterization of cDNA coding for beta1, 2N-acetylglucosaminyltransferase I (GlcNAc-TI) from *Nicotiana tabacum*. Glycobiology, 9:779-785

Sturm A and Chrispeels MJ, 1986. The high mannose oligosaccharide of phytohemagglutinin is attached to asparagine 12 and the modified oligosaccharide to asparagine 60. Plant Physiol, 80:320-322

Sturm A, Johnson KD, Szumilo T, Elbein AD and Chrispeels MJ, 1987a. Subcellular localization of glycosidases and glycosyltransferases involved in the processing of the N-linked oligosaccharides. Plant Physiol, 85:741-745

Sturm A, van Kuik JA, Vliegenthart JFG and Chrispeels MJ, 1987b. Structure, position, and biosynthesis of the high mannose and the complex oligosaccharide side chains of the bean storage protein phaseolin. J Biol Chem, 262:13392-13403

Sturm A, Bergwerff AA and Vliegenthart JFG, 1992. 1H-NMR structural determination of the N-linked carbohydrate chains on glycopeptides obtained from the bean lectin phytohemagglutinin. Eur J Biochem, 204:313-316

Szumilo T, Kaushal GP and Elbein AD, 1986a. Purification and properties of glucosidase I from mung bean seedlings. Arch Biochem Biophys, 247:261-271

Szumilo T, Kaushal GP, Hori H and Elbein AD, 1986b. Purification and properties of a glycoprotein processing α-mannosidase from mung bean seedling. Plant Physiol, 81:383-389

Tezuka K, Hayashi M, Ishihara H, Akazawa T and Takahashi N, 1992. Studies on synthesis of xylose-containing N-linked oligosaccharides deduced from substrate specificities of the processing enzymes in sycamore cells (*Acer pseudoplatanus* L.). Eur J Biochem, 203:401-413

Trombetta SE, Bosch M and Parodi AJ, 1989. Glucosylation of glycoproteins by mammalian, plant, fungal and trypanosomatid protozoa microsomal membranes. Biochemistry 28, 8108-8116

van Die I, Gomord V, Kooyman FNJ, van den Berg TK, Cummings RD and Vervelde L, 1999. Core α1-3fucose is a common modification of N-glycans in parasitic helminths and constitutes an important epitope for IgE from *Haemonclus contortus* infected sheep. FEBS Lett, 463:189-193

van Ree R and Aalberse RC, 1993. Pollen-vegetable food crossreactivity: serological and clinical relevance of crossreactive IgE. J Clin Immunoassay, 16:124-130

van Ree R, Cabanes-Macheteau M, Akkerdaas J, Milazzo JP, Loutelier-Bourhis C, Rayon C, Villalba M, Koppelman S, Aalberse R, Rodriguez R, Faye L and Lerouge P, 2000. β (1, 2)-xylose and α(1, 3)-fucose residues have a strong contribution in IgE binding to plant glycoallergens. J Biol Chem, 275:11451-11458

Vitale A and Chrispeels MJ, 1984. Transient N-acetylglucosamine in the biosynthesis of phytohemagglutinin : attachment in the Golgi apparatus and removal in protein bodies. J Cell Biol, 99:133-140

von Schaewen A, Sturm A, O'Neill J and Chrispeels MJ, 1993. Isolation of a mutant *Arabidopsis* plant that lacks N-acetylglucosaminyltransferase I is unable to synthesize Golgi-modified complex N-linked glycans. Plant Physiol, 102:1109-1118

Wee EQ, Sherrier DJ, Prime TA and Dupree P, 1998. Targeting of active sialyltransferase to the plant Golgi apparatus. Plant Cell, 10:1759-1768

Wenderoth I and von Schaewen A, 2000. Isolation and characterization of plant N-acetylglucosaminyltransferase I (GnT I) cDNA sequences. Functional analyses in the *Arabidopsis* cgl mutant and in antisense plants. Plant Physiol, 123:1097-1108

Wilson IBH, Harthill JE, Mullin NP, Ashford DA and Altmann F, 1998. Core α 1, 3-fucose is a part of the epitope recognised by antibodies reacting against plant N-linked oligosaccharides and is present in a wide variety of plant extracts. Glycobiology 8:651-661

Wright A and Morrison SL, 1997. Effect of glycosylation on antibody function: implications for genetic engineering. Trends Biotechnol, 15:26-32

Zeitlin L, Olmsted SS, Moench TC, Co MS, Martinell BJ, Paradkar VM, Russell DR, Queen C, Cone RA and Whaley KJ, 1998. A humanized monoclonal antibody produced in transgenic plants for immunoprotection of the vagina against genital herpes. Nat Biotechnol, 16:1361-1364

Zeng Y, Bannon G, Thomas Hayden V, Rice K, Drake R and Elbein A, 1997. Purification and specificity of b1, 2-xylosyltransferase, an enzyme that contributes to the allergenicity of some plant proteins. J Biol Chem, 272:31340-31347

Chapter 5

PRODUCTION AND COMMERCIALIZATION OF BIOPHARMACEUTICALS FROM MILK

M.G.A. Peters[1] and M.F. Brink[2]
[1]*Pharming Group N.V., 2333 CA, Leiden, The Netherlands;* [2]*Pharming Technologies B.V., 2333 CA, Leiden, The Netherlands*

Keywords: transgenic cattle, bioreactors, microinjection, nuclear transfer, lactoferrin, fibrinogen

Abstract The ability to produce recombinant proteins is one of the major successes of biotechnology. Transgenic animals are excellent bioreactor candidates for the production of such proteins as they possess the biochemical/cellular machinery required for the complex post-translational modifications typical of most proteins of therapeutic significance. Recent developments in gene transfer techniques with animals have greatly enhanced the feasibility and attractiveness of this approach which is being harnessed to produce valuable proteins in tissues such as milk, egg white, blood, urine, and seminal plasma. Pharming has developed a transgenic cattle platform for such production and has successfully applied this technology for the synthesis of human lactoferrin and fibrinogen.

1. INTRODUCTION

Thanks to the progress in biotechnology and medical science, therapies for hitherto untreatable diseases can be developed. Today's technological advances open the way for new and improved pharmaceutical products. However, there is increasing pressure on (bio)pharmaceutical manufacturers to restrain the costs of their products, leading to an increasing need for more cost-effective production methods. One possible answer to this paradox of innovation on the one hand and cost-restriction on the other is offered by transgenic milk technology, i.e. the production of biopharmaceuticals in the milk of transgenic

L. Erickson et al. (eds.),
Molecular Farming of Plants and Animals for Human and Veterinary Medicine, 111–125.

animals. This technology enables the production of complex biopharmaceuticals in high volumes, resulting in both innovative and cost-effective healthcare products.

2. THE RISING COST OF PHARMACEUTICALS

Over the last decade, pharmaceutical costs have been on the rise. According to IMS Health, a leading provider of information solutions to the pharmaceutical and healthcare industries, drug spending increased 18.8% in 1999 alone. The largest part of this percentage can be attributed to increased utilization of medicines for prevention and treatment of serious conditions. Medicines are the least invasive form of treatment for many diseases. In many cases, drug therapies have replaced more traditional medical therapies that involved hospitalizations, surgery, or other invasive procedures (Pharmaceutical Research and Manufacturers of America, 2000a).

Due to ongoing efforts in new medicine discovery and development, more conditions can be treated with medicines, which is another reason for the increase in pharmaceutical expenditures. As scientific knowledge and technology advance, researchers are able to target more complex diseases, providing treatments for conditions that could not be treated before. The prices for these new drugs are usually relatively high, which can be explained by looking at the different factors that contribute to the high development costs of new drugs. According to the US pharmaceutical industry association, PhRMA, US drug companies spent more than $22 billion on research and development in 2000, a 10 percent increase over the previous year. According to drug manufacturers, it normally takes an average of 12-15 years and $500 million in operational costs to discover and develop a new drug. In addition, only one in ten drugs under development makes it through the complicated and expensive clinical trial process to the market. The candidate drugs that do make it have to recover not only their own costs, but also the costs of all the "dry holes", i.e. the drugs that never make it.

Containment of the increasing cost of medicines has been the subject of debate for quite some time now among manufacturers, health care providers, and politicians. Pressure is building to develop new lower cost medicines that, at the same time, represent medical advances. New medicines are expected to produce clear benefits over their predecessors. They will need to be more effective and safer than the older medicines they replace.

In addition to these "replacement" medicines there is the growing group of new treatments for diseases that were considered untreatable. Scientific and medical progress has resulted in elucidating the causes of many 'untreatable' diseases and everyday new medical puzzles are unravelled. In many cases, this explains why effective treatment could not be developed in the past. However, for some diseases treatment has been possible for some time, but an adequate production process for that particular medicine was lacking. Either there was no viable production platform to produce sufficiently high volumes of the active ingredients or the production costs were simply too high.

Developments in the biotechnology industry make it possible to meet today's healthcare demands: innovative, safer and more effective medicines that provide long term cost-effectiveness benefits through high-volume production platforms.

3. ADVANCES IN BIOTECHNOLOGY

Drug manufacturers have for some time been working on developing medicinal products that are either identical to or similar to naturally occurring substances found in the human body that counter disease. These products are called biopharmaceuticals, or "biotechnology medicines".

In the past, two main problems were associated with the therapeutic use of these drugs: supply and safety. Often, the only source for such products was human cadavers. The amount of available cadavers was not sufficient to satisfy the demand, which automatically made biopharmaceuticals more expensive.

The other problem was that products derived from human sources possess a higher probability of containing contaminants that are detrimental to human health when compared to traditional pharmaceuticals (Institut Pasteur, 2000)

With the advent of genetic engineering in the 1970s, a new era for the production of biopharmaceuticals started. Two breakthroughs formed the basis of the modern pharmaceutical biotechnology industry: the interspecies transplantation of genetic material, and the fusion of tumor cells and certain leukocytes. The cells resulting from such fusion - hybridomas - replicate endlessly and can be geared to produce specific antibodies in bulk (Steinberg and Raso, 1998). This new technology made it possible to eliminate the two problems mentioned above. Contamination issues can be eliminated when a drug is produced in microbial, e.g. *Eschericia coli* (*E. Coli*) or mammalian (cell) systems and because these production systems are more plentiful, costs can be reduced.

Modern pharmaceutical biotechnology encompasses gene cloning and recombinant DNA technology. Gene cloning comprises isolating a DNA-molecule segment that corresponds to a single gene and synthesizing ("copying") the segment. Recombinant DNA technology, or gene splicing, comprises altering genetic material outside an organism—for example, by inserting into a DNA molecule a segment from a very different DNA molecule—and making the altered material (recombinant DNA) function in living things. Most biopharmaceuticals are proteins and are directly coded for by DNA. The gene that codes for the biopharmaceutical of interest is removed from the genome of the animal. This gene is then inserted into the genome of a microbial organism, such as *E.coli*, mammalian cell cultures, such as Chinese Hamster Ovary (CHO) cells, or another animal, e.g. cows, rabbits, goats and sheep.

According to data from PhRMA, the organization representing the Pharmaceutical Research and Manufacturers of America, 369 medicines that meet the definition of "biotechnology medicines" are at various stages in the development pipeline (Figure 20). All together, the biotechnology medicines in the pipeline target more than 200 diseases. Nearly half the medicines, 175, are for cancer. Medicines under development target, among many other diseases, infectious diseases, autoimmune diseases, heart disease, asthma, and cystic fibrosis. New developments will increase the number of targets at which medicines can be aimed. The Human Genome Project is an attempt to sequence each of the approximately 100,000 genes that dictate just about every physical characteristic of a human being. Once the sequence of the genes is known, scientists can begin to figure out what protein the gene produces and what that protein does in the body. This knowledge will help them determine how some diseases are triggered. More targets and an understanding of how they are involved in the development of human disease will allow more precise treatments and the prevention of more diseases. Debilitating symptoms will not have to be seen before diagnosis can occur and treatment begins. This will eventually lead to more and better medicines (Pharmaceutical Research and Manufacturers of America, 2000b).

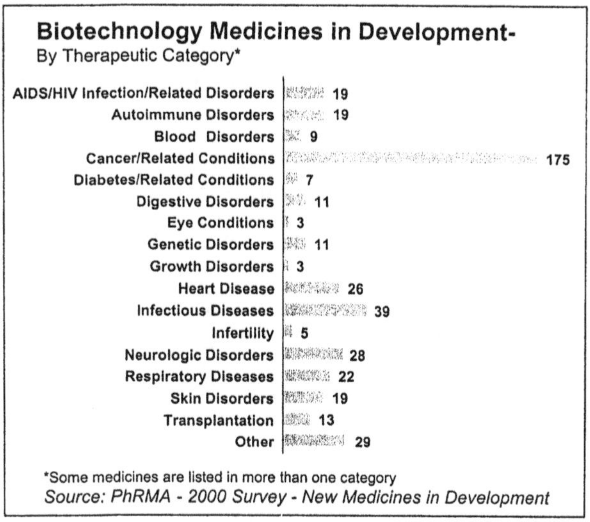

Figure 20. Biotechnology medicines in development.

4. LIVESTOCK VERSUS BACTERIAL FERMENTATION

As mentioned above, there are several technologies for producing biopharmaceuticals. A more traditional method is through bacterial fermentation systems. Genes are isolated, cloned, and inserted into prokaryotic bioreactors for mass production.There are several critical factors that make livestock preferable. The genetic code in itself does not make a functional protein. Much of the pharmaceutically significant structure of a protein is actually due to post-translational chemical and architectural modifications carried out by the cellular machinery on the particular protein molecule coded for by the inserted gene. The problem with bacterial fermentation systems is that prokaryotes (bacterial cells that do not have a separate nucleus) do not have the same mechanics for glycosylation, folding, amidation, and other post-translational modifications as eukaryotic (nucleated) cells. Glycosylation, for example, must be carried out after fermentation in separate chemical reactions. This adds considerable expense and additional problems for purification and evaluation of products (Houdebine, 2000).

Another problem with fermentation systems is that of purification. Even using specific affinity techniques, nonsecreted, intracellular proteins are always more difficult to purify than extracellular proteins. This is not due only to the mass action effect of such a large number of other proteins present in cells rather than in the culture medium. Cellular debris and enzymes released when cells and organelles are ruptured can bind or denature pharmaceutical proteins during purification cycles. Losses due to these mass-action effects and problems of contamination due to the presence of unknown numbers of physically similar molecules inevitably occur. Using whole animals as bioreactors gets around these problems. Normally, nonsecreted proteins can be targeted to secretory organs such as mammary glands. These proteins are "dumped" as added constituents into the normal secretory fluid, providing an easy source for harvesting and purification (Houdebine, 2000).

Other benefits of choosing animals as bioreactors over biofermentors are those of storage, manufacturing, and purification costs. Once an appropriate founder population is established, no high-tech processes are necessary to reproduce the system; the animals merely do what comes naturally. Cell fermentation technology works well enough when relatively small volumes of protein are needed. However, for large-scale production, fermentation is not a cost-effective platform. To implement a sizeable, 100 kg pure protein fermentation plant, a capital investment of US $ 100 million-plus is required. A transgenic dairy animal system requires only about a third of that investment. Also, dairy herds are much more flexible. If you start out with 20 animals and the market proves to be bigger than imagined, you can easily increase the herd size. It is not so easy to increase the capacity of a factory.

5. TRANSGENESIS

Transgenesis can be defined as the alteration of genomic information with the intent to modify a specific physical trait of an animal. First applied in the early 1980s in mice, when genomic insertion (and subsequent expression) of the gene encoding human growth hormone resulting in enhanced growth, the addition or disruption of a host of different genes has yielded a flood of invaluable information regarding biomedical and developmental issues. Once transgenic mice could be generated with relative ease via pronuclear micro-injection of recombinant DNA, attempts to modify the genome of other animal species soon followed (Houdebine, 2000). Since 1989, Pharming has been on the forefront of these developments, as evidenced by the birth of Herman, the world's first transgenic bull.

6. TRANSGENIC MILK TECHNOLOGY

One of the most intriguing applications of transgenesis is the genetic modification of animals for production of (human) recombinant proteins in milk. Particularly for (human) proteins that are scarce, production via transgenic animals would constitute a breakthrough (Houdebine, 2000). For example, the current supply of human blood clotting factors is still largely dependent on purification from donor blood, a source limited in availability and under constant threat of viral contamination. Due to the latter, frequent product recalls are not unusual. Although production of recombinant clotting factors in mammalian cell culture systems provides a safer alternative, the production capacity of this approach is limited. As mentioned before, (mammalian) cell culture systems are generally expensive, which is illustrated by the fact that therapy for hemophilia can amount to a hundred thousand dollars per patient annually.

For high-value and small volume products Pharming uses transgenic rabbits as a production platform. One of the characteristics of the transgenic rabbit production system is that rabbits are efficient breeders and will produce milk containing the desired protein within one year after the start of the project. Rabbits can produce up to 10 liters of milk a year and expression levels of the transgenic protein can be as high as 20 grams per litre. For small and medium sized indications, such as orphan (rare) diseases, the rabbit system is ideal to produce up to 50 kg of protein per year.

Our transgenic rabbits are generated through micro-injection. The procedure starts with the collection of fertilized oocytes, after which the DNA construct encoding the protein of choice is injected directly into the nucleus. The injected embryos are transferred to foster mothers, where they will develop into healthy rabbits. The first batches of milk containing the protein are available after six months, when the female transgenic rabbits start lactating. Sperm of the male transgenic rabbits is used for the (rapid) generation of a transgenic production colony (Figure 21).

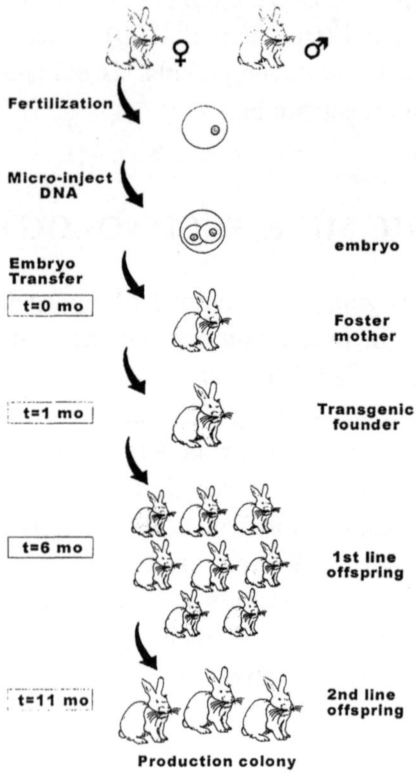

Figure 21. Transgenic rabbit system.

For higher volume production of biopharmaceuticals, farm animals are preferred as a production platform. Whereas 'small' farm animals such as pigs, goats and sheep produce between 300-900 L of milk annually, the yield of a single dairy cow can amount to ~10,000 L. The production potential of a livestock platform is therefore formidable. Since the expression levels of recombinant protein in milk of transgenic animals routinely exceed 1 g/L, for every protein of therapeutic value, a limited number of animals would be required to meet patient needs. Although micro-injection is still a very viable method for generating smaller animals, Pharming is now working on generating transgenic cattle through nuclear transfer.

7. GENERATING TRANSGENIC ANIMALS THROUGH MICRO-INJECTION

Until recently, micro-injection of recombinant DNA into the pronucleus of a fertilized oocyte remained the method of choice to generate a transgenic animal. The basic procedure for generating a transgenic animal is similar for most species. After oocyte maturation and fertilization, DNA constructs encoding the recombinant protein of choice are injected directly into the pronucleus. Subsequently, the injected embryos are transferred to foster mothers.

Combined data from a number of different laboratories show that the overall efficiency of micro-injection as a technique to introduce recombinant DNA into the genome varies between species. Generation of transgenic animals is a very labour intensive and costly matter requiring large numbers of animals and dedicated facilities. This is particularly so for cattle, where only 1 or 2 injected embryos can be transferred to a single recipient heifer. Nevertheless, given their potential to produce large quantities of recombinant protein, Pharming has performed several such large-scale embryo transfer (ET) trials to generate transgenic (founder) cattle. The transgenesis rate was somewhat disappointing, as it reached only 4.7% out of 342 pregnancies established in all our trials combined. These results were nearly identical to the rate reported in other studies.

In addition to studies aimed at improvement of transgenesis rates, alternative ways were explored to use the available resources more effectively. For example, improved cattle management resulted in increased pregnancy rates. The number of recipient heifers available for new transfers was maximized, and due to the use of a new strategy for 'early transgene detection' that allowed for early exclusion of non-transgenic pregnancies, all calves born were transgenic.

Not every transgenic founder animal is suitable for the establishment of a production herd. Several factors can contribute to the transgene being essentially inactive. Determination of the recombinant protein concentration in milk is the only reliable method for selection of suitable founder animals. No more than 25% of the offspring of transgenic founder animals (i.e., the transgenic females) will be suitable to join a production herd. The transgene transfer rate from founder to offspring often is lower than 25%.

8. NEW DEVELOPMENTS: NUCLEAR TRANSFER TECHNOLOGY

The relative inefficiency of micro-injection as a tool to generate transgenic (farm) animals has sparked the development of many different approaches to distinguish transgenic from non-transgenic individuals. Such attempts included the co-injection of selectable marker genes or pre-selection of transgenic embryos via the polymerase chain reaction (PCR). In addition, increased transgenesis rates have been achieved by alternative transgene delivery systems. Many of these methods are still under development and their success has been limited. Whether some of them will find their way to commercial application remains to be seen, as transgenic farm animal technology is being revolutionized by what promises to be a very powerful strategy: genetic manipulation of totipotent cells and generation of transgenic animals via nuclear transfer (NT).

NT (Figure 22) involves the introduction of the nucleus from a totipotent donor cell into a matured oocyte of which the nucleus has been removed. The resulting embryo is transferred to a surrogate mother for development into a live calf. Key to the overall success of NT is the requirement for totipotency; defined here as the potential to initiate and direct normal development. Although many of the factors controlling totipotency have yet to be elucidated, blastomeres and various cell types from fetal and adult origin have been successfully used to clone a number of different species, including farm animals such as sheep, goats, and cattle.

For the purpose of generating transgenic cattle, the critical breakthrough for an NT-based strategy lies in the pre-selection and expansion of characterized transgenic cells, prior to the generation of transferable embryos. For generation of transgenic cattle, any cell used for NT must undergo genetic manipulation and selection procedures combined with prolonged culturing in vitro, while remaining totipotent. In general, the number of times the cell has been passed and the number of days it has been maintained in culture are two of the many factors that can affect totipotency. By minimizing the impact of these factors the success of an embryo transfer program can be increased. The percentage of transgenic animals born (100%), using NT technology, is in sharp contrast to a micro-injection based approach which yields on average no more than 5%

transgenic births.

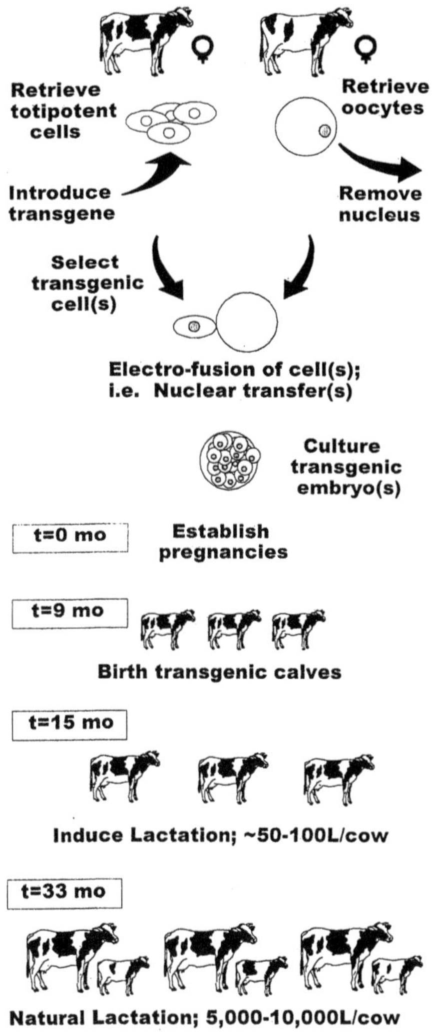

Figure 22. Nuclear transfer system.

Due to the use of female totipotent cells for genetic manipulation and the subsequent selection of transgenic cells before nuclear transfer, all calves born will be female, transgenic, and non-mosaic. These advantages alone already constitute a major breakthrough compared to micro-injection. An additional reason to apply NT lies in the fact that a time consuming breeding process is not required to establish a production herd. Instead, herds consisting of genetically identical transgenic cows can be established instantly. For the development of therapeutic proteins, this is of significant importance, since it increases product consistency and reduces the "time-to-market". Given the advantages of preselection of transgenic embryos, genetic consistency, and reduced time of product development, it is not surprising that NT is rapidly becoming the industry's technology of choice to generate transgenic animals for the production of biopharmaceuticals in milk. In fact, the combination of large protein quantities, safety, lower costs, and the reduced time of development, has made the transgenic animal platform even more attractive for production of therapeutic proteins.

9. PHARMING'S TRANSGENIC CATTLE PLATFORM

Pharming generates transgenic cattle in collaboration with Infigen Inc. of DeForest, Wisconsin, U.S.A., one of the leaders in the field of nuclear transfer. In order to ensure appropriate care of our transgenic cattle in compliance with FDA guidelines, we have constructed a unique, state-of-the-art cattle facility, Vienna Pharms in Vienna, Wisconsin. The first in its kind, Vienna Pharms can house up to eighty transgenic calves.

In January 1999, we reported the generation of our first-ever transgenic female calves. This was a promising and practical application of nuclear transfer technology – that is, the generation of a herd of transgenic cows capable, on reaching maturity, of producing a medicine in their milk. This was also the first time that nuclear transfer was applied to generate female calves carrying a transgene of pharmaceutical significance. Companies are now exploring the possibility of using nuclear transfer for xenotransplantation, for example, where donor organs are grown in transgenic pigs, and to develop transgenic animals as models for the study of human diseases.

10. EFFECTIVENESS OF THE TRANSGENIC PLATFORM

10.1 Human Lactoferrin

Pharming produces recombinant human lactoferrin (rhLF) in the milk of transgenic cattle. Human lactoferrin is a naturally-occurring protein with antibiotic, probiotic, anti-inflammatory and immuno-modulating properties. Human lactoferrin is present in substantial quantities in mother's milk and plays an important role in the defense system of infants. The protein is also present in various body fluids and, also later in life continues to be an important agent against a wide range of bacterial, fungal and viral pathogens.

A phase I clinical study with intravenously administered rhLF demonstrated that it is very well tolerated up to very high doses, which offers opportunities for developing various parenteral applications. Pharming is, among other applications, concentrating on systemic bacterial infections. This type of infection is difficult to treat otherwise due to the continuously increasing resistance to more traditional antibiotics.

In a collaborative project with the Department of Infectious Diseases of the University Hospital of Leiden, the Netherlands, recombinant hLF was shown to be highly effective in vivo against a number of pathogens, including multiple resistant *Staphylococcus aureus* strains. In addition, several proprietary recombinant hLF-derived small peptides have recently been synthesized, which have shown potent antibiotic activity against a wider range of bacteria.

The excellent safety profile of intravenously administered rhLF, combined with the data generated in animal model studies make rhLF a promising candidate as a therapy for resistant systemic infections.

10.2 Human Fibrinogen

In an exclusive alliance with the American Red Cross Biomedical Services, Pharming is developing human fibrinogen (hFIB)-based wound sealants. HFIB is a naturally-occurring soluble plasma protein, pivotal in generating an insoluble fibrin matrix, which is formed by the human body to stop internal or external bleeding. Because of its key role in this process, hFIB is the main component of a biological tissue sealant that can be used to control bleeding in trauma and surgery patients.

The interaction between fibrinogen and thrombin, another clotting protein, provides the biochemical basis for a variety of fibrin-based tissue sealant products, such as spray-on, bandage and foam products, all currently under development. In all these products, fibrinogen and thrombin are mixed at the site of the injury to form fibrin, which prevents excessive bleeding and facilitates the tissue repair process. There are several important advantages of fibrin sealants over conventional wound sealing techniques. A biodegradable clot is achieved in seconds to minutes, without leaving a scar. Fibrin sealants can be used to close surgical areas that are difficult to close with sutures or staples. The risk of wound inflammation is reduced and the sealant can be dosed accurately and consistently.

As is the case for the majority of plasma proteins, human fibrinogen is scarce. At present hFIB is derived from pooled donor blood. As mentioned before, there are several significant drawbacks to using pooled human blood: there is the risk of (viral) disease transmission and the supply depends on the number of blood-donating volunteers. In vitro cell culture technology provides a way to circumvent the safety concerns, but the supply problems remain. Transgenic milk technology, as employed by Pharming, offers a virtually unlimited supply of plasma proteins like hFIB.

Recently, several independent hFIB founder animals, generated through the use of nuclear transfer, carrying multiple copies of each fibrinogen gene, were born. Analysis of the first milk demonstrated expression of hFIB in the gram per litre range and according to preliminary results, both the structural and functional characteristics of the purified recombinant hFIB are virtually identical to the plasma derived hFIB. The development of a large scale purification process and characterization methods is in progress.

11. CONCLUSION

Pharming's achievements so far have demonstrated that the use of transgenic animals offers a viable production platform for (large volumes) of complex biopharmaceuticals. Technological innovations, such as recent developments in nuclear transfer, will continue to improve the timelines and cost-effectiveness of this platform. Therefore, we believe that transgenic technology will become increasingly important in meeting the growing demand for innovative therapies and medical solutions.

REFERENCES

Houdebine LM, 2000. Transgenic animal bioreactors. Transgenic Res, 9:305-320

Institut Pasteur, 2000. Pharmaceuticals and viral safety: a global view. Euroconference Abstracts, March, 2000

Pharmaceutical Research and Manufacturers of America (PhRMA), 2000a. Survey 2000: Why do prescription drugs cost so much? 1100 Fifteenth St, NW Washington, DC, 20005

Pharmaceutical Research and Manufacturers of America, 2000b.Biotechnology: New Medicines in Development: A 2000 Survey. 1100 Fifteenth St, NW Washington, DC, 20005

Steinberg F and Raso J, 1998. Biotech pharmaceuticals and biotherapy: an overview, J Pharm Sci, 1: 48-59

Stokes R, 1999. Commercialization of 'Milked Medicines', Pharmaceutical Technology Europe, March 1999

Chapter 6

A SEED-DERIVED ORAL SUBUNIT VACCINE FOR HUMAN CYTOMEGALOVIRUS

E.S. Tackaberry[1,2], K. E. Wright[2], A.K. Dudani[1,3], I. Altosaar and P.R. Ganz[1,2]

[1]Bureau of Biologics and Radiopharmaceuticals, Therapeutic Products Programme, Health Canada, Ottawa, K1A 0L2, Canada; [2]Department of Biochemistry, Microbiology and Immunology and [3]Department of Cellular and Molecular Medicine, University of Ottawa, K1N 6N5, Canada

Keywords: transgenic tobacco, seeds, human cytomegalovirus, glycoprotein B, vaccine

Abstract Transgenic plants offer many potential advantages as production systems for making human biopharmaceuticals. Subunit vaccines are especially promising candidates for this technology since it may also be possible to administer the recombinant protein vaccinogen orally, in edible plant products. In order to study this promising approach we have established a model system to express the major glycoprotein, gB, of human cytomegalovirus (HCMV) in the seeds of tobacco plants. A novel expression vector was developed incorporating regulatory sequences of glutelin, the major rice seed storage protein, to direct synthesis of recombinant glycoprotein B to seeds. Following transformation, selected calli and subsequently mature plants were regenerated. Screening of genomic DNA by PCR amplification and Southern blotting showed that 71% harboured the gB coding sequence. Specific immunoassays were established using neutralizing and non-neutralizing monoclonal antibodies specific for gB. These revealed that protein extracts from seeds of positive plants produced antigenic gB at levels ranging from 70 - 146 ng/mg extracted protein. In addition, similarity with native gB produced in HCMV-infected cells was demonstrated by inhibition of immunofluorescence on HCMV-infected human fibroblasts. Although several mammalian proteins have been expressed in tobacco, localization of the proteins within transgenic tissues has not been extensively examined. In order to identify the site(s) of recombinant gB deposition in mature tobacco seeds, we used immunogold labelling and electron microscopy. We observed specific labelling for gB in the endosperm of transgenic seeds, with gB localized almost exclusively in protein storage vesicles. This was found equally in seeds that were freshly harvested and in seeds that had been stored at room temperature for several

L. Erickson et al. (eds.),
Molecular Farming of Plants and Animals for Human and Veterinary Medicine, 127–153.

months. Collectively, our results demonstrate that this large and complex viral glycoprotein can be expressed in plants in a highly tissue-specific manner. They also show that biologically significant structural features, including viral neutralizing epitopes, are retained in this heterologous system, and that the gB is sorted within seeds in a manner similar to plant storage proteins. These data provide further evidence to support the suitability of plants for producing recombinant proteins of potential clinical relevance.

1. PLANTS AS EXPRESSION SYSTEMS FOR BIOTHERAPEUTICS

Over the past decade increasing attention has been directed to the feasibility of developing oral vaccines for the control of infectious diseases. While some endeavours have focused on diseases caused by enteric pathogens, others have concentrated on the large number of non-enteric infectious agents which nonetheless may gain entry to their hosts at mucosal surfaces (Service, 1994; Manganaro et al., 1994; Shalaby, 1995). A corollary of this research is the goal of developing subunit vaccines in edible plant products - that is, of using transgenic plants as expression systems for producing the recombinant protein vaccinogens of interest - for both human and veterinary use.

The idea of using plants as expression systems for therapeutic proteins is highly attractive (reviewed in Pen, 1996; Cramer et al., 1999). Potential advantages include: products that are safer than recombinant proteins made in mammalian cell culture systems, since they are unlikely to harbour viral contaminants pathogenic to humans; products that can be produced cheaply without highly sophisticated and expensive scaled-up production facilities; and products that can be made in quantities sufficient to meet worldwide demand, with increased cultivation being a ready alternative if expression levels or product yields are low. Another advantage is that, unlike bacteria, plants can effect post-translational modifications such as glycosylation and terminal processing or assembly steps often required for authentic structure and function of many complex eukaryotic proteins.

An additional opportunity provided by plant-based systems is the potential for oral delivery of the therapeutic protein if it is produced in an edible plant or plant product. This option is being actively pursued for oral vaccines. Most vaccines currently approved for human use are administered by injection. This not only increases the cost of vaccine programs, a significant factor in developing countries, but decreases people's willingness to get vaccinated and comply with booster regimens, compromising the effectiveness of vaccination campaigns. Edible vaccines offer a way of circumventing these difficulties and foster the goals of the international Children's Vaccine Initiative which are dedicated to maximizing protection against infectious diseases by developing new technologies, including oral vaccines, to increase vaccine availability, safety and effectiveness (Mitchell et al., 1993).

Two recent reviews provide an excellent summary of progress in this area (Richter and Kipp, 1999; Walmsley and Arntzen, 2000). In 1992 Mason et al., reported that potential vaccinogens could be expressed in transgenic plant systems, and soon thereafter the immunogenicity of several plant-derived recombinant antigens was demonstrated in experimental animals (Thanavala et al., 1995; Haq et al., 1995; Mason et al., 1996). Most recently, the B subunit of *E. coli* enterotoxin (Tacket et al., 1998) and the Norwalk virus capsid protein (Tacket et al., 2000) have been delivered orally via raw potato tubers to human volunteers. These plant-derived vaccinogens proved to be immunogenic in most recipients, with variable antibody titres in serum and stool samples. Protective immunity to challenge with pathogens has not yet been reported.

Plant seeds are particularly well suited for the targeted synthesis of heterologous proteins. Unlike proteins synthesized in vegetative plant tissues, seed storage proteins are compartmentalized in specialized vacuoles in the mature seed called protein bodies or protein storage vesicles (PSV) (Adeli and Altosaar, 1984; Vitale and Raikhel, 1999; Miller and Anderson, 1999). The PSV provide a dry and stable environment devoid of significant enzymatic activity prior to germination (Bewley and Black, 1994). A number of specific promoters of seed storage protein genes have been identified and characterized including glutelin (Gt), the major reserve endosperm protein in rice seeds. Glutelin is encoded by a small multigene family with subfamilies designated Gt1, Gt2, Gt3, etc. (Okita et al., 1989). The glutelin promoters have been shown to direct the expression of various reporter genes in transgenic plant systems, resulting in gene expression that is tissue specific and developmentally regulated (Leisy et al., 1990; Zhao et al., 1994). This prompted our laboratory to investigate glutelin promoters for directing the synthesis of heterologous proteins of clinical relevance in seeds.

2. HUMAN CYTOMEGALOVIRUS

Human cytomegalovirus (HCMV) is a widely distributed member of the herpesvirus family. The virus is responsible for severe, sometimes fatal disease in patients who are immunocompromised thus posing a particular threat to transplant patients on immunosuppressive drugs, AIDS patients and low weight premature infants. HCMV also causes intrauterine infection which may result in fatal systemic disease or may be limited to the central nervous system. The frequency of congenital infection in Western countries is estimated at 1% of live births, with about 10% of these infants developing clinical symptoms (Ho, 1991; Plotkin, 1999).

Since the cytomegaloviruses are species-specific, humans are the only reservoir for HCMV and the virus will only productively infect human cells. In addition to transplacental infection, transmission may occur via body fluids such as cervical secretions, breast milk, urine, saliva and blood. In otherwise healthy individuals HCMV infection is usually asymptomatic. But, as for all herpesviruses, primary HCMV infection is followed by a latent and persistent infection for the lifetime of the host. It may reactivate from latency under suitable circumstances (usually associated with reduced immune function) causing clinical symptoms ranging from pneumonia (typically in transplant recipients) to retinitis and gastrointestinal manifestations (typically in AIDS patients) (Ho, 1991). The site of latent infection remains uncertain, but hematopoietic progenitor cells in bone marrow have been implicated (Crapnell et al., 2000), and certainly hematopoietic complications in recipients of bone marrow transplants are a significant clinical problem (Stocchi et al., 1999).

HCMV is a large virus and encodes over 200 proteins including several glycoprotein complexes situated on the outer surface envelope of the virus. Of these, glycoprotein B (gB, UL55) has been identified as one of the prime candidates for inclusion in any HCMV subunit vaccine (Landini, 1992; Tackaberry et al., 1993; Britt and Mach, 1996; Tackaberry et al., 1997; Speckner et al., 1999). This is based on numerous studies which demonstrate, for example, that over half the neutralizing antibodies developing after natural HCMV infection are directed towards gB, that high serum titres of anti-gB diminish the severity of HCMV disease in renal transplant patients, and that monoclonal antibodies (mAb) against gB will neutralize the infectivity of

HCMV in vitro (see Britt, 1996; Plotkin, 1999).

Both humoral and cell-mediated immune responses to many different HCMV antigens occur in immunocompetent individuals with each component playing a different role. It appears that cell-mediated immunity is essential for recovering from active infection and maintaining latency, whereas the ability to mount a robust humoral immune response is critical for controlling primary infection and for reducing the gravity of disease once it occurs (Snydman, 1990; Fowler et al., 1992; Riddell et al., 1992). Data from numerous studies support the probability that an HCMV vaccine could provide at least limited protection against infection and serious disease, and is thus a high priority goal for the target populations most at risk (Fowler et al., 1992; Britt, 1996; Plotkin, 1999).

Nonetheless, the development of an effective and safe vaccine for HCMV has eluded researchers for many years. A recent review (Plotkin, 1999) summarizes some of the current approaches including a live attenuated (Towne) vaccine (Adler et al., 1995; Wang et al., 1996), live recombinant vaccines (Adler et al., 1999), DNA vaccines (Pande et al., 1995; Schleiss et al., 2000) and recombinant subunit vaccines (Frey et al., 1999; Pass et al., 1999). Adoptive immunotherapy with antigen-specific cytotoxic T lymphocytes has also been used to reconstitute HCMV cellular immunity in recipients of allogeneic bone marrow transplants (Walter et al., 1995; Lucas et al.,2000).

3. CONCERNS AND CHALLENGES

Despite the considerable promise of plant-derived biotherapeutics, there are numerous unresolved biochemical and immunological issues which await clarification before the safety and efficacy of such products, including vaccinogens, can be assured. For example, it is known that plants process glycoproteins differently than mammalian cells and that several plant glycans are not present in mammalian glycoproteins (see Lerouge et al., 1998 and chapter 4, this volume). Depending on the use of the plant-derived therapeutic, for example the route of recipient exposure, an undesirable immune response may be initiated such that the product becomes ineffective and/or triggers an adverse immune response (Jenkins et al., 1996; Chrispeels and Faye, 1996). Other issues relate to whether plant systems are capable of expressing heterologous genes in a manner yielding recombinant proteins that are sufficiently authentic in structure and function. For example, like gB of

HCMV, many immunodominant viral antigens are transmembrane glycoproteins that require numerous and specific glycosidases, proteases, chaperonins, etc. for achieving their correct structure and function. Whether a plant host system can adequately mimic these molecular processing events to produce structurally authentic viral glycoproteins has not yet been determined.

4. PREVIOUS WORK

As a model system to investigate these issues we are using cDNA encoding the immunodominant gB of HCMV, with expression directed to plant seeds. Previous reports from our laboratory have described the introduction into tobacco of 432 bp of cDNA encoding the human cytokine granulocyte-macrophage colony stimulating factor (GM-CSF) (Ganz et al., 1995; Ganz et al., 1996; Sardana et al., 1997). Our studies showed that GM-CSF was expressed in seeds of selected tobacco transformants. The seed-derived GM-CSF displayed immunological reactivity with both monoclonal and polyclonal antibodies as demonstrated by Western blotting and specific ELISAs, and also exhibited biological activity in vitro (Ganz et al., 1996).

Here we describe a plasmid vector containing the cDNA sequence encoding mature HCMV gB, transformation of tobacco tissues and analysis of the resulting transgenic tobacco plants for recombinant gB, characterization of the plant-derived gB for antigenic comparability with native gB and identification of the subcellular location of this heterologous protein in the transgenic tobacco seeds.

5. TRANSFER OF THE HCMV Gb GENE TO TOBACCO

5.1 Vector Constructs and Transfer to Plants

A plasmid vector previously designed in our laboratory was modified (Tackaberry et al., 1999) to enable synthesis of the mature gB protein without any associated amino acid residues of the glutelin protein (Ganz et al., 1996; Sardana et al., 1997). This new construct contained the 980 bp glutelin Gt3 promoter and associated 72 bp Gt3 signal peptide sequence ligated in-frame to the DNA coding sequence for serine, the first amino acid of the cDNA of the mature gB protein of HCMV Towne prototype strain without its signal peptide Cranage et al., 1988), and the NOS transcription termination sequence The construct was designated pPH2/gB/NOS-T. It was subsequently subcloned into pRD400 (Datla et al., 1992), a binary vector carrying the kanamycin resistance gene to enable transformation of tobacco (Figure 23A, 1B). DNA sequencing was carried out on ligation products to insure correct orientation and fidelity of ligation junctions.

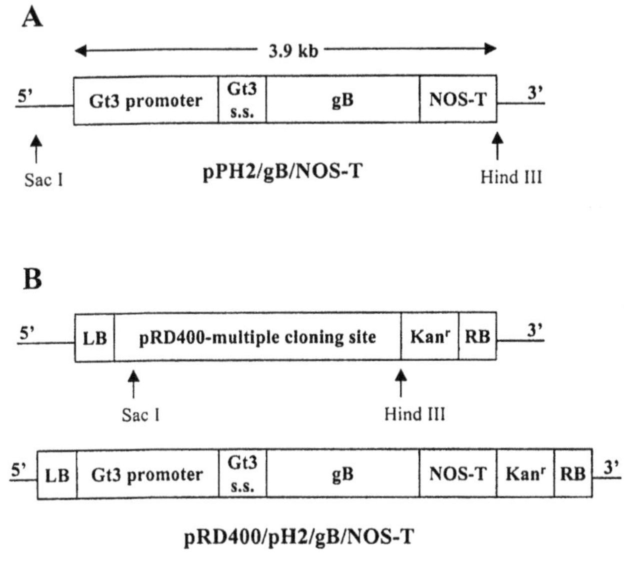

Figure 23. Engineering of pRD400/pPH2/gB/NOS-T, an expression vector designed to target synthesis of the heterologous viral glycoprotein gB to the seeds of tobacco plants (Tackaberry et al. 1999). A, the rice glutelin 3 (Gt3) promoter and signal peptide sequence were ligated directly upstream of, and in-frame with, cDNA for the mature gB gene, from which the sequences for the signal peptide had been previously removed. B, After inclusion of NOS-T, a termination signal, pPH2/gB/NOS-T was subcloned into pRD400, a binary vector carrying the NPTII gene for kanamycin resistance and including right border (RB) and left border (LB) regulatory sequences of T-DNA to effect transfer into *A. tumefaciens*.

Tobacco (*N. tabacum* cv. Xanthi) leaf disk tissue was transformed via *Agrobacterium* (strain LBA4404) that carried the binary vector pRD400/pPH2/gB/ NOS-T (Horsch et al., 1985; Zambryski, 1988). Kanamycin incorporated into tissue culture media allowed for selection of transformed callus tissue from which shoots and thereafter mature tobacco plants were regenerated over a period of about six months. Tobacco shoots were regenerated in tissue culture and selected plants grown to maturity in soil (Robert et al., 1989).

5.2 Molecular Analysis of Plants Containing the HCMV gB Gene

Young leaves for PCR and Southern blotting analyses were collected from the tops of plants, quickly frozen in liquid N_2 and stored at -85°C. Self-pollination of flowers was assured by using pollination bags (Applied Extrusion Technologies Inc., Middleton, DE). Unless otherwise specified, seeds were collected near maturity at approximately day 24 post-pollination, quickly frozen in liquid N_2 and stored at -85°C. Genomic DNA was extracted from tobacco leaves (DNeasy Kit, Qiagen, Chatsworth, CA) and DNA concentrations estimated spectrophotometrically (Sambrook et al., 1989). PCR analysis of genomic DNA extracted from leaf tissue verified that the transgene was present in 20 of 28 plants tested (71%) (Tackaberry et al., 1999).

Genomic DNA extracted from selected PCR-positive plants was analyzed by Southern blotting to gain information about organization of the transgene. Ten μg of each genomic DNA sample was digested with Hind III and Sac I, restriction enzymes with target cleavage sites flanking the 3.9 kb pPH2/gB/NOS-T construct, electrophoresed through 0.8% agarose and then alkali blotted onto a Pall Biodyne B membrane (Canadian Life Technologies, Burlington, Ontario). Pre-hybridization (QuickHyb, Stratagene, La Jolla, CA) was followed by DNA hybridization using a random primed ^{32}P-dATP labelled probe (Ready-To-Go DNA Labelling Kit, Pharmacia Biotech) generated from the full 3.9 kb pPH2/gB/NOS-T construct. Washings were carried out according to manufacturer's directions and the membrane then exposed to Kodak XOMAT XAR film with 2 intensifying screens at -85°C.

As seen in Figure 24, there was a variety of different patterns of hybridizing fragments, with a band at 3.9 kb in plants A2, A6, A7, A11, A19, A20, A23, A24 and A26 corresponding to the size of the intact pPH2/gB/NOS-T transgene. One plant (A27) had no band visible at 3.9 kb but did show at least 6 other specific bands both greater and smaller than 3.9 kb. DNA extracted from non-transformed control tobacco did not reveal any hybridizing fragments.

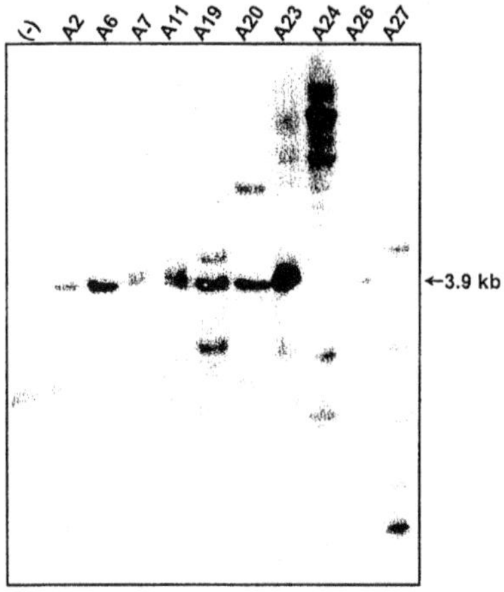

Figure 24. Southern blot analysis of genomic DNA. Ten µg of genomic DNA extracted from young leaves of tobacco plants was digested with Sac I and Hind III, electrophoresed through agarose and transferred to a membrane. A 32P-dATP probe generated from the full 3.9 kb construct pPH2/gB/NOS-T was hybridized to the immobilized DNA, then exposed to film for 47 h. Lanes A2 - A27 represent DNA extracted from tobacco plants transformed with pPH2/gB/NOS-T; lane (-) represents pooled DNA extracted from non-transformed tobacco plants.

6. Immunological Detection of gB and Quantification in Seed Protein Extracts

6.1 Extraction of Soluble Seed Proteins

Previously harvested and frozen seeds were thawed in 3 vol of ice cold extraction buffer (50 mM Tris-HCl pH 7.5 containing 50 mM NaCl, 5 mM EDTA, 0.1% Triton X-100, 1% (w/v) ascorbic acid, 1% (w/v) insoluble PVP, 1 mM PMSF, 20 mM DTT) and set on ice for 5-30 min (Sardana et al., 1997). The seeds were homogenized (Brinkmann Polytron, 12 mm generator probe), the homogenate centrifuged at 3000 x g for 20 min at 4°C and the resulting supernatant further clarified by centrifugation at 20,000 x g for 10 min at 4°C. This extract was frozen in liquid N_2 and stored at -85°C. Proteins were extracted from tobacco leaves in a similar manner.

6.2 gB-Specific ELISA

Volumes of 50 μl were generally used throughout and wells were washed 3x between each step (PBS pH 7.2 containing 0.02% Tween 20 and 0.02% NaN_3). Wells of Immulon II microtiter plates (Dynatech Laboratories) were coated with the $F(ab')_2$ fragment of CMVB1, a gB-specific murine mAb with viral neutralizing activity against HCMV (Tackaberry et al., 1993). Wells were blocked for 2 h with 2% (w/v) bovine serum albumin (BSA) in PBS, followed by addition of test sample or control for 1 h. Prior to assay the test samples (seed extracts) were adjusted so that they were at comparable protein concentrations and none was greater than that of the control non-transformed seed extracts. Where specified, a supplementary control of protein extracts from seeds producing human GM-CSF (Ganz et al., 1996) was also included. The GM-CSF seeds were derived from tobacco plants transformed with a plasmid vector similar to pPH2/gB/NOS-T except that the gene encoding GM-CSF was substituted for the gB gene (Ganz et al., 1996). A pool of gB-specific mAb (CMVB1 as above and #13-127-001 and #13-128-001, ABI, Columbia, MD) in 2% BSA/ PBS was used as the primary antibody with Fc-specific goat anti-mouse IgG conjugated to alkaline phosphatase (Jackson Laboratories, Westgrove, PA and ICN Biomedicals Inc., Aurora, OH) in 3% BSA/PBS used as the secondary antibody. This was followed by phosphatase substrate (1 mg/mL p-nitrophenyl phosphate, Sigma, St. Louis, MO) and subsequent measurement of absorbance at 405 nm. Sample wells were blanked against wells in which buffer was substituted for test sample and all assays were performed in triplicate. Raw ELISA data were converted to ng gB/mg total extracted protein by reference to an ELISA standard curve constructed with recombinant gB produced in Chinese hamster ovary (CHO) cells (Austral Biologics, San Ramon, CA) (Tackaberry et al., 1999).

Soluble proteins extracted from mature tobacco seeds of selected plants yielded extracts with protein concentrations ranging from 1.3 - 7.4 mg/mL. For ELISA testing, each seed extract was diluted such that none exceeded the protein concentration of that in the non-transformed tobacco seed extract control. Samples were analyzed for gB immunologic reactivity by an ELISA utilizing a pool of gB-specific murine mAb which recognize both neutralizing and non-neutralizing epitopes of gB, with quantification by reference to recombinant gB derived from mammalian CHO cells. Representative data shown in Figure 25 demonstrate that extracts from plants A2 - A27 contained immunologically reactive gB, at concentrations of 70 - 146 ng/mg total extracted protein, with the upper value corresponding to 658 ng gB/g dry seed material. For comparison purposes, seed extracts from A22 (positive in PCR and Southern analysis) and A28 (negative in PCR and Southern analysis) (Tackaberry et al., 1999) were assayed but neither contained significant amounts of gB. Additional controls consisting of protein extracts from tobacco seeds producing the cytokine GM-CSF and protein extracts from leaves from gB positive plants were also tested by ELISA. No gB was detected in these samples.

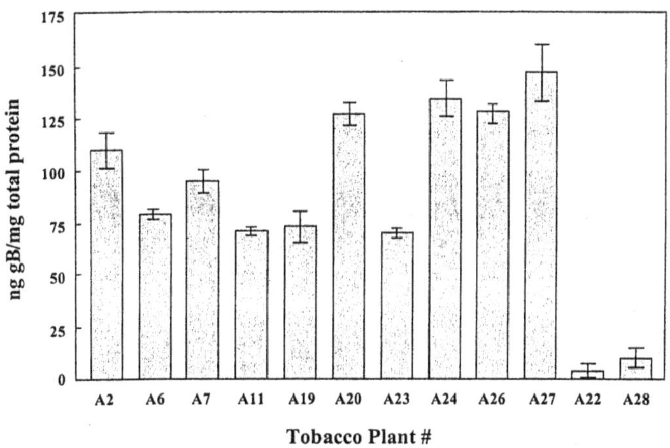

Figure 25. Detection and quantification of gB protein in soluble extracts of tobacco seeds. Seeds were collected approximately 24 days after pollination from plants transformed with pPH2/gB/NOS-T (A2 - A28) or negative controls. Seed extracts were tested for gB by ELISA with sample wells blanked against wells in which buffer was substituted for test sample. As negative controls seed extracts from non-transformed plants or from plants producing an unrelated heterologous protein (human GM-CSF) were assayed at a comparable protein concentration and each yielded ELISA OD405 values of approximately 0.150. Raw ELISA data were converted to ng gB/mg total protein by reference to a standard curve constructed with recombinant gB produced in CHO cells. All data represent the mean of triplicate values. A2 - A28, protein extracts from seeds of transformed tobacco plants.

7. Antigenic Comparability of Seed-Derived gB vs. gB Produced in Infected Human Cells

To evaluate the antigenic comparability between gB derived from seeds and gB produced in HCMV-infected cells, we investigated whether seed-derived gB could inhibit the immunofluorescence normally observed when a gB-specific mAb with neutralizing activity bound to HCMV-infected human fibroblasts. Positive results (inhibition) would indicate that the seed-derived gB expressed the target epitope of the mAb with sufficient fidelity to preclude any residual mAb binding to this gB neutralization epitope in the infected cells.

An immunofluorescent assay (IFA) specific for gB of HCMV was performed using human foreskin fibroblasts infected with the Towne strain of HCMV (Tackaberry et al., 1993) . The IFA was performed by blocking wells with 10% normal goat serum followed by application of the gB-specific murine mAb CMVB1. Binding of CMVB1 to the infected cells was visualized by addition of fluorescein isothiocyanate (FITC)-conjugated goat antiserum to mouse immunoglobulins IgG, IgA, IgM (Cappel/Organon Teknika Corp., West Chester, PA) diluted in 3% BSA/PBS containing Evans Blue. For inhibition of the IFA, the protocol was modified as follows: (i) mAb CMVB1 was diluted in 2% BSA/PBS to 1/16,000, the minimum concentration producing strong fluorescence; and (ii) while blocking was underway, diluted mAb CMVB1 was mixed with an equal volume of transgenic tobacco seed extract (or suitable control) and incubated at 37°C. After 1 h, residual binding of mAb CMVB1 to gB in infected cells was assayed by applying the reaction mixture to the blocked wells for 20 min at 37°C. The assay was then continued as usual. Controls included substituting diluent or extract from non-transformed tobacco seeds into the reaction mixture, instead of transgenic tobacco seed extract. In all experiments, seed extracts from both transformed and non-transformed plants were adjusted to the same protein concentration of 4.2 mg/mL (Tackaberry et al., 1999).

As illustrated in Figure 26a - c, under conditions of no inhibition, binding of mAb CMVB1 resulted in a pattern of fluorescence typical of antibodies directed against late HCMV antigens (Larose et al., 1991; Tackaberry et al., 1993) with a prominent perinuclear inclusion body in the cytoplasm (Figure 26). However, complete inhibition of fluorescence was observed when the mAb was pre-incubated with seed extract derived from transformed plants, in this case #A27 (Figure 26b). The same concentration of seed extract derived from non-transformed tobacco plants had no inhibitory effect (Figure 4c).

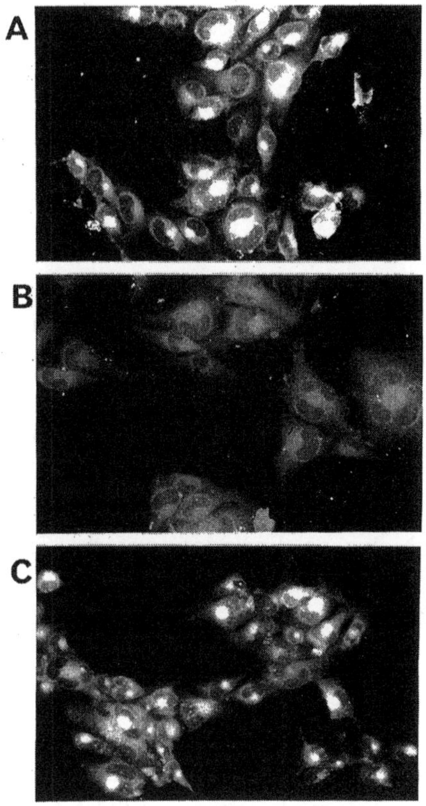

Figure 26. Comparability of gB derived from seeds and gB, in HCMV-infected cells, by inhibition of immunofluorescence. Protein samples extracted from seeds of transgenic tobacco plants were tested for their ability to inhibit binding of the gB-specific murine mAb CMVB1 to human fibroblasts infected with HCMV, by a gB-specific IFA. mAb CMVB1 was diluted in BSA/PBS to the minimum concentration producing strong fluorescence, mixed with an equal volume of tobacco seed protein extract (or suitable control) and incubated for 1 h. Residual mAb CMVB1 binding to gB was then measured by applying this reaction mixture to blocked wells followed by FITC-conjugated goat antiserum to mouse immunoglobulins (IgG, IgA, IgM). (a) conditions of no inhibition, with mAb CMVB1 pre-incubated with an equal volume of BSA/PBS; (b) mAb CMVB1 pre-incubated with seed extract derived from transformed plant A27; (c) mAb CMVB1 pre-incubated with seed extract derived from non-transformed tobacco plants, used at the same protein concentration as the A27 extract shown in (b). Scale bar (in c) represents 100 μm in all panels.

8. Subcellular Localization of gB in Endosperm Tissue of Mature Transformed Seeds

In order to extend our understanding of gB expression in plants, we wished to determine whether gB expression occurred in seed endosperm, and if so, to identify where in the endosperm cells this heterologous protein would accumulate. To this end, we undertook immunogold labelling of ultra-thin sections of seed endosperm from transformed tobacco plants, followed by transmission electron microscopy.

Seeds from transgenic plants expressing gB and control non-transformed plants were harvested when mature (day 25-30 post-pollination). For preparation of seed samples two conditions of fixation were used with no apparent differences: 2% paraformaldehyde/1.5% glutaraldehyde or 0.3% paraformaldehyde/0.8% glutaraldehyde, both prepared in 0.2 M sodium cacodylate buffer, pH 7.4. The endosperm and seed coat were dissected away from embryo tissue, which was discarded, and the fixed tissues gradually dehydrated (50%, 70% and 95% ETOH). Tissues were infiltrated with LR white resin (Merivac, Canada) by incubation overnight at RT. Polymerization was then carried out in fresh resin for a minimum of 24 h at 55°C. Ultra-thin sections (85 nm) of tissue were collected on formvar and carbon coated grids (Wright et al., 2001).

For immunogold labelling, grids were incubated for 1 h in 3% BSA/0.1% gelatin in Tris buffered saline with Tween 20 (TBST: 0.05 M Tris, 0.75 M NaCl, 0.25% Tween 20), followed by an additional 1 h in a pool of gB-specific murine mAb (as for the ELISA, above). The mAb pool was pre-adsorbed for 1 hr with fixed and dehydrated endosperm from mature seeds of non-transgenic plants. The grids were washed, incubated in anti-mouse IgG (Jackson Research Laboratories) and finally incubated in 10 nm protein-A gold (AuroProbe Protein A G10, Amersham Life Sciences). Grids were post-stained in uranyl acetate and lead citrate (Wright et al., 2001). Controls included identical preparation and labelling of endosperm from non-transgenic tobacco seeds, and treatment of transgenic seeds with rabbit anti-mouse IgG and protein-A gold ,or with protein-A gold alone in the absence of anti-gB mAbs.

Results of this study are illustrated in Figures 27 and 28. They reveal that gB is present in the endosperm of mature seeds from transgenic tobacco plants, consistent with published reports that the Gt3 promoter can drive expression of non-plant proteins in the seed endosperm of plants other than rice (Leisy et al., 1990; Zhao et al., 1994).

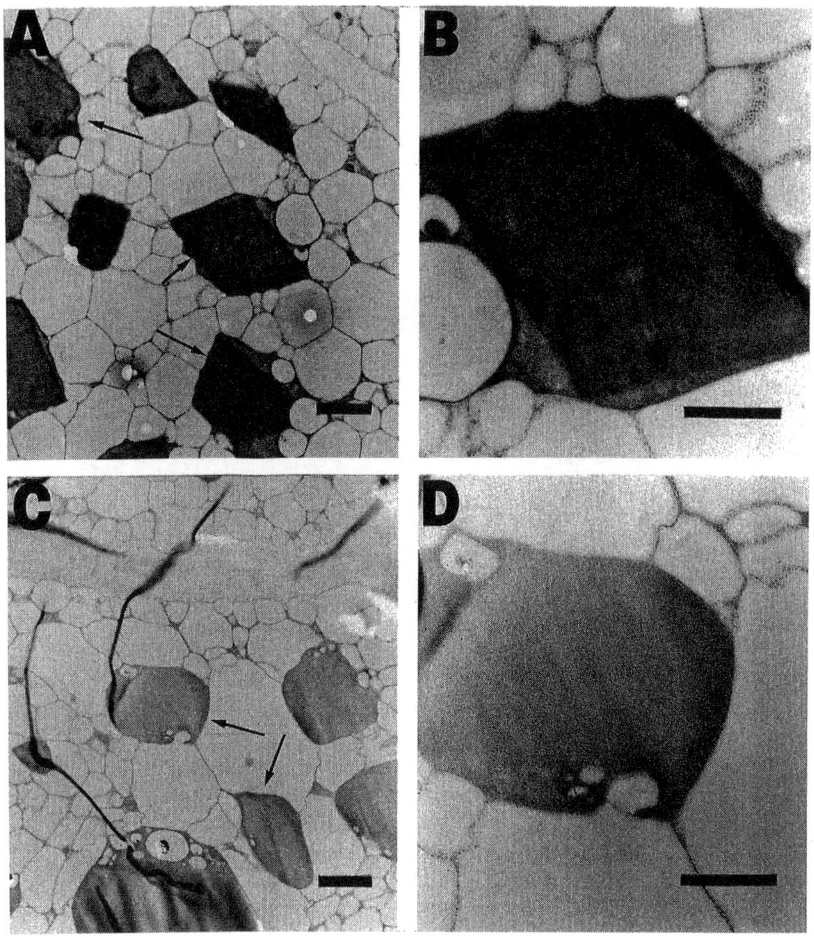

Figure 27. Immunogold labelling of endosperm from freshly harvested mature seeds. Seeds from transformed (A, B) and non-transformed (C, D) seeds were harvested at day 26 post-pollination and fixed directly. Endosperm was labelled with a pool of mouse mAb to gB, followed by rabbit anti-mouse IgG and protein-A gold (10 nm diameter) as described under Materials and Methods. Two magnifications of each section are presented: in A and C the magnification bar represents 2 microns; in B and D the magnification bar represents 1 micron. Arrows indicate some of the PSV.

Within the endosperm, gold was observed almost exclusively in protein storage vesicles (PSV), most of which were largely crystalline (Figure 27A, B). We also occasionally observed gold accumulating in smooth vesicles with diameters of 300-1000 nm in close proximity to PSV (Figure 28A). Seeds that had been stored for several months at room temperature (Figure 28A) labelled with as much gold as seeds that were processed immediately after harvest (Figure 27A, B). Specificity of the immunogold labelling for gB was confirmed by the relative absence of gold particles on sections of endosperm from non-transformed seeds (Figure 27C, D), as well as on sections of the same transgenic seed seen in Figure 28A but treated with rabbit anti-mouse IgG and protein-A gold in the absence of gB specific mAbs (Figure 28B) or treated with protein-A gold alone (not shown).

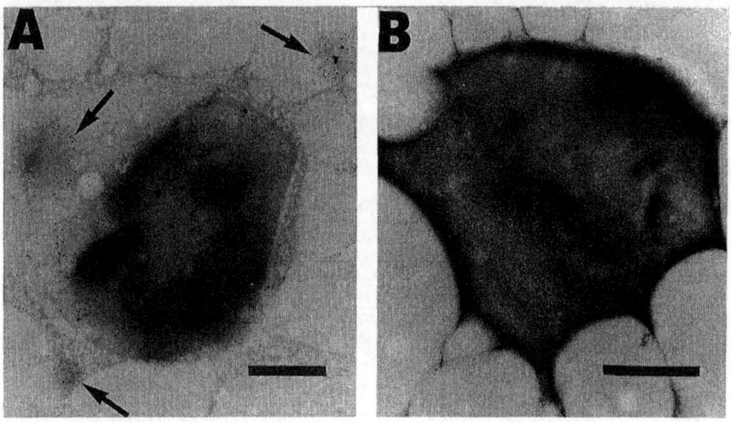

Figure 28. Immunogold labeling of endosperm from stored mature seeds. (A) Seeds from transformed tobacco that had been stored for several months at room temperature were fixed and the seed endosperm was labelled with a pool of mouse mAb to gB, followed by rabbit anti-mouse IgG and protein-A gold (10 nm diameter) as described under Materials and Methods. In (B), a section from the same seed was labeled with rabbit anti-mouse IgG and protein-A gold in the absence of gB specific mAbs. Arrows indicate smooth vesicles. The magnification bar represents 1 micron.

9. GENERAL DISCUSSION

9.1 Cloning, Transfer and Expression of gB Protein

The development of plant-derived, orally delivered subunit vaccines appears to be a feasible goal. In this chapter we describe our studies with the potential vaccinogen gB, an immunodominant envelope glycoprotein of the herpesvirus HCMV: its cloning and expression in transgenic tobacco seeds, immunological characterization and subcellular localization in mature seed endosperm.

Regulatory sequences of the rice seed storage protein glutelin were used to direct expression of the gB coding sequence to seeds. In contrast to other plant tissues, seeds provide a stable environment for protein synthesis and storage. Seeds may also be practical as delivery vehicles for targeting orally administered vaccines to the mucosal lymphoid tissue of the gut, since protein antigens must be protected from denaturation until reaching the ileum of the small intestine. Construction of plasmid pPH2/gB/NOS-T was designed so that transcription of the full gB sequence of approximately 2.7 kb was directed by the glutelin Gt3 promoter. In addition, to facilitate sorting of gB polypeptides to the secretory pathway of plant cells, the 72 bp signal peptide sequence of gB was removed and replaced with the 72 bp signal peptide sequence of Gt3. After transformation of tobacco and regeneration of mature plants, genomic DNA extracts from leaves of selected transformants was screened for the transgene. Results demonstrated that the transgene was carried in the majority of plants tested. Southern analysis further revealed that most of the plants exhibited a hybridizing band of DNA at 3.9 kb, the same size as the transgene DNA. But, as is typical of plants transformed with binary vectors via Agrobacterium species, transfer and subsequent integration of the pPH2/gB/NOS-T construct into the tobacco genome was variable and there was considerable transgene heterogeneity, with restriction fragments of many sizes identified by a probe specific for sequences of pPH2/gB/ NOS-T. These results suggest different transgene copy numbers and integration of the transgene into different sites. Multiple hybridization fragments may also result from tandem inserts, elimination of restriction sites flanking pPH2/gB/NOS-T, rearrangement or partial deletion of the insert, or scrambling of transferred DNA by recombination during or after insertion (Scott et al., 1988). Each of these events could impact on synthesis and yields of the recombinant product (Albright et al., 1987; Iglesias et al., 1997) and emphasize the need to generate a considerable number of integration events in order to recover plant lines with appropriate levels of stable transgene expression.

Using a capture ELISA, our data show that gB expression was targeted to the seeds of transformed tobacco plants and that the recombinant gB retained four unique gB epitopes of the wild-type molecule (unpublished observations). Three of the epitopes are neutralizing and at least one is conformational. The amount of recovered gB accumulated to an appreciable level with a maximum yield of 146 ng gB /mg total extracted protein. Furthermore, some of the transformed tobacco plant lines have now been producing gB at steady levels for over 36 months, indicative of a degree of transgene stability.

Further demonstration of the structural authenticity of gB epitopes was shown by the ability of seed-derived gB to completely inhibit all binding of a neutralizing gB-specific mAb to HCMV-infected cells, visualized by immunofluorescence. If this neutralization epitope on seed-derived gB was even slightly altered from gB produced in HCMV-infected cells, then some degree of residual binding by the anti-gB mAb would likely have been observed.

9.2 Sub-cellular Localization of gB in Seed Endosperm

When storage proteins from other plants are expressed in tobacco seeds, they appear to be correctly glycosylated, oligomerized, post-translationally processed and transported to the PSV (see Herman and Larkins, 1999; Cramer et al., 1999). However, similar studies of heterologous proteins from animal viruses in plant seeds have not been reported. To extend our studies, we carried out experiments to investigate whether gB expression occurred in the seed endosperm and, if so, to identify where in the endosperm cells the protein accumulated. We approached this by performing immunogold labelling of ultra-thin sections of seed endosperm from transformed tobacco plants and examining the sections by transmission electron microscopy.

Our results indicate that gB synthesized in tobacco seeds is transported to and stored in PSV as are plant storage proteins. In plants, an N-terminal signal peptide is necessary for cotranslational insertion of storage proteins into the membrane of the rough endoplasmic reticulum (RER). After insertion, the signal peptide is cotranslationally cleaved, an event that may be a requirement for correct folding and oligomerization (Dickinson et al., 1987; Coleman et al., 1995). From the RER, proteins are transported to PSV by one of two routes, both of which require specific sorting determinants. Some, for example many prolamins, exit the ER by budding directly into smooth dense vesicles. Others, such as the glutelins and globulins, are transported to the Golgi and subsequently bud from the trans-Golgi network into dense vesicles indistinguishable from those associated with the ER (Hohl et al., 1996; Miller and Anderson, 1999). These vesicles range in size from 130 to 300 nm in diameter and appear to fuse with existing vacuoles to form PSV (Herman and Larkins, 1999; Vitale and Raikhel, 1999; Miller and Anderson, 1999).

9.3 Sorting of gB in Endosperm Cells

gB is a transmembrane protein of 907 amino acids for the prototype Towne strain (Spaete et al., 1988), which is initially synthesized in infected cells as a 105 kDa non-glycosylated polypeptide. Cotranslational glycosylation, cleavage of the N-terminal 24 amino acid signal peptide, oligomerization and folding take place in the ER of the cell, where it is transiently associated with a membrane-bound chaperonin (Yamashita et al., 1996). This results in transport of a 150 kDa gB precursor to the Golgi complex where further carbohydrate modifications occur and the polypeptide is proteolytically cleaved to yield products of 116 kDa and 58 kDa which are disulphide linked (Spaete et al., 1988; Britt and Vugler, 1989; Kari et al., 1990). Both species are targets for neutralizing and non-neutralizing antibodies, each representing both continuous and discontinuous epitopes (Kari and Gehrz, 1991; Qadri et al., 1992; Speckner et al., 1999). A recently identified phosphorylation site is located in the cytoplasmic tail (Norais et al., 1996) and may be important for correct intracellular trafficking (Fish et al., 1998). By analogy with other mammalian glycoproteins that have been expressed in tobacco plants (Salmon et al., 1998; Cabanes-Macheteau et al., 1999) we believe that gB travels from the ER to the Golgi. However, we have not yet been able to confirm this as the Golgi cannot be easily visualized in unosmicated tissues embedded under the conditions we used (Hoffman et al., 1988; Wright et al., 2001). From the Golgi, gB is probably transported via smooth dense vesicles to the PSV like plant storage proteins. At this time we cannot be sure that the smooth vesicles we observed specifically labelling for gB are transport vesicles, as they are larger than the dense vesicles that have been described by others (Hoffman et al., 1988; Miller and Anderson 1999).

This direct sorting of gB to PSV is quite unlike the events occurring in mammalian cells during viral infection. Mature gB is localized both to the plasma membrane (Stinski et al., 1979) and to a cytoplasmic compartment (Tooze et al., 1993; Sanchez et al., 2000). The current evidence supports a model whereby gB is transported from the Golgi to the plasma membrane, and movement to the cytoplasmic vacuoles occurs after internalization from the surface by endocytosis (Radsak et al., 1996; Tugizov et al., 1998; Fish et al., 1998). The cytoplasmic tail of gB contains several potential sorting motifs for entry into the endocytic pathway in mammalian cells (Tugizov et al., 1998), but whether these motifs are also essential for targeting to cytoplasmic compartments has not been established.

Nonetheless, our data indicate that gB must contain a PSV sorting signal that is recognized in the plant cell, whatever the route taken to the storage site. Storage proteins exiting the Golgi and destined for PSV either possess a non-consensus cleavable C-terminal sorting signal or they contain sorting determinants that involve conformations or exposed sequences within the mature protein (see Neuhaus and Rogers, 1998; Vitale and Raikhel, 1999; Miller and Anderson, 1999). We think it unlikely that the endocytic targeting motifs of this viral protein would function in plants, favouring instead the hypothesis that mature gB possesses a cryptic signal responsible for sorting to the PSV since no plant sorting sequences were incorporated into the construct used to transform tobacco.

10. SUMMARY

Collectively, these data show that the synthesis of heterologous proteins can be targeted successfully to a particular plant tissue, in this case the endosperm of seeds, with authentic reproduction of biologically relevant, immunoreactive structural features. They also extend our previous studies with the cytokine GM-CSF (Ganz et al., 1996; Sardana et al, 1997), which at 18 kDa is considerably smaller than gB, regarding the capability and utility of plant systems for producing large proteins of biopharmaceutical interest. Nonetheless, it is unlikely that gB or other potential subunit vaccine candidates made in transgenic plant systems are identical to their native counterparts. How potential differences, however slight, may affect the clinical safety and/or effectiveness of the products is unknown and extensive pre-clinical evaluation will certainly be required. In this regard the present study provides a platform for ongoing and future investigations in our laboratory, including expressing gB in plants more appropriate for human consumption, such as rice, and more detailed biochemical and immunological characterization of the plant-derived gB products.

ACKNOWLEDGEMENTS

The authors wish to thank Dr. G. Marshall (University of Louisville, KY) and Dr. T. W. Okita (Washington State University, WA) for providing the gB plasmid pSVL/CMVgB and the glutelin promoter plasmid pJH18, respectively. They are also indebted to Ms. F. Prior and Ms. M. Tocchi for outstanding technical assistance, and to Dr. J. Austin and Mr. G. Sanders for invaluable support with the electron microscopy.

REFERENCES

Adeli K and Altosaar I, 1984. Compartmentalization of seed reserve proteins -- proposed discriminatory factors. FEBS Lett, 178:193-198

Adler SP, Starr SE, Plotkin SA, Hempfling SH, Buis J, Manning ML and Best AM, 1995. Immunity induced by a primary human cytomegalovirus infection protects against secondary infection among women of childbearing age. J Infect Dis, 171:26-32

Adler SP, Plotkin SA, Gonczol E, Cadoz M, Meric C, Wang JB, Dellamonica P, Best AI, Zahnradnik J, Pincus S, Berencsi K, Cox WI and Gyulai Z, 1999. A canarypox vector expressing cytomegalovirus (CMV) glycoprotein B primes for antibody responses to a live attenuated CMV vaccine (Towne). J Infect Dis, 180:843-846

Albright LM, Yanofsky MF, Leroux B, Ma DQ and Nester EW, 1987. Processing of the T-DNA of *Agrobacterium tumefaciens* generates border nicks and linear, single-stranded T-DNA. J Bacteriol, 169:1046-1055

Bewley JD and Black M, 1994. Seed development and maturation. In *Seeds: Physiology of development and germination* . Eds. Bewley JD and Black M. Plenum Press, New York, pp 35-115

Britt WJ, 1996. Vaccines against cytomegalovirus: Time to test. Trends in Microbiol, 4: 34-38

Britt WJ and Vugler LG, 1989. Processing of the gp55-116 envelope glycoprotein complex (gB) of human cytomegalovirus. J Virol, 63: 403-410

Britt WJ and Mach M, 1996. Human cytomegalovirus glycoproteins. Intervirol, 39:401-412

Cabanes-Macheteau M, Fischette-Lainé A-C, Loutier-Bourhis, C, Lange C, Vine ND, Ma JKC, Lerouge P and Fay L, 1999. N-Glycosylation of a mouse IgG expressed in transgenic tobacco plants. Glycobiology, 9:365-372

Chrispeels MJ and Faye L, 1996. The production of recombinant glycoproteins with defined non-immunogenic glycans. In *Transgenic plants: A production system for industrial and pharmaceutical proteins.* Eds. Owen MRL and Owen P. John Wiley and Sons Ltd., New

York, pp 99-113

Coleman CE, Lopes MA, Gillikein JW, Boston RS and Larkins BA, 1995. A defective signal peptide in the maize high-lysine mutant floury-2. Proc Natl Acad Sci USA, 92, 6828-6831

Cramer CL, Boothe JG and Oishi KK, 1999. Transgenic plants for therapeutic proteins: Linking upstream and downstream strategies. Curr Top Microbiol Immunol, 240:95-118

Cranage MP, Smith GL, Bell SE, Hart H, Brown C, Bankier AT, Tomlinson P, Barrell BG and Minson TC, 1988. Identification and expression of a human cytomegalovirus glycoprotein with homology to the Epstein-Barr virus BXLF2 product, varicella-zoster virus gpIII, and herpes simplex virus type I glycoprotein H. J Virol, 62:1416-1422

Crapnell K, Zanjani ES, Chaudhuri A, Ascensao JL, St Jeor S and Maciejewski JP, 2000. In vitro infection of megakaryocytes and their precursors by human cytomegalovirus. Blood, 95: 487-493

Datla RSS, Hammerlindl JK, Panchuk B, Pelcher LE and Keller W, 1992. Modified plant transformation vectors with the wild-type gene encoding NPTII. Gene, 122:383-384

Dickinson CD, Floener L, Lilley G and Nielson N, 1987. Self-assembly of proglycinin and hybrid proglycinin synthesized in vitro from cDNA. Proc Natl Acad Sci USA, 84:5525-5529

Fish KN, Soderberg-Naucler C and Nelson JA, 1998. Steady-state plasma membrane expression of human cytomegalovirus gB is determined by the phosphorylation state of Ser_{900}. J Virol, 72:6657-6664

Fowler KB, Stagno D, Pass RF, Britt WJ, Boll TJ and Alford CA, 1992. The outcome of congenital cytomegalovirus infection in relation to maternal antibody status. N Engl J Med, 326:663-667

Frey SE, Harrison C, Pass RF, Yang E, Boken D, Sekulovich R, Percell S, Izu A, Hirabayashi S, Burke RL and Duliege A-M, 1999. Effects of antigen dose and immunization regimens on antibody responses to a cytomegalovirus glycoprotein B subunit vaccine. J Infect Dis, 180:1700-1703

Ganz PR, Dudani A, Tackaberry E, Sardana R, Sauder C, Cheng X and Altosaar I, 1995. Plants as factories for producing human blood proteins. Can Soc Trans Med Bull, 7:112-121

Ganz PR, Dudani AK, Tackaberry ES, Sardana R, Sauder C, Cheng X and Altosaar I, 1996. Expression of human blood proteins in transgenic plants: the cytokine GM-CSF as a model protein. In *Transgenic plants: A production system for industrial and pharmaceutical proteins.* Eds. Owen MRL and Owen P. John Wiley and Sons Ltd., New York, pp 281-97

Haq TA, Mason H, Clements JD and Arntzen CJ, 1995. Oral immunization with a recombinant bacterial antigen produced in transgenic plants. Science, 268:714-717

Herman E and Larkins B, 1999. Protein storage bodies and vacuoles. Plant Cell, 11:601-613

Ho M, 1991. *Cytomegalovirus: Biology and infection.* 2nd Ed. Plenum Medical Book Co., New York, USA

Hoffman LM, Donaldson DD and Herman EM, 1988. A modified storage protein is synthesized, processed, and degraded in the seeds of transgenic plants. Plant Mol Biol, 11:717-729

Hohl I, Robinson DG, Chrispeels MJ and Hinz G, 1996. Transport of storage proteins to the vacuole is mediated by vesicles without a clathrin coat. J Cell Sci, 109:2539-2550

Horsch RB, Fry JE, Hoffmann, NL, Eichholtz D, Rogers SG and Fraley RT, 1995. A simple and general method for transferring genes into plants. Science, 227:1229-1231

Iglesias VA, Moscone EA, Neuhuber F, Michalowski S, Phelan T, Spiker S, Matzke M and Matzke AJ, 1997. Molecular and cytogenic analyses of stably and unstably expressed transgenic loci in tobacco. Plant Cell, 9:1251-1264

Jenkins N, Parekh RB and James DC, 1996. Getting the glycosylation right: Implications for the biotechnology industry. Nat Biotech, 14:975-981

Kari B, Liu Y-N, Lussenhop N, Stinski M and Gehrz R, 1990. Structure and composition of a family of human cytomegalovirus glycoprotein complexes designated gC-1. J Gen Virol, 71:2673-2680

Kari B and Gehrz R, 1991. Biochemical and immunological analysis of discontinuous epitopes in the family of human cytomegalovirus glycoprotein complexes designated gC-I. J Gen Virol, 72:1975-1983

Kusnadi AR, Nikolov ZL and Howard JA, 1997. Production of recombinant proteins in transgenic plants: Practical considerations. Biotech and Bioengin, 56:473-484

Landini MP, 1992. Antibody responses to human cytomegalovirus proteins. Rev Med Virol, 2:63-72

Larose Y, Tackaberry ES and Brodeur BR, 1991. Human monoclonal antibodies to cytomegalovirus recognize viral epitopes on the surface of viral-infected cells. Human Antibod Hybridomas, 2:65-73

Leisy DJ, Hnilo J, Zhao Y and Okita TW, 1990. Expression of a rice glutelin promoter in transgenic tobacco. Plant Mol Biol, 14:41-50

Lerouge P, Cabanes-Macheteau M, Rayon C, Fischette-Lainé A-C, Gomord V and Faye L, 1998. N-glycoprotein biosynthesis in plants: Recent developments and future trends. Plant Mol Biol, 38:31-48

Lucas KG, Sun Q, Burton RL, Tilden A, Vaughan WP, Carabasi M, Salzman D and Ship A, 2000. A phase I-II trial to examine the toxicity of CMV- and EBV-specific cytotoxic T lymphocytes when used for prophylaxis against EBV and CMV disease in recipients of CD34-selected /T cell-depleted stem cell transplants. Hum Gene Ther, 11:1453-1493

Manganaro M, Ogra PL. and Ernst PB, 1994. Oral immunization: Turning fantasy into reality. Intl Arch Allergy Immunol, 103:223-233

Mason HS, Lam DM and Arntzen CJ, 1992. Expression of hepatitis B surface antigen in transgenic plants. Proc Natl Acad Sci USA, 89:11745-11749

Mason HS, Ball JM, Shi J-J, Jiang X, Estes MK and Arntzen CJ, 1996. Expression of Norwalk virus capsid protein in transgenic tobacco and potato and its oral immunogenicity in mice. Proc Natl Acad Sci USA, 93:5335-5340

Mason HS, Haq TA, Clements JD and Arntzen CJ, 1998. Edible vaccine protects mice against *Escherichia coli* heat-labile enterotoxin (LT): potatoes expressing a synthetic LT-B gene. Vaccine, 16:1336-1348

Miller EA and Anderson MA, 1999. Uncoating the mechanisms of vacuolar protein transport. Trends Plant Sci, 4:46-48

Mitchell VS, Philipose NM, Sanford JP, 1993. *The children's vaccine initiative - Achieving the vision.* Institute of Medicine, National Academy Press, Washington.

Moffat AS, 1995. Exploring transgenic plants as a new vaccine source. Science, 268:658-660

Neuhaus JM and Rogers JC, 1998. Sorting of proteins to vacuoles in plant cells. Plant Mol Biol, 38:127-144

Norais N, Hall JA, Gross L, Tang D, Kaur S, Chamberlain S, Burke RL and Marcus F, 1996. Evidence for a phosphorylation site in cytomegalovirus glycoprotein gB. J Virol, 70:5716-5719

Okita TW, Hwang YS, Hnilo J, Kim WT, Aryan AP, Larson R and Krishnan HB, 1989. Structure and expression of the rice glutelin multigene family. J Biol Chem, 264:12573-12581

Pande H, Campo K, Tanamachi B, Forman S and Zaia J, 1995. Direct immunization of mice with plasmid DNA encoding the tegument protein pp65 (ppUL83) of human cytomegalovirus induces high levels of circulating antibody to encoded protein. Scand J Infect Dis, 99:117-120

Pass RF, Duliege A-M, Boppana S, Sekulovich R, Percell S, Britt W and Burke RL, 1999. J Infect Dis, 180:970-975

Pen J, 1996. Comparison of host systems for the production of recombinant proteins. In *Transgenic plants: A production system for industrial and pharmaceutical proteins.* Eds. Owen, MRL and Owen P. John Wiley and Sons Ltd., New York, pp.149-167

Plotkin SA, 1999. Cytomegalovirus vaccine. Am Heart J, 138:S484-487

Qadri I, Navarro D, Paz P and Pereira L, 1992. Assembly of conformation-dependent neutralizing domains on glycoprotein B of human cytomegalovirus. J Gen Virol, 73:2913-2921

Radsak K, Eickmann M, Mockenhaupt T, Bogner E, Kern H, Eis-Hübinger A and Reschke M, 1996. Retrieval of human cytomegalovirus glycoprotein B from the infected cell surface for virus envelopment. Arch Virol, 141:557-572

Richter L and Kipp PB, 1999. Transgenic plants as edible vaccines. Curr Top Microbiol Immunol, 240:159-176

Riddell SR, Watanabe KS, Goodrich JM, Li CR, Agha ME and Greenberg PD, 1992. Restoration of viral immunity in immunodeficient humans by the adoptive transfer of T cell clones. Science, 257:238-241

Robert LS, Donaldson PA, Ladaique C, Altosaar I, Arnison PG and Fabijanski SF, 1989. Antisense RNA inhibition of β-glucuronidase gene expression in transgenic tobacco plants.

Plant Molec Biol, 13:399-409

Salmon V, Legrand D, Slomianny MC, Yazidi IE, Spik G, Gruber V, Bournat P, Olagnier B, Mison D, Theisen M and Merot B, 1998. Production of human lactoferrin in transgenic plants. Prot Exp Pur, 13:127-135

Sambrook J, Fritsch EF and Maniatis T , 1989. *Molecular cloning: A laboratory manual.* 2nd Ed. Cold Spring Harbor, New York

Sanchez V, Greis KD, Sztul E and Britt WJ, 2000. Accumulation of virion tegument and envelope proteins in a stable cytoplasmic compartment during human cytomegalovirus replication: characterization of a potential site of virus assembly. J Virol, 74:975-986

Sardana R, Ganz PR, Dudani A, Tackaberry ES, Cheng X and Altosaar I, 1997. Synthesis of recombinant human cytokine GM-CSF in the seeds of transgenic tobacco plants. In *Methods in Biotechnology*, Vol 3. Eds. Cunningham C and Porter A. Humana Press Inc, Totowa, New Jersey, pp 77-87

Schleiss MR, Jensen BN, Bravo NJ and Bernstein DI, 2000. Immunogenicity evaluation of DNA vaccines that target guinea pig cytomegalovirus proteins glycoprotein B and UL83. Viral Immunol, 13:155-167

Scott R, Draper J, Jefferson R, Dury G and Jacob L, 1988. Analysis of gene organization and expression in plants. In *Plant genetic transformation and gene expression: A laboratory manual.* Eds. Draper J, Scott R and Armitage P. Blackwell Scientific Publications, Oxford, pp 263-339

Service RF, 1994. Triggering the first line of defense. Science, 265:1522-1524

Shalaby WSW, 1995. Development of oral vaccines to stimulate mucosal and systemic immunity: barriers and novel strategies. Clin Immunol Immunopath, 74:127-134

Snydman DR, 1990. Cytomegalovirus immunoglobulins in the prevention and treatment of cytomegalovirus disease. Rev Infect Dis, 12:S839-848

Spaete RR, Thayer RM, Probert WS, Masiarz FR, Chamberlain SH, Rasmussen L, Merigan TC and Pachl C, 1988. Human cytomegalovirus strain Towne glycoprotein B is processed by proteolytic cleavage. Virology, 167:207-225

Speckner A, Glykofrydes D, Ohlin M and Mach M, 1999. Antigenic domain 1 of human cytomegalovirus glycoprotein B induces a multitude of different antibodies which, when combined, results in incomplete virus neutralization. J Gen Virol, 80:2183-2191

Stinski MF, Mocarski ES, Thomsen DR and Urbanowski ML, 1979. Membrane glycoprotein and antigens induced by human cytomegalovirus. J Gen Virol, 43:119-129

Stocchi R, Ward KN, Fanin R, Baccarani M and Apperley JF, 1999. Management of human cytomegalovirus infection and disease after allogeneic bone marrow transplantation. Haematologica, 84:71-79

Tackaberry ES, Hamel J, Larose Y and Brodeur BR, 1993. Anti-idiotypic mimicry of a neutralizing epitope on the glycoprotein B (gB) complex of human cytomegalovirus. J Virol, 67:6815-6819

Tackaberry ES, Hamel J and Brodeur B, 1997. Anti-idiotypic studies of the glycoprotein B (gB, UL55) complex of human cytomegalovirus: Mimicry of a neutralizing epitope and development of diagnostics. In *Idiotypes in medicine: Autoimmunity, infection and cancer*. Eds. Shoenfeld Y, Kennedy R and Ferrone S. Elsevier Science Publications, Amsterdam, pp 259-68

Tackaberry ES, Dudani AK, Prior F, Tocchi M, Sardana R, Altosaar I and Ganz PR, 1999. Development of biopharmaceuticals in plant expression systems: cloning, expression and immunological reactivity of human cytomegalovirus glycoprotein B (UL55) in seeds of transgenic tobacco. Vaccine, 17:3020-3029

Tacket CO, Mason HS, Losonsky G, Clements JD, Levine MM and Arntzen CJ, 1998. Immunogenicity in humans of a recombinant bacterial antigen delivered in a transgenic potato. Nat Med, 4:607-609

Tacket CO, Mason HS, Losonsky G, Estes MK, Levine MM and Arntzen CJ, 2000. Human responses to a novel Norwalk virus vaccine delivered in transgenic potatoes. J Infect Dis, 182:302-305

Thanavala Y, Yang Y-F, Lyons P, Mason HS and Arntzen C, 1995. Immunogenicity of transgenic plant-derived hepatitis B surface antigen. Proc Natl Acad Sci USA, 92:3358-3361

Tooze J, Hollinshead M, Reis B, Radsak K and Kern H, 1993. Progeny vaccinia and human cytomegalovirus particles utilize early endosomal cisternae for their envelopes. Eur J Cell Biol, 60:163-178

Tugizov S, Maidji E, Xiao J, Zheng Z and Periera L, 1998. Human cytomegalovirus glycoprotein B contains autonomous determinants for vectorial targeting to apical membranes of polarized epithelial cells. J Virol, 72:7374-7386

Vitale A and Raikhel NV, 1999. What do proteins need to reach different vacuoles? Trends Plant Sci, 4:149-155

Walmsley AM and Arntzen CJ, 2000. Plants for delivery of edible vaccines. Curr Opin Biotechnol, 11:126-129

Walter EA, Greenberg PD, Gilbert MJ, Finch RJ, Watanabe KS, Thomas ED and Riddell SR, 1995. Reconstitution of cellular immunity against cytomegalovirus in recipients of allogeneic bone marrow by transfer of T-cell clones from the donor. N Engl J Med, 333:1038-1044

Wang JB, Adler SP, Hempfling S, Burke RL, Duliege A-M, Starr SE and Plotkin SA, 1996. Mucosal antibodies to human cytomegalovirus glycoprotein B occur following both natural infection and immunization with human cytomegalovirus vaccines. J Infect Dis, 174:387-392

Wright KE, Prior F, Sardana R, Altosaar I, Dudani AK, Ganz PR and Tackaberry ES, 2001. Sorting of glycoprotein B from human cytomegalovirus to protein storage vesicles in seeds of transgenic tobacco. Transgenic Res,10:177-181

Yamashita Y, Shimokata K, Mizuno S, Daikoku T, Tsurumi T and Nishiyama Y, 1996. Calnexin acts as a molecular chaperone during the folding of glycoprotein B of human cytomegalovirus. J Virol, 70:2237-2246

Zambryski P, 1988. Basic processes underlying *Agrobacterium* mediated DNA transfer to plant cells. Annu Rev Genet, 22:1-30

Zhao Y, Leisy DJ and Okita TW, 1994. Tissue-specific expression and temporal regulation of the rice glutelin Gt3 gene are conferred by at least two spatially separated *cis*-regulatory elements. Plant Mol Biol, 25:429-436

Chapter 7

USE OF PLANT VIRUS-BASED EXPRESSION SYSTEMS FOR THE PRODUCTION OF HIV VACCINES

G.G. Zhang
Genesis Research and Development Corporation Ltd., PO Box 50, Auckland, New Zealand

Keywords: antigenicity, chimeric virus particle, coat protein, epitope presentation, human immunodeficiency virus type 1 (HIV-1), immunogenicity, plant virus, vaccine antigen, viral vector

Abstract The number of patients carrying acquired immune deficiency syndrome (AIDS) keeps increasing as new infections with human immunodeficiency virus (HIV) continue to occur. Although anti-retroviral drug therapies are capable of reducing viral load of infected patients, it is clear that the effective and ultimate solution to control AIDS is vaccination against HIV infection. Recent development of HIV vaccines has shown great promise and different stages of HIV vaccine clinical trials in humans are currently being conducted worldwide. However, vaccines produced using traditional production systems are costly and future products may be unaffordable in developing counties, where most HIV infections occur. Several plant-based systems have been developed and are being optimized for cost effective production of vaccines. RNA plant viruses have attracted great interest because of their utility in over expressing vaccine antigens. In addition, plant virus-derived products have been shown to be effective in inducing immunoresponses when administered to animals. Results from these studies support the further development of plant virus-based expression systems for the production of cost effective HIV vaccines. This chapter focuses on the recent development of RNA plant virus-based expression systems and the immunogenicity and potential clinical applications of plant virus-derived HIV vaccines.

L. Erickson et al. (eds.),
Molecular Farming of Plants and Animals for Human and Veterinary Medicine, 155–177.

1. INTRODUCTION

Acquired immune deficiency syndrome (AIDS) is progressively threatening human life throughout the world. As of the end of 1998, the number of people living with AIDS increased to 33.4 million (Cohen, 2000). Unfortunately, no cures are available once one is infected by the human immunodeficiency virus (HIV), which is responsible for the development of AIDS. Although current anti-retroviral drugs are able to reduce patient viral loads, such therapeutic agents are simply not affordable in developing countries where most HIV infections occur. Even in industrialized countries where the drugs are available, long-term effectiveness is questionable because of poor drug tolerance and the emergence of drug-resistant viral strains. Consequently, there is an urgent need to find effective means to prevent HIV infections. To meet this need, HIV vaccines are currently being developed as an effective solution for prevention of HIV infections.

Live attenuated simian immunodeficiency virus (SIV) vaccines have proven to induce protective immunity against SIV in monkeys far more consistently than any other type of retroviral vaccine. However, the occurrence of AIDS in monkeys inoculated with live-attenuated SIV vaccines demonstrated possibility of recovery of the virulent virus. As HIV-1 mutates extremely rapidly, accumulating genetic alterations in a non-pathogenic virus used as a vaccine might lead eventually to the virus regaining its pathogenic potential (Letvin, 1998). As reversion to a virulent form is always of clinical concern, no live-attenuated HIV has been used in clinical trials, although such a form has been proven to be effective in monkeys. Inactivated HIV is a safer approach and is capable of inducing an immunoresponse. However, it has not been widely pursued because of related technical difficulties. With the increased understanding of HIV and human immunology, subunit vaccines consisting of macromolecules derived from the virus are currently under development. However, the cost of subunit vaccines is high because of the use of traditional production systems. To increase the availability of future vaccines, development of cost-effective production systems are needed.

With recent advances in plant molecular genetics, several commercially valuable and clinically important recombinant proteins have been expressed in plants. For example, the hepatitis B surface antigen, Norwalk virus capsid protein, heat-labile enterotoxin B, and cholera B subunit oligomers have been produced from transgenically-engineered tobacco and potato (Mason et al., 1992; 1996; Arakawa et al., 1997; 1998). When the plant-derived proteins were purified and introduced into animals, they were able to elicit immunological responses (Thanavala et al., 1995; Arakawa et al., 1997), thus supporting the use of plant-produced antigens for the development of vaccines. More interestingly, recent studies have shown that antigens produced in plants are able to induce immunoresponse in animals that are directly fed antigen-producing plant tissues, thereby supporting the feasibility of edible vaccine (Haq et al., 1995; Tacket et al., 1998; Arakawa et al., 1998). Production of vaccines in transgenic plants have been reviewed extensively (Mason and Arntzen, 1995; Featherstone, 1996; Meloen et al., 1998; Palmer et al., 1999; Ma and Vine, 1999).

As an alternative to stable expression via transgenic plants, several RNA plant viruses have been investigated as expression vectors for transient production of vaccine antigens in plants (Beachy et al., 1996; Jagadish et al., 1996; Porta et al., 1996; Spall et al., 1997, Palmer et al., 1999; Ma and Vine, 1999). Compared with the expression of transgenes, very high levels of gene expression can be easily achieved using plant viral vectors due to the nature of autonomous replication of plant viruses. The yield of plant virus-based vaccines was reported to reach levels as high as milligrams per gram fresh plant tissues (see Ma and Vine, 1999). Since some vaccine proteins might be detrimental to plant growth or toxic to plants, an added advantage of viral vectors is that infection of plants can be delayed until they reach an appropriate developmental stage or size to maximize the yield of proteins.

2. CANDIDATE PEPTIDES AND PROTEINS FOR HIV VACCINES

The HIV-1 genome contains three major genes, the group-specific antigen gene (*gag*), the polymerase gene (*pol*) and the envelope gene (*env*). Both the *gag* and *env* encode precursor polyproteins, which are cut into smaller structural proteins during post-translational processing. The p55 encoded by *gag* gives rise to p17, p24, p7 and p6. The *env*-encoded gp160 is processed into two glycoproteins, gp120 and gp41 (Figure 29). Both are located on the surface of the HIV particle, with gp120 forming the exterior domain and gp41 forming the transmembrane domain. These two glycoproteins play important roles in attachment of HIV to target cells and subsequent penetration (Wyatt and Sodroski, 1998).

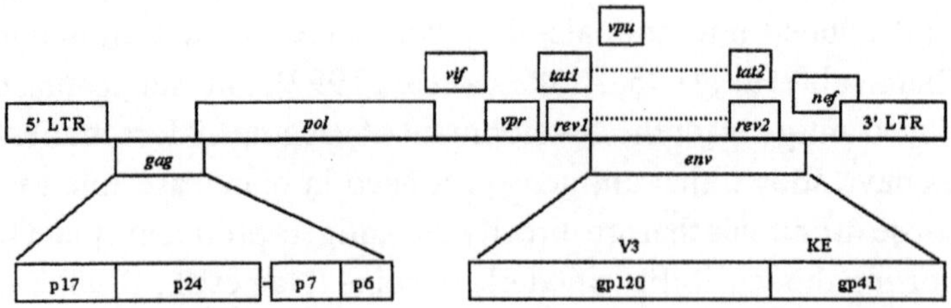

Figure 29. HIV-1 genome organization The genome, shown as a horizontal line, contains 9 genes (depicted as boxes) including gag, pol, vif, vpr, tat, rev, vpu, env and nef. gag, the group specific antigen gene, encodes a 55 kDa precursor polyprotein which is cleaved into 17 (P17), 24 (p24), 7 (P7) and 6 kDa (P6) structural proteins; env, the envelope gene, encodes a 160 kDa precursor polyprotein that is processed into 120 kDa (gp120) and 41 kDa (gp41) structural proteins. Shaded regions indicate the approximate location of V3 from gp120 and the Kennedy epitope (KE) from gp41.

The location of gp120 and gp41 on the surface of HIV-1 makes them the primary target for antibodies that could potentially block the key functions of these proteins. It has been shown that gp120 and gp41 induce specific antibodies against HIV-1 in AIDS patients. Some of these antibodies have viral neutralizing activity and therefore are able to prevent attachment of HIV to cells and subsequent penetration (Shaw et al., 1984; Weiss et al., 1985). The p24 is a capsid protein forming the core capsid. It is an important early marker of HIV infection and has been shown to induce cellular and humoral immune responses (Martin et al., 1993; Benson et al., 1999). In addition, p24 has been widely used in blood screening and clinical diagnoses (Lefrere et al., 1992; Stiehm et al., 2000). Several HIV proteins including gp120, gp41 and p24 have been used as components in subunit vaccines in current preventive as well as therapeutic HIV vaccine trials (Table 4).

Characterization of gp120 and gp41 showed that the neutralizing antibody response is directed against certain regions or epitopes of these surface glycoproteins (Weiss, 1993; Tsoukas and Bernard, 1994). Four seroreactivity regions have been identified in patients' sera (Broliden et al., 1992). Two are involved in the early stages of virus entry and have been localized on gp120, the V3 loop region and the CD4 binding site (Ho et al., 1991; Weiss, 1993; Moore et al., 1994). The other two regions are located on the transmembrane envelope glycoprotein gp41. One is situated on the extracellular domain of gp41 at residues 647 to 671 (Muster et al., 1993; Sattentau et al., 1995; Chen and Dierich, 1996). The other is in the intracellular location at residues 731 to 751, referred to as the Kennedy epitope (Kennedy et al., 1986; Modrow et al., 1987; Gallagher, 1992)

Table 4. Candidate HIV/AIDS vaccines in development and clinical trials*

Alpha Vax	
-gag only	in development
-gag, env, pol	in development
Aventis	
-therapore-p24	in development
Aventis Pasteur	
-gp 160	in phases I/II
Chiron	
-p24	in phase I
-gp120	in phase II Thailand

*Data from the National Institute of Allergy and Infectious Disease Division for AIDS

The V3 loop consists of 35 amino acids (SVQINCTRPNYNKRKRIHIG PGRAFYTTKNIIGTIRQAHCNIS) and has been shown to elicit virus-neutralization antibodies (Rencher et al., 1995). Within the V3 loop, a 13 amino acid peptide (pep13; KSIHIGPGRAFYT) has been identified as the central motif of the V3 region (Arendrup et al., 1993). This peptide is the principal neutralization determinant and antibodies against this peptide have binding capacity to HIV-1 (Putney et al., 1991). The Kennedy epitope contains 22 amino acid residues (PRGPDRPEGIEEEGGERDRDRS) of HIV-1 gp41 and is shown to induce cross-reactive neutralizing antibodies in humans (Kennedy et al., 1986; Modrow et al., 1987; Gallagher, 1992). Due to their small size, the V3 loop and Kennedy epitope have been used to develop plant virus-based vaccine production systems for HIV. In addition, expression of a full-length p24 was also reported (Zhang et al., 2000).

3. EXPRESSION OF HIV VACCINES USING RNA PLANT VIRUS-BASED EXPRESSION VECTORS

RNA plant viruses are very efficient in replication and movement throughout host plants. Their genetic information is effectively stored in compact genomes, and many nucleotide sequences serve multiple functions. Also, each viral protein plays a key role in the life cycle including replication, symptom development, cell-to-cell movement, virus particle assembly and long-distance movement in transducting tissues throughout plants. Proteins responsible for replication and movement usually cannot be altered. However, the coat protein of most viruses, which is synthesized in large quantities, can be modified to a certain extent. The subgenomic promoter which drives the synthesis of a smaller viral subgenomic mRNA encoding coat protein is usually very active and yields high levels of coat protein mRNA. Therefore, it is advantageous to place sequences encoding foreign proteins under the control of the coat protein promoter. For a few viruses, the coat protein is dispensable for whole plant infection; therefore, the coat protein open reading frame (ORF) can be replaced with genes of interest. However, most viruses require assembly into virus particles prior to systemic movement. In this case, short peptides can be inserted into a coat protein region or fused to the coat protein at the N or C terminus so that the peptide will be displayed on the surface of the assembled virus particle. However, there is usually a size constraint for these peptide inserts. If the insert is too long, the virus is unable to assemble and fails to move systemically. Strategies for cloning an HIV sequence into a virus largely depend on the flexibility of a virus and the size of peptides or proteins to be expressed. In most studies, defined peptide sequences corresponding to antigenic sites from HIV-encoded proteins are chosen, such as the V3 loop from gp120 and the Kennedy epitope from gp41. Several viral expression systems used to produce HIV vaccines will be discussed below.

3.1 Presentation of the Kennedy Epitope from HIV-1 gp41 on the Surface of Cowpea Mosaic Virus

Cowpea mosaic virus (CPMV) is the type member of the comovirus. The bipartite CPMV genome consists of two separately encapsidated positive-strand

RNA molecules, a 5889 nucleotide RNA1 and a 3481 nucleotide RNA2. Each RNA molecule contains a single ORF that encodes precursor polyproteins. Post-translational processing of the polyproteins results in the formation of functional proteins. RNA1 encodes the RNA dependent RNA polymerases and therefore is able to self replicate in plant cells in the absence of RNA2. RNA2 encodes two (one large and one small) coat proteins as well as a movement protein and is required for systemic invasion of whole plants (Lomonossoff and Hamilton, 1999).

The structure of CPMV has been determined (Lomonossoff and Johnson, 1991). CPMV capsids contain 60 copies of large and 60 copies of small coat proteins, forming icosahedral virus particles. The coat protein of CPMV is essential for systemic movement, and replacement of the coat protein with a heterologous sequence disables systemic spread of the virus (Usha et al., 1993). Analysis of amino acid sequences and three-dimensional structures revealed a βB-βC loop of the small coat protein where short peptides could be inserted and displayed on the surface of the virus particle upon assembly (Figure 30).

Usha et al. (1993) first used CPMV coat protein as a carrier to express a foreign peptide of the major antigenic site of foot-and-mouth virus. The sequence encoding the peptide was inserted immediately adjacent to the βB strand next to threonine 19. While the chimeric virus remained infectious, with the ability to systemic spread, the entire insert was deleted due to the presence of direct repeats flanking the insert (Usha et al., 1993). Porta et al. (1994) optimized the cloning strategy by inserting sequences of interest adjacent to proline 23 at the center of the βB-βC loop without direct repeats flanking the insert. Using this strategy, a nucleotide sequence encoding the 22 amino acid Kennedy epitope from HIV-1 gp41 was introduced into CPMV coat protein. Transcripts of the chimeric virus were used to inoculate cowpea plants and inoculated leaves developed typical chlorotic lesions. The modified small coat protein remained functional and successfully assembled into virus particles. The chimeric virus particles were fully competent for cell-to-cell movement as well as long distance transport. Systemic leaves showed typical mosaic symptoms and RT-PCR analysis of viral RNA confirmed the presence of chimeric RNA2 carrying the epitope. After ten serial passages, the chimeric virus remained stable with no sign of deletion of the insert. Chimeric virus particles were isolated and purified from infected cowpea plants. The yield of virus particles reached 1.2 to 1.5 mg per gram of fresh tissue (Porta et al., 1994). Antigenic analysis using Western blotting, immuno-labelling, antigen-coated plate and double-antibody sandwich methods using a monoclonal antibody against the Kennedy epitope confirmed the antigenicity of the chimeric CPMV-based product (Port et al., 1996).

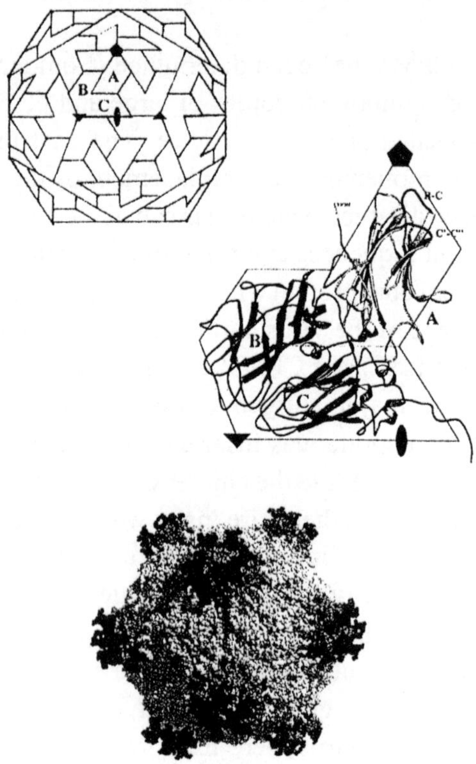

Figure 30. Presentation of foreign peptides on the surface of the CPMV particle. (A) The icosahedral asymmetry of the virus particle. One of the 60 asymmetry units is outlined, which is composed of a small [A] and a large [B and C] coat protein. (B) An enlarged asymmetry unit showing detailed arrangement of the three-barrels [A, B and C]. The B-C and C'-C" loops highlighted in black are tolerable to insertions of epitopes. (C) Space filling model of a CPMV carrying foreign epitopes. The large coat protein subunit is shown in light grey, the small coat protein subunit is shown in dark grey. The inserted epitope is shown in black attached to the dark grey small coat protein subunit (Adapted with permission from Spall et al., 1997).

In subsequent studies, purified CPMV chimeric particles containing the

Kennedy epitope were injected subcutaneously with alum adjuvant into adult mice. Antisera collected reacted with a free synthetic peptide corresponding to the inserted epitope, demonstrating that the epitope incorporated into the chimera virus particle was immunogenic. Further characterization in an assay based on the inhibition of syncytium formation in a HIV-1 susceptible T cell line demonstrated the neutralization ability of the virus particles (McLain et al., 1995). An injection of 10 μg of chimeric virus particles containing 170 ng HIV-1 epitope gave 99% neutralization of HIV-1 strains IIIB in C8166 mouse cells at a serum dilution of 1/2000. The longevity of the neutralizing antibody response increased as the immunogen doses decreased. While re-stimulation of the mice with the same immunogen has no apparent effect on the neutralization titres, a third injection improved the longevity of the antibody response (McLain et al., 1996). Interestingly, wild type virus particles were found to specifically stimulate the neutralizing activity (McLain et al., 1995).

CPMV chimeric virus particles are capable of inducing peptide-specific antibodies in the absence of adjuvant. However, inclusion of adjuvants was found to improve the immunogenecity of CPMV chimeric epitope. For instance, adjuvents such as alum, complete Freund's adjuvant, Quil A, AdjuPrime and Ribi were reported to elicit strong antibody responses to CPMV displaying the epitope. Among these, Quil A elicit the highest and most consistent responses to the peptide as reflected in both ELISA titres with immobilized peptide and in HIV-1-neutralizing antibody (McInerney, 1999).

To control pandemic HIV-1 infection, specific mucosal immunity is required to protect the genital regions through which transmission often occurs. To develop HIV vaccines that stimulate a disseminated mucosal and systemic protective immune response, the chimeric CPMV particle was used to stimulate HIV-1-specific and CPMV-specific mucosal and serum antibodies. Although less effective when mice were immunized orally, all mice immunized intranasally with two doses of 10 ug of chimeric CPMV particles produced both HIV-1-specific IgA in faeces as well as higher levels of specific, predominantly IgG2a, serum antibody. Thus there was a predominantly T helper 1 cell response. All mice also responded strongly to CPMV epitopes (Durrani et al., 1998).

In addition to potential development of an HIV vaccine, CPMV chimeras were employed to characterize the immunogenicity of HIV epitopes. For instance, neutralizing antisera from mice immunized with the CPMV antisera were used to resolve the controversy over of the neutralization site of Kennedy epitope. Using the flock house virus antigen-presenting system, the neutralizing response that is uniquely directed against a conformational epitope was mapped to the ^{746}ERDRD^{750} portion of the sequence. It was also demonstrated that the major antibody response is against the ^{740}IEEE^{743} sequence, but these antibodies do not have any neutralizing ability. Uses of plant virus-based HIV vaccine products in basic immunological studies support the possible broad utility of plant virus-based HIV vaccine antigens (Buratti et al., 1998). Despite the size constraint that limits the use of chimeric CPMV as carriers for large peptides or full-length proteins, the epitope presentation of vaccine on the surface of chimeric CPMV is a successful example of using plant viruses as HIV vaccine expression vectors.

3.2 Expression of V3 Loop from Hiv-1 Gp120 Using Tobacco Mosaic Virus

Tobacco mosaic virus (TMV) is a single-stranded positive sense RNA virus. The genome of TMV encodes a 126 kDa and a 183 kDa protein responsible for virus replication, a 30 kDa protein for movement and a 17.6 kDa coat protein. The virus is rod-shaped and composed of about 2,100 coat proteins arranged in an elongated helix. Both N and C termini are exposed on the surface of the virus particle. Compared with spherical viruses, rod-shaped viruses are usually more flexible in terms of packaging constraints. With its broad host range, TMV is a good candidate virus for foreign gene expression. In fact, TMV was the first virus that was genetically modified to carry a foreign epitope (Haynes et al., 1986). Several different strategies have been developed for the expression of HIV-1 antigens based on the length of inserts.

3.2.1 Presentation of a 13 Amino Acid Epitope from V3 Loop of Hiv-1 Gp120 on the Surface of Tobacco Mosaic Virus

Sugiyama et al. (1995) used an in-frame fusion strategy to express the 13 amino acid residue (pep13) derived from the V3 loop of HIV-1 gp120 (Putney et al., 1991). To clone pep13 to the C terminus of TMV coat protein, the coat protein stop codon UAA was mutated into an amber stop codon UAG. A 39-nucleotide sequence encoding the pep13 was fused to the amber stop codon UAG, followed by UAA and the original 3' sequence. Therefore, two translation products were expected. Wild type coat protein was produced when translation stopped at UAG, and read-through of the leaky UAG stop codon yielded a mutant coat protein containing pep13.

Despite the extension of a foreign sequence, inoculation of tobacco plants with the transcripts resulted in mosaic symptoms and subsequent systemic spread. Presence of pep13 in the coat protein did not appear to interfere with the virus particle assembly. The chimeric virus was stable and no loss of foreign insert was observed after three serial passages. Chimeric viral particles were isolated and purified from infected tobacco leaves. Western blot analysis, using antiserum against pep13, revealed accumulation of both wild type coat protein and the mutant coat protein carrying the pep13.

To confirm the location of the inserted pep13, purified virus particles were subjected to trypsin treatment, which digests peptide sequences located on the surface. Western blotting was used to examine proteins of virus particles treated with trypsin. Antibodies were no longer bound to the chimeric virus after trypsin treatment, suggesting that inserted pep13 was located on the surface of the chimeric virus particle.

The yield of virus particles reached 1.6 mg per gram of fresh tobacco leaves, which is comparable to that of CPMV obtained in cowpea leaves. However, pep13 epitope was not displayed on all coat proteins. The expression of the pep13 largely depended on the rate of read-through, which is usually low. Therefore, the actual yield of coat proteins carrying pep13 may be limited. To improve the efficiency of HIV epitope expression, Beachy et al. (1996) inserted several HIV peptides into the coat protein between amino acid 154 and 155 (Figure 31). This strategy allowed the accumulation solely of modified coat protein in both infected and systemic leaves. In infected leaves, virus particle levels reached 10 to 40% that of total leaf protein (Beachy et al., 1996).

While fusion to the C terminus has been shown to be a good strategy to produce short peptides, it is believed that more than 25 amino acid residues fused to TMV coat protein leads to failure of virus assembly. In one study, full-length reporter genes were expressed by replacement of the coat protein (Takamatsu et al., 1987). As in CPMV, the coat protein of TMV is required for efficient long-distance movement; thus replacement of the coat protein with genes of interest would disable the mobility of the virus (Dawson et al., 1988).

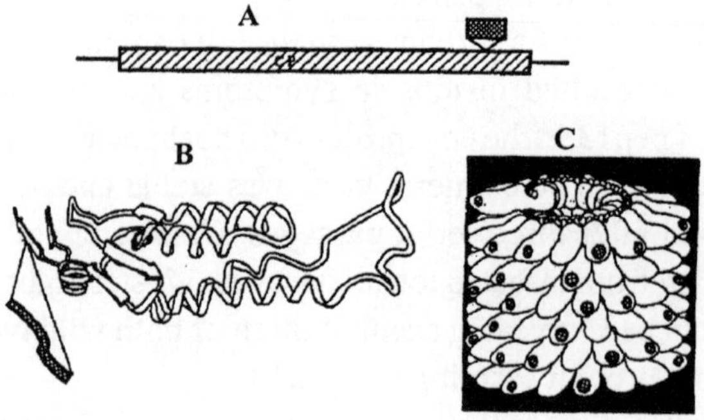

Figure 31. Presentation of vaccine epitopes on the surface of TMV. (A) Linear diagram of the coat protein of TMV showing the approximate position at which the epitopes are inserted. (B) Structure of TMV coat protein derived from X-ray crystallography and presented as a ribbon diagram. Epitope (hatched region) is inserted between amino acid 154 and 155; (C) Diagrammatic representation of TMV rod showing where the epitope is displayed (Adapted from Beachy et al, 1996, with permission).

3.2.2 Expression of V3 Loop from Hiv-1 Gp120 Using a Chimeric Tobacco Mosaic Virus with an Alfalfa Mosaic Virus Coat Protein Amplicon

To expand the use of TMV, particularly for expression of longer peptides, Yusibov et al. (1997) used alfalfa mosaic virus (AMV) coat protein as the carrier molecule to express the gp120 V3 loop of HIV-1 MN isolate from a chimeric TMV/AMV virus. The coat protein of AMV has great flexibility in protein-protein interactions. This is indicated by the fact that coat proteins form particles of different sizes in a range of 20-60 nm with different shapes such as spherical, ellipsoid, and bacilliform depending on the length of encapsidated RNA. The N terminus of the AMV coat protein is located on the surface of the virus particles and does not appear to interfere with virus assembly (Bol et al., 1971).

To take advantage of the flexibility of AMV coat protein, Yusibov et al. (1997) engineered an AMV coat protein amplicon, a coat protein subgenomic expression unit. A nucleotide sequence coding for 47 amino acids (38 residues for the V3 loop plus linker) was inserted upstream of the AMC coat protein ORF. The fusion protein ORF contains flanking 5' and 3' non-coding regions of AMV coat protein including the AMV origin of assembly. The expression unit was placed under the control of a second TMV coat protein promoter (Figure 32).

Infection of *Nicotiana benthamiana* plants with transcripts of the TMV/AMV chimera resulted in symptom development and systemic spread. Electron microscopy revealed two types of virus particles. In addition to the wild type TMV rods, spherical virus particles were observed, indicating that the recombinant subgenomic RNA was successfully packaged with the modified AMC coat protein. Analysis of coat protein using SDS-PAGE showed accumulation of a 28.9 kDa AMV coat protein from the chimeric virus particles compared with that of 24 kDa from the wild type AMV, suggesting the presence of theV3 loop at the extruding N terminus.

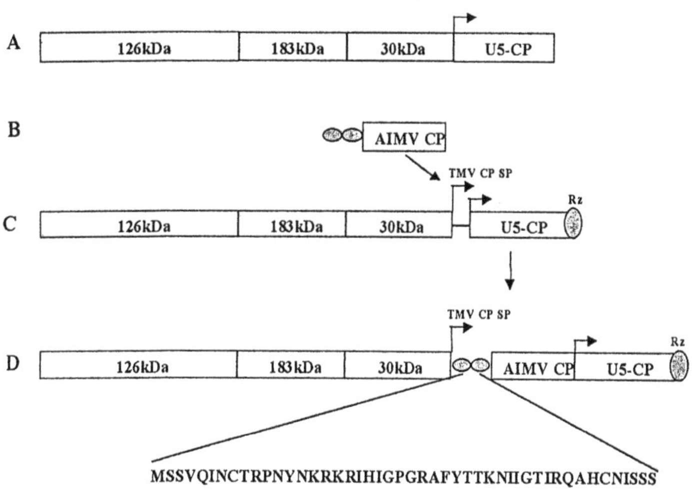

Figure 32. Expression of V3 loop from HIV-1 gp120 using TMV with an AMV coat protein amplicon. (A) Schematic representation of wild type TMV genome. The 126 and 183 kDa proteins are required for TMV replication. The 30 kDa protein is the viral movement protein. The arrow indicates the subgenomic promoter of TMV coat protein. The U5-CP is viral coat protein. (B) The V3 loop, as indicated by the two ellipsoids, fused to the 5' end of the AMV coat protein ORF. (C) TMV genome with an additional coat protein subgenomic promoter to drive the AMV coat protein with V3 loop. Rz, ribozyme for self-cleavage. (D) Chimeric TMV with an AMV amplicon. The amino acid sequence containing the V3 loop is shown under the genome (Adapted from Yusibov et al. 1997 with permission).

The recombinant virus is stable and retained the fusion protein during systemic movement throughout the plant tissues. This strategy appears to be more promising for longer peptides. The only concern with this approach is possible homologous recombination of the chimeric virus because of the presence of the duplicated sequence of the subgenomic promoter. However, this did not seem to be problematic, as the chimeric virus remained stable in systemically infected leaves. A closely related viral subgenomic promoter may be a better choice for further development.

Chimeric AMV particles isolated from plant tissues were introduced intraperitoneally into mice. Antibodies specific for the V3 loop of HIV MN isolate were detected in sera. Sera from immunized mice resulted in up to 80% neutralization of an HIV-1 MN isolate. Use of the complete Freund's adjuvant appeared to have no apparent effect on the serum antibody response in this case (Yusibov et al., 1997).

3.3 Expression of an Epitope from V3 Loop of Hiv-1 Gp120 and the Full-length P24 Using Tomato Bushy Stunt Virus

TBSV, the prototype member of the genus *Tombusvirus*, has a single stranded, positive sense RNA genome of 4776 nucleotides and encodes five functional ORFs (Figure 33). The 5'-proximally encoded products, p33 and p92, are translated directly from the genome and are essential for viral RNA replication (Hearne et al., 1990; Oster et al., 1998). Products encoded more 3' in the genome are expressed from two subgenomic mRNAs that are synthesized during infections (Hearne et al., 1990). The 41 kDa viral coat protein is translated from the larger 2.1 kb subgenomic mRNA1 whereas p22 and p19, which facilitate cell-to-cell and systemic spread of the infection, are translated from the 0.9 kb subgenomic mRNA2 (Hearne et al., 1990). The TBSV particle is icosahedral and composed of 180 capsid protein subunits. The C terminus portion of coat protein consists of a protruding domain which is located on the surface of virus particles (Harrison et al., 1978; Olson et al., 1983). The location of the protruding domain provides an opportunity to construct chimeric virus particles with peptides expressed on the surface as fusions to the carboxyl end (Olson et al., 1983). The coat protein of TBSV is dispensable for long-distance movement of the virus throughout host plants. Therefore, the coat protein can be replaced with genes of interest without affecting systemic spread of the virus (Scholthof et al., 1993). Two strategies have been used to express HIV vaccines using TBSV.

3.3.1 Presentation of Pep13 Epitope on the Surface of Tomato Bushy Stunt Virus as a C-terminal Extension of Coat Protein

Using epitope presentation, Joelson et al. (1997) fused a 16 amino acid peptide including the 13 amino acid epitope pep13 from V3 loop of HIV-1 gp120 (LaRosa et al., 1990) to the C-terminus of the TBSV capsid protein (Figure 33).

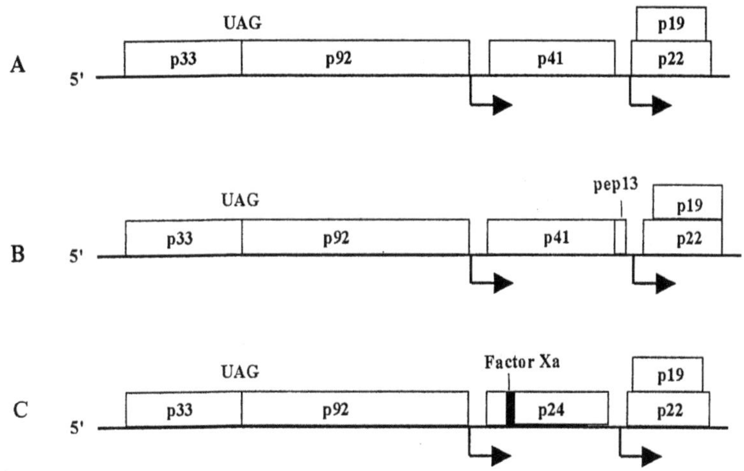

Figure 33. Expression of pep13 and p24 using TBSV. (A) Schematic representation of wild type TBSV. The genome is shown as a horizontal line, with coding regions depicted as boxes which include the approximate molecular mass values (in thousands of Daltons) of the encoded proteins. Arrows indicate the approximate positions of the initiation sites for subgenomic mRNA synthesis within the genome. (B) In-frame fusion of pep13 to the coat protein at the C terminus. (C) In-frame fusion of p24 to the N terminus of coat protein. A protease recognition site (Factor Xa) was inserted between the N-terminus of TBSV coat protein and p24.

The resulting mutant induced typical local lesions on *Chenopodium amaraanticolar* and systemic symptoms in *N. benthamiana*. Immunosorbant electronmicroscopy revealed the successful assembly of the virus particle. The chimeric virus particles were purified from infected tissues of *N. benthamiana* with an average yield of 0.9 mg per gram leaf tissue comparable to 1.1 mg per gram in wild type TBSV. Western analysis showed that the chimeric coat protein is recognized by a monoclonal antibody raised against the central motif of the V3 region in HIV-1 gp120. To test whether the chimeric TBSV particles function as antibody traps in directed ELISA, microtitre plate wells were coated with purified chimeric particles and then incubated with human sera from different stages of HIV infected patients. Positivity of sera correlated with development of AIDS in patients, demonstrating that pep13 was expressed on the surface of the virus to react with antibodies. Injection of mice with purified virus elicited a viral specific primary antibody response.

3.3.2 Expression of p24 Protein Using Tomato Bushy Stunt Virus

As mentioned earlier, a unique feature of TBSV is that the coat protein is not essential for systemic movement of the TBSV genome (Scholthof et al., 1993). This feature makes it possible to use the virus to express full-length vaccine antigens using a coat protein replacement strategy. Previous studies with reporter genes showed that the coat protein gene could tolerate inserts up to 1.9 kb (Scholthof et al., 1993). To express HIV-1 p24, the p24 ORF was fused in-frame with a small 5'-terminal portion of the TBSV coat protein ORF (Zhang et al., 2000). Therefore, p24 was under the control of the coat protein subgenomic promoter. Additionally, the Factor Xa sequence, which can be recognized and cleaved by Factor Xa protease, was inserted between the N terminal peptide of TBSV coat protein and the p24 sequence (Figure 33).

Inoculation of *N. benthamiana* with the chimeric TBSV transcripts showed that replacement of the viral coat protein ORF with that of p24 had no apparent effect on the accumulation of progeny genomic RNA. However, the levels of subgenomic RNA2, which encodes the movement proteins, decreased significantly compared with the wild type TBSV due to the presence of intervening sequences in the p24 ORF. As a result, systemic spread of the chimeric virus was delayed. Progeny viral RNAs were detected subsequently in systemic leaves with partial deletions of the p24 insert. Despite reduced fitness of the chimeric virus, the plant-based p24 fusion product from inoculated leaves reacted with mouse anti-HIV-1 p24 monoclonal antibody, confirming the maintenance of key antigenic determinants. Further analysis using ELISA showed that p24 protein accumulated up to 5 to 8% of soluble leaf protein 7 days post inoculation. However, no immunogenic data are yet available (Zhang et al., 2000).

4. SUMMARY AND FUTURE PROSPECTS

In this chapter, development of several RNA plant virus-based expression vectors for the production of HIV subunit vaccines was discussed. Due to the fact that most viruses require their coat proteins for systemic movement throughout the host plants and that insertion or fusion of large peptides interferes with the assembly of virus particle, a common strategy used was epitope presentation on the surface of chimeric virus particles. This has been a successful strategy for all viruses employed for HIV vaccine production discussed above.

Expression systems for full-length proteins require further development. While a larger peptide of up to 47 amino acids from V3 loop of HIV-1 gp120 was displayed on the surface of chimeric AMV particles using AMV amplicon in TMV, it would be difficult to display full-length proteins at the extruding terminus. Alternatively, larger peptides and full-length proteins could be expressed using the coat protein replacement strategy in AMV coat protein subgenomic amplicons. High levels of proteins are expected although no AMV subgenomic virus particles will be formed. Furthermore, viruses such as TBSV with a dispensable coat protein represent good candidates.

Vaccine production using plants has been found to be a safe alternative as the pathogenic agents of concern in mammalian cell culture are not found in plants. There have been no reports of the pathogenic nature of plant viruses in humans. In addition, chimeric virus particles, such as CPMV articles, are stable at temperatures up to 65°C and survive exposure to a protease and to pH values as low as 1 (Xu et al., 1996). These features make it possible to use plant virus-based products as edible vaccines, which make the product even more affordable. As virus-like particles are easy to purify, down stream processing would be inexpensive, if required. However, insertion of a protease recognition site at the point of fusion would be required if epitopes are to be isolated.

Plant viruses can also be used for the production of HIV antibodies for the prevention of HIV infection as well as passive immunotherapy. It has been shown that HIV-1 infection could be prevented in monkeys and chimpanzees using antibody therapy (Putkonen et al., 1991; Prince et al., 1991; Emini et al., 1992). Production of complex antibodies have been successfully demonstrated in transgenic plants (Hiatt et al., 1989; Ma et al., 1995). High yield of HIV antibodies could potentially be achieved using plant viruses.

Recent studies have shown the great potential of plant viruses as expression vectors for the production of HIV vaccines, therapeuticals and diagnostic reagents. It is hoped that success in the current human clinical trials and global demand for HIV vaccines will prompt commercial production of cost-effective products. Plant virus-based production systems have a clear competitive advantage for the development of such marketable and cost-effective products.

ACKNOWLEDGEMENTS

The author wants to thank Dr. K. Andrew White, Dr. Grant Munro and Ms. Catherine Bryant for their helpful comments and suggestions.

REFERENCES

Arakawa T, Chong DKX, and Langridge WHR, 1998. Efficacy of a food plant-based oral cholera toxin B subunit vaccine. Nat Biotechnol, 16:292-297

Arakawa T, Chong DKX, Merritt JL and Langridge WHR, 1997. Expression of cholera toxin B subunit oligomers in transgenic potato plants. Transgenic Res, 6:403-413

Arendrup M, Sonnerborg A, Svennerholm B, Akerblom L, Nielsen C, Clausen H, Olofsson S, Nielsen JO and Hansen J-ES, 1993. Neutralizing antibody response during human immunodeficiency virus type 1 infection: type and group specificity and viral escape. J Gen Virol, 74:855-863

Beachy RN, Fitchen JH and Hein MB, 1996. Use of plant viruses for delivery of vaccine epitopes. Ann N Y Acad Sci, 792:43-49

Benson EM, Clarkson J, Law M, Marshall P, Kelleher AD, Smith DE, Patou G, Stewart GJ, Cooper DA and French RA, 1999. Therapeutic vaccination with p24-VLP and zidovudine augments HIV-specific cytotoxic T lymphocyte activity in asymptomatic HIV-infected individuals. AIDS Res Hum Retroviruses, 15:105-113

Bol JF, van Vloten-Doting L and Jaspars EM, 1971. A functional equivalence of top component RNA and coat protein in the initiation of infection by alfalfa mosaic virus. Virology, 46:73-85

Broliden PA, Von Gegerfelt A, Clapham P, Rosen J, Fenyo E-M Wahren B and Brodien K, 1992. Identification of human neutralization-inducing regions of the human immunodeficiency virus type 1 envelope glycoproteins. Proc Natl Acad Sci USA, 89:461-465

Buratti E, McLain L, Tisminetzky S, Cleveland SM, Dimmock NJ and Baralle E, 1998. The neutralizing antibody response against a conserved region of human immunodeficiency virus type 1 gp41 (amino acid residues 731-752) is uniquely directed against a conformational epitope. J Gen Virol, 79:2709-2716

Chen YH and Dierich MP, 1996. Identification of a second site in HIV-1 gp41 mediating binding to cells. Immunol Lett, 52:153-156

Cohen J, 2000. Ground zero: AIDS research in Africa. Science, 288:2150-2153

Dawson WO, Bubrick P and Grantham GL, 1988. Modification of the tobacco mosaic virus coat protein gene affecting replication, movement, and symptomatology. Phytopathology, 78:783-789

Durrani Z, McInerney TL, McLain L, Jones T, Bellaby T, Brennan FR and Dimmock NJ, 1998. Intranasal immunization with a plant virus expressing a peptide from HIV-1 gp41 stimulates better mucosal and systemic HIV-1-specific IgA and IgG than oral immunization. J Immunol Methods, 220:93-103

Emini EA, Schleif WA, Nunberg JH, Conley AJ, Eda Y, Tokiyoshi S, Putney SD, Matsushita S, Cobb KE, Jett CM, Eichberg JW and Murphy KK, 1992. Prevention of HIV-1 infection in chimpanzees by gp 120 V3 domain-specific monoclonal antibody. Nature, 355:728-730

Featherstone C, 1996. Vaccine by agriculture. Mol Med Today, 2:278-281

Gallagher DM, 1992. HIV/AIDS: the epidemic continues. Imprint, 39: 64-65

Haq TA, Mason HS, Clements JD and Arntzen CJ, 1995. Oral immunization with a recombinant bacterial antigen produced in transgenic plants. Science, 268:714-715

Haynes JR, Cunningham J, von Seefried A, Lennick M, Garvin T and Shen S, 1998. Development of a genetically-engineered, candidate polio vaccine employing the self-assembling properties of the tobacco mosaic virus coat protein. Bio/Technology, 4:637-641

Harrison SC, Olson AJ, Schutt CE, Winkler FK and Bricogne G, 1978. Tomato bushy stunt virus at 2.9 Å resolution. Nature, 276:368-373

Hearne PQ, Knorr DA, Hillman BI and Morris TJ, 1990. The complete genome structure and synthesis of infectious RNA from clones of tomato bushy stunt virus. Virology, 177:141-151

Hiatt AC, Cafferkey R and Bowdish K, 1989. Production of antibodies in transgenic plants. Nature, 342:76-78

Ho DD, McKeating JA, Li XL, Moudgil T, Daar ES, Sun NC and Robinson JE, 1991. Conformational epitope on gp120 important in CD4 binding and human immunodeficiency virus type 1 neutralization identified by a human monoclonal antibody. J Virol, 65:489-493

Jagadish MN, Edwards SJ, Hayden MB, Grusovin J, Vandenberg K, Schoofs P, Hamilton RC, Shukla DD, Kalnins H, McNamara M, Haynes J, Nisbet IT, Ward CW and Pye D, 1996. Chimeric potyvirus-like particles as vaccine carriers. Intervirology, 39:85-92

Joelson T, Akerblom L, Oxelfelt P, Strandberg B, Tomenius K and Morris TJ, 1997. Presentation of a foreign peptide on the surface of tomato bushy stunt virus. J Gen Virol, 78:1213-1217

Kennedy RC, Henkel RD, Pauletti D, Allan JS, Lee TH, Essex M and Dreesman GR, 1986. Antiserum to a synthetic peptide recognised the HTLV-III envelope glycoprotein. Science,

231:1556-1559

LaRosa GJ, David JP, Weinhold K, Waterbury JA, Profy AT, Lewis JA, Langlois AJ, Dreesman GR, Boswell NR, Shadduck P, Holley HL, Karplus M, Bolognesi DP, Matthews TJ, Emini EA and Putney SD, 1990. Conserved sequence and structural elements in the HIV-1 principal neutralizing determinant. Science, 249:932-935

Lefrere J-J, Courouce A-M, Rouger P, Duedari N and Elghouzzi M-H, 1992. P24 antigen and HIV screening. Lancet, 339:999-1000

Letvin NL, 1998. Progress in the development of an HIV-1 vaccine. Science, 280:1875-1880

Lomonossoff GP and Hamilton WD, 1999. Cowpea mosaic virus-based vaccines. Curr Top Microbiol Immunol, 240:177-189

Lomonossoff GP and Johnson JF, 1991. The synthesis and structure of comovirus capsids. Prog Biophys Mol Biol, 55:107-137

Lomonossoff GP and Johnson JE, 1996. Use of macromolecular assemblies as expression systems for peptides and synthetic vaccines. Curr Opin Struct Biol, 6:176-182

Ma JK, Hiatt A, Hein M, Vine ND, Wang F, Stabila P, van Dolleweerd C, Mostov K and Lehner T, 1995. Generation and assembly of secretory antibodies in plants. Science, 268:716-719

Ma J K-C and Vine ND, 1999. Plant expression systems for the production of vaccines. Curr Top Microbiol Immunol, 236:275-292

Martin SJ, Vyakarnam A, Cheingsong-Popov R, Callow D, Jones KL, Senior JM, Adams SE, Kingsman AJ, Mater P, Gotch FM, McMichael AJ, Roitt IM and Weber JN, 1993. Immunization of human HIV-seronegative volunteers with recombinant p17/p24:Ty virus-like particles elicits HIV-1 p24-specific cellular and humoral immune responses. AIDS, 7:1315-1323

Mason HS and Arntzen CJ, 1995. Transgenic plants as vaccine production systems. TIBTech, 13:388-392

Mason HS, Ball JM, Shi J-J, Jian X, Estes MK and Arntzen CJ, 1996. Expression of Norwalk virus capsid protein in transgenic tobacco and potato and its oral immunogenicity in mice. Proc Natl Acad Sci USA, 93:5335-5340

Mason HS, Lam DM-K and Arntzen CJ, 1992. Expression of hepatitis B surface antigen in transgenic plants. Proc Natl Acad Sci USA, 89:11745-11749

McInerney TL, Brennan FR, Jones TD and Dimmock NJ, 1999. Analysis of the ability of five adjuvants to enhance immune responses to a chimeric plant virus displaying an HIV-1 peptide. Vaccine, 17:1359-1368

McLain L, Durrani Z, Wisniewski A, Port C, Lomonossoff GP and Dimmock NJ, 1996. Stimulation of neutralizing antibodies to human immunodeficiency virus type 1 in three strains of mice immunized with a 22-mer amino acid peptide expressed on the surface of a plant virus. Vaccine, 14:799-810

McLain L, Durrani Z, Dimmock NJ, Wisniewski LA, Port C and Lomonossoff GP, 1996. A plant virus-HIV-1 chimera stimulates antibody that neutralized HIV-1. In *Vaccine 96*, Eds. Brown

F, Burton DR, Collier J, Mekalanos J and Norrby E. Cold Spring Harbor Laboratory Press, Cold Spring Harbor, pp 311-316

McLain L, Porta C, Lomonossoff GP, Durrani Z and Dimmock NJ, 1995. Human immuno-deficiency virus type 1 neutralizing antibodies raised to a gp41 peptide expressed on the surface of a plant virus. AIDS Res Hum Retroviruses, 11:327-334

Meloen RH, Hamilton WD, Casal JI, Dalsgaard K and Langeveld JP, 1998. Edible vaccines. Vet Q, 20 suppl 3:S92-S95

Modrow S, Hahn BH, Shaw GM, Gallo RC, Wong-Staal F and Wolf H, 1987. Computer assisted analysis of envelope protein sequences of seven human immunodeficiency virus isolates: prediction of antigenic epitopes in conserved and variable regions. J Virol, 61:570-578

Montroni M, Silvestri G, Butini L, Bartocci C, Regnery C and Danieli G, 1992. p24 antigenaemia as a predictor of good immunological responsiveness to zidovudine therapy in asymptomatic HIV infection (letter). AIDS, 6:338-339

Moore JP, Sattentau QJ, Wyatt R and Sodroski J, 1994. Mapping the topology of the human immunodeficiency virus glycoprotein gp120 with a panel of monoclonal antibodies. J Virol, 68:469-484

Muster T, Steindl F, Purtscher M, Trkola A, Klima A, Himmler G, Ruker F and Katinger H, 1993. A conserved neutralizing epitope on gp41 of human immunodeficiency virus type 1. J Virol, 67:6642-6647

Olson AJ, Bricogne G and Harrison SC, 1983. Structure of tomato bushy stunt virus. IV. The virus particle at 2.9 Å resolution. J Mol Biol, 171:61-93

Oster SK, Wu B and White KA, 1998. Uncoupled expression of p33 and p92 permit amplification of tomato bushy stunt virus RNAs. J Virol, 72:5845-5851

Palmer KE, Arntzen CJ, and Lomonossoff GP, 1999. Antigen delivery systems: transgenic plants and recombinant plant viruses. In *Mucosal immunity, 2nd Edition*, Eds. Ogra PI, Mestecky J, Lamm ME, Stober W, McGhee JR and Bienenstock J. Academic Press, Orlando, pp 793-807

Port C and Lomonossoff GP, 1996. Use of viral replicons for the expression of genes in plants. Mol Biotechnol, 5:209-221

Porta C, Spall VE, Lin T, Johnson JE and Lomonossoff GP, 1996. The development of cowpea mosaic virus as a potential source of novel vaccines. Intervirology, 39:79-84

Porta C, Spall VE, Loveland J, Johnson JE, Baker PJ and Lomonossoff GP, 1994. Development of cowpea mosaic virus as a high-yielding system for the presentation of foreign peptides. Virology, 202:949-955

Prince AM, Reesink H, Pascual D, Horowitz B, Hewlett I, Murphy KK, Cobb KE and Eichberg JW, 1991. Prevention of HIV infection by passive immunization with HIV immunoglobulin. AIDS Res Hum Retroviruses, 7:971-973

Putney SD, LaRosa GJ, Profy AT, Silver S, Scott CF, Javaherian K, Robinson J, Langlois AJ, Matthews TJ, Bolognesi DP, Lewis JA and Emini EA, 1991. The HIV-1 principal neutralization determinant elicits broadly neutralizing antibodies. In *Vaccines 91*. Eds.

Chanock R, Ginsberg H, Brown F and Lerner R. Cold Spring Harbor Laboratory Press, Cold Spring Harbour, NY, pp 9-13

Putkonen P, Thorstensson R, Ghavamzadeh L, Albert J, Hild K, Biberfeld G and Norrby E, 1991. Prevention of HIV-2 and SIVsm infection by passive immunization in cynomolgus monkeys. Nature, 352:436-438

Reddy MM, Winger EE, Hargrove D, McHugh T, McKinley GF and Grieco MH, 1992. An improved method for monitoring efficacy of anti-retroviral therapy in HIV-infected individuals; a highly sensitive HIV p24 antigen assay. J Clinic Lab Anal, 6:125-129

Rencher SD, Slobod KS, Dawson DH, Lockey TD and Hurwits JL, 1995. Does the key to a successful HIV type 1 vaccine lie among the envelope sequences of infected individuals? AIDS Res Hum Retroviruses, 11:1131-1133

Robert-Guroff M, Brown M and Gallo RC, 1985. HTLV-III-neutralizing antibodies in patients with AIDS and AIDS-related complex. Nature, 316:72-74

Sattentau QJ, Zolla-Pazner S and Poignard P, 1995. Epitope exposure on functional, oligomeric HIV-1 gp41 molecules. Virology, 206:713-717

Scholthof HB, Morris TJ and Jackson AO, 1993. The capsid protein gene of tomato bushy stunt virus is dispensable for systemic movement and can be replaced for localized expression of foreign genes. Mol Plant-Microbe Interact, 6:309-322

Scholthof HB, Scholthof KB and Jackson AO, 1996. Plant virus gene vector for transient expression of foreign genes in plants. Ann Rev Phytopathol, 34:299-323

Shaw GM, Broder S, Essex M and Gallo RC, 1984. Human T-cell leukemia virus: its discovery and role in leukemogenesis and immunosuppression. Adv Intern Med, 30:1-27

Sherry B, Mosser AG, Colonno RJ and Ruecker RR, 1986. Use of monoclonal antibodies to identify four neutralisation immunogens on a common cold picornavirus, human rhinovirus 14. J Virol, 57:246-257

Spall VE, Porta C, Taylor KM, Lin T, Johnson JE and Lomonossoff GP, 1997. Antigen expression on the surface of a plant virus for vaccine production. In *Engineering crop plants for industrial end uses*. Eds. Shewry PR, Napier JA and Davis PJ. Portland, London, pp 35-46

Spector SA, Kennedy C, McCutchan JA, Straube RG, Connor JD and Richman DD, 1989. The antiviral effect of zidovudine and ribavirin in clinical trails and the use of p24 antigen levels as a virologic marker. J Infect Dis, 159:822-828

Stiehm ER, Fletcher CV, Mofenson LM, Palumbo PE, Kang M, Fenton T, Sapan CV, Meyer WA, Shearer WT, Hawkins E, Fowler MG, Bouquin P, Purdue L, Sloand EM, Nemo GJ, Wara D, Bryson YJ, Starr SE, Petru A and Burchett S, 2000. Use of human immunodeficiency virus (HIV) human hyperimmune immunoglobulin in HIV type 1-infected children (Pediatric AIDS clinical trials group protocol 271). J Infect Dis, 181:548-554

Sugiyama Y, Hamamoto H, Takemoto S, Watanabe Y and Okada Y, 1995. Systemic production of foreign peptides on the particle surface of tobacco mosaic virus. FEBS Lett, 359:247-250

Tacket CO, Mason HS, Lonsonsky G, Clements JD, Levine MM and Arntzen CJ, 1998. Immunogenicity in humans of a recombinant bacterial antigen delivered in a transgenic potato. Nat Med, 4:607-609

Takamatsu N, Ishidawa M, Meshi T and Okada Y, 1987. Expression of bacterial chloramphenicol acetyltransferase gene in tobacco plants mediated by TMV-RNA. EMBO J, 6:307-311

Thanavala Y, Yang YF, Lyons P, Mason HS and Arntzen CJ, 1995. Immunogenicity of transgenic plant-derived hepatitis B surface antigen. Proc Natl Acad Sci USA, 92:3358-3361

Tsoukas CM and Bernard NF, 1994. Markers predicting progression of human immunodeficiency virus-related disease. Clinical Microbiol Rev, 7:14-28

Usha R, Rohll JB, Spall VE, Shanks M, Maule AJ, Johnson JE and Lomonossoff GP, 1993. Expression of an animal virus antigenic site on the surface of a plant virus particle. Virology,197:366-374

Weiss RA, 1993. How does HIV cause AIDS? Science, 260:1273-1279

Weiss RA, Clapham PR, Cheingsong-Popov R, Dalgleish AG, Carne CA, Weller IV and Tedder RS, 1985. Neutralization of human T-lymphotropic virus type III by sera of AIDS and AIDS-risk patients. Nature, 316:69-72

Wyatt R and Sordroski J, 1998. The HIV envelope glycoproteins: fusogens, antigens, and immunogens. Science, 280:1884-1888

Yusibov V, Modelska A, Steplewski K, Agadjanyan M, Weiner D, Hooper DC and Koprowski H, 1997. Antigens produced in plants by infection with chimeric plant viruses immunize against rabies virus and HIV-1. Proc Natl Acad Sci USA, 94:5784-5788

Yusibov V, Shivprasad S, Turpen TH, Dawson W and Koprowski H, 1999. Plant viral vectors based on tobamoviruses. Curr Top Microbiol Immunol, 240:81-94

Xu F, Jones TD and Rodgers PB, 1996. Potential of chimeric plant virus particle as novel stable vaccine. Dev Biol Stand, 87:201-205

Zhang G, Slowinski V and White KA, 1999. Subgenomic mRNA regulation by a distal RNA element in a (+)-strand RNA virus. RNA, 5:550-561

Zhang G, Leung C, Murdin L, Rovinski B and White KA, 2000. In planta expression of HIV-1 p24 protein using an RNA plant virus-based expression vector. Mol Biotechnol, 14:99-107

Chapter 8

SUPPRESSION OF AUTOIMMUNE DIABETES BY THE USE OF TRANSGENIC PLANTS EXPRESSING AUTOANTIGENS TO INDUCE ORAL TOLERANCE

S. Ma and A.M. Jevnikar
Lawson Health Research Institute, Robarts Research Institute, Multi Organ Transplant Program, London Health Sciences Centre and the University of Western Ontario, London, Ontario, Canada.

Keywords: transgenic plants, protein antigens, oral immune tolerance

Abstract The use of transgenic plants expressing autoantigens to induce protective oral immune tolerance represents a novel approach for antigen-specific therapy of autoimmune disease. A transgenic plant-based strategy offers several attractive advantages for oral tolerance, including relatively inexpensive production of large amounts of soluble protein antigens essential for the clinical application of oral tolerance, and absence of potential pathogens and endotoxins which are issues in the production of therapeutic proteins from animal tissues or bacterial cells. Plants also offer a simple and efficient system for the direct oral delivery of human proteins for immune modulation which in many cases will not require extensive and expensive protein purification steps. Here we discuss strategies to prevent autoimmune diabetes by oral delivery of diabetes-associated autoantigens such as glutamic acid decarboxylase (GAD) and insulin in transgenic plants. These principles and approaches may be relevant to other clinical indications such as transplant rejection, allergy and the delivery of bio-therapeutics in inflammatory bowel disease.

1. INTRODUCTION

Type 1 (insulin-dependent) diabetes mellitus (IDDM) is a chronic disease caused by the progressive immunological destruction of insulin-secreting

L. Erickson et al. (eds.),
Molecular Farming of Plants and Animals for Human and Veterinary Medicine, 179–196.

pancreatic beta cells by autoreactive T lymphocytes. A number of islet cell target autoantigens have been identified including glutamic acid decarboxylase (GAD), insulin, and tyrosine phosphatase-like IA-2 which play a role in the initiation and maintenance of the pathogenic events that lead to IDDM (Atkinson and Maclaren, 1994; Bosi and Botazzo, 1995). In the non-obese diabetic (NOD) mouse model, which closely resembles human IDDM, overt diabetes is preceded by a period in which there is a characteristic lymphocytic infiltration of pancreatic islets termed insulitis. This is a period in which intervention can alter disease and results in this mouse model have suggested that there may be a potential for immunological intervention and prevention of IDDM in clinical practice.

Although immunosuppression with specific drugs, such as cyclosporine A, can alter T-cell immune responses and thus inhibit the development of diabetes in NOD mice and humans, therapies that more specifically target the destructive immune responses which cause organ specific disease but without causing global immunosuppression would be much safer and therefore more desirable to patients. The adverse effects of all general immunosuppressive drugs include increased rates of infection and cancer, as the immune system is critical for surveillance against such "dangers". Interestingly, it has been demonstrated that specific peripheral immune responses can be reduced to orally administered proteins leaving the immune system intact for responses to pathogens. The mechanisms involved in this endogenous immune regulatory system have been collectively referred to as "oral immune tolerance".

An approach using oral immune tolerance to islet cell autoantigens including glutamic acid decarboxylase (GAD) and insulin, has been shown by our laboratory and others to prevent the spontaneous onset of diabetes in NOD mice. This method of treatment may therefore represent a convenient and potentially effective antigen-specific therapeutic modality for human autoimmune diabetes. However, to a large extent, the clinical success of this approach as a therapy may be determined by practical issues such as the availability of suitably large amounts of soluble protein antigens at low cost, the efficiency of oral tolerance induction by proteins versus peptides and the establishment of a simple delivery system. Recently, we have explored transgenic plants as a potential cost-effective expression system for production and delivery of the diabetes-associated autoantigen, mouse GAD67. We have shown that transgenic plants can express GAD in an immunologically active form, and that feeding transgenic plant tissues containing GAD can protect

NOD mice from diabetes (Ma et al., 1997). This first demonstration of the use of transgenic plants to induce protective oral tolerance may have an important impact on future treatment strategies for diabetes and many other autoimmune disorders, allergy and organ transplant rejection. Here we discuss some of the critical aspects of the plant-based approaches for production of diabetes-associated β-cell autoantigens as well as their application in the induction of oral immune tolerance to the prevention of animal diabetes. This new technology may one day play an important role in the therapy and cure of human Type 1 diabetes. A discussion is also provided of our recent progress and new developments in the area.

2. ORAL TOLERANCE AND THE TREATMENT OF AUTOIMMUNE DISEASES

Oral tolerance refers to the oral administration of protein antigens, which induces a state of systemic non-responsiveness specific for the fed antigen. The phenomenon of orally induced tolerance was first described by Wells in 1911 (Wells, 1911). In these early studies, Guinea pigs fed hen proteins or dinitrochlorobenzene (DNB) did not develop responses to subsequent systemic challenges with the same antigen. Later, it was shown that the ingestion of a wide range of proteins can lead to attenuated or absent systemic immune responses to those particular antigens (Mowat, 1987). It is now well established that oral tolerance is an important natural physiological property of the immune system, whereby the host can avoid dangerous reactions such as DTH (delayed-type hypersensitivity) to "non-dangerous" antigens and other substances encountered in our diets (Mowat, 1987).

A milestone in the area of oral tolerance was the recognition that self-antigens could be administered to prevent and treat autoimmune diseases. Induction of oral tolerance in animal models has been used in the prevention of several experimental autoimmune diseases, such as rodent (NOD) diabetes (Bergerot et al., 1994; Zhang et al., 1991; Bergerot et al., 1994), uveo-retinitis (Nussenblatt et al., 1990), experimental autoimmune encephalomyelitis (EAE) (Higgins and Weiner, 1988), and rheumatoid arthritis (RA) (Thompson and Staines, 1986). Induction of oral tolerance in humans was also demonstrated using the nominal antigen keyhole limpet hemocyanin (KLH). Human volunteers, when orally given KLH, showed decreased T cell proliferative responses and DTH reactivity specific to KLH (Husby et al., 1994). These early studies have led to clinical trials that test oral antigen administration in human autoimmune diseases. Trials have been conducted in multiple sclerosis, rheumatoid arthritis, juvenile rheumatoid arthritis, inflammatory uveitis, autoimmune thyroiditis, and insulin-dependent diabetes mellitus. Importantly no toxicity or exacerbation of disease has been observed following oral administration of autoantigen (Weiner, 1997), although this remains a potential concern. Although improvement has been noted in some patients treated with oral antigen, predictable clinical improvement has not been observed in controlled trials. However, it is clear that the choice of antigen (requirement for purified form of antigen), dose of antigen (low-dose versus high-dose tolerance), and timing of antigen administration (before or during early stages of disease) have an effect on the therapeutic efficacy of oral tolerance. Although there are still many studies required, it is likely that the major effectiveness of oral tolerance as a therapy will be in the prevention of disease in those with an identified increased risk. This includes diseases in which the relevant antigens have been well defined in animal and human studies, such as in the case of Type 1 diabetes with clear genetic linkage to HLA loci and demonstrable early evidence of immune reactivity to antoantigens such as GAD. This disease also has the benefit of having the NOD model which is spontaneous, and is genetically very similar at several key loci, to human Type I autoimmune diabetes.

Although there are several mechanisms identified in oral tolerance, a primary factor that determines efficacy is the dose of antigen administered. Tolerance to high doses of antigen is mediated by the inactivation or clonal deletion of T helper (Th1) cells characterized by the production of interleukin-2 (IL-2) and interferon-γ (IFN-γ). Th1 autoimmune T cells have been associated with the pathogenesis of autoimmune diabetes (Weiner et al., 1994). In contrast, tolerance in response to "lower dose" antigen involves the induction of regulatory T cells which can produce Th2 type cytokines (interleukin-4 (IL-4) and interleukin-10 (IL-10)) or Th3 type cytokines such as transforming growth factor-β (TGF-β) in response to antigen (Weiner et al., 1994). Both Th2 and Th3 cytokines may function by preventing Th1 reactivity rather than having a direct beneficial effect. Oral tolerance can also be enhanced by the use of mucosal adjuvants. The most commonly used mucosal adjuvant is cholera toxin B subunit (CTB) (Czerkinsky et al., 1996; Holmgren et al., 1993; Sun et al., 1994). Cholera toxin (CT) consists of a toxic 27 kDa A subunit (CTA) having

ADP ribosyl transferase activity and a non-toxic pentamer comprised of 11.6 kDa B subunits (CTB) that bind to the A subunit and facilitates its entry into intestinal epithelial cells. CTB given simultaneously with a variety of antigens orally can enhance oral tolerance and allows a reduction in the antigen dose and dosing schedule (frequency of antigen feeding) required for tolerance induction (Sun et al., 1996; Bergerot et al., 1997). One mechanism of such enhancement is that CTB bound to antigen, by virtue of its capacity to bind to eukaryotic cell surfaces *via* GM1 ganglioside receptors present on intestinal epithelial cell surfaces, focuses or concentrates antigen in specialized mucosal cells with antigen presentation capacity and thus facilitates the development of immune tolerance. However, responses may be limited as the immune system will have neutralizing effects on any components of infectious agents which may represent "danger." Th2 cytokines have been examined for possible effect in the enhancement of oral tolerance, because of their known roles in differentiating precursor Th0 cells toTh2 lineages (Mosmann and Coffman, 1989). Parenterally administered IL-4, oral IL-4, and oral IL-10 can enhance oral tolerance (Weiner et al., 1994).

3. UNIQUE ADVANTAGES OF USING TRANSGENIC PLANTS FOR ORAL TOLERANCE INDUCTION

The induction of oral immune tolerance requires the ingestion of large amounts of soluble protein antigens. To a large extent the clinical application of oral tolerance induction as a potential therapeutic strategy may depend on the availability of the target proteins in sufficient quantities and at an economical cost. Conventional protein production systems, while capable of producing such amounts of protein, are unlikely to be cost effective enough to allow clinical studies. Therefore, an important factor for our group in considering the application of oral tolerance therapeutically was to devise a production system capable of generating large quantities of immunologically and, in some cases, biologically active proteins at low cost. Presently, several heterologous protein production systems are available, such as bacteria, yeast, and mammalian and insect cell cultures. In these conventional production systems, sufficiently high expression levels of natively folded mammalian proteins are often limited by the inability to form disulfide bridges or to add glycans to the recombinant proteins. Also, there is a high cost associated with the use of sophisticated equipment (i.e. fermentors) and sterile culture media, absolute exclusion of pathogenic viruses such as HIV and hepatitis B, and removal of other harmful substances such as bacterial endotoxin. All recombinant proteins derived from these systems need to be purified before clinical use, which further adds expenses to the production cost.

The use of transgenic plants as an alternative option to conventional cell culture-based systems has recently attracted much interest (Goddijn and Pen, 1995). Transgenic plant systems have the highest potential for large-scale and cost-effective production of many recombinant proteins. Plants can be contained, grown easily and inexpensively in large quantities, can be harvested and processed with available agronomic infrastructures, and scaling up is simple. Plants, as higher eukaryotes, also have a crucial advantage over bacteria in being able to perform many of the complex protein processing steps, such as isoprenylation, oligomerization, disulfide bridge formation and proteolytic cleavage as in a mammalian system. This factor may be particularly important in producing recombinant proteins that have *in vivo* activities identical to their animal counterparts. Plants do not harbor infectious agents such as viruses and prions harmful to humans as these agents cannot replicate in plants. Safety is a primary concern when any therapeutic proteins for human use are prepared from animal tissues or bacterial cells, regardless of the route of administration. More importantly, as the production of antigenic proteins for oral tolerance can be targeted to edible transgenic plants, there is no need for extensive purification resulting in a major cost reduction. In the last few years it has been shown that plants are capable of synthesizing a wide range of valuable recombinant products, such as monoclonal antibodies (Hiatt et al., 1989; Ma et al., 1995), biopharmaceuticals such as enkephalin, α-interferon, human serum albumin (HAS), vaccines (Mason et al., 1992; McGarvey et al., 1995) and costly drugs used in the "orphan disease" group of rare diseases: human glucocerebrosidase (Cramer et al., 1996) and murine granulocyte-macrophage colony-stimulating factor (Lee et al., 1997). Recently there have been two successful human trial demonstrations of prototype plant "edible vaccines" for the prevention of infectious diarrheal illness. Potatoes containing the binding subunit of the heat-labile enterotoxin of *E.coli* (LT-B) were fed uncooked to volunteers and serum and secretory antibodies specific for LT-B were induced (Tacket et al.,1998). In separate clinical studies, volunteers who ate uncooked potatoes containing the Norwalk virus capsid protein (causal agent of epidemic gastroenteritis) developed both serum and secretory antibodies specific to the capsid protein

(Tacket et al., 2000). Both these human trials as well as earlier animal trials (Haq et al., 1995; Mason et al., 1996; Arakawa et al., 1998; Gomez et al., 2000; Richter et al., 2000; Sandhu et al., 2000), suggest that recombinant antigens expressed in transgenic plants can be delivered as food into the gut mucosa, and are capable of inducing immune responses. It is also possible that autoantigens delivered in transgenic plants will be capable of altering harmful immune responses.

4. TRANSGENIC PLANTS EXPRESSING GAD

Foreign gene expression in plants can be achieved either by transformation of the nuclear genome by stable integration of a foreign DNA using *Agrobacterium*-mediated transformation, direct gene transfer *via* particle bombardment and electroporation, or by transient expression using genetically modified plant viruses (Porta and Lomonossoff, 1996). *Agrobacterium*-mediated transformation is the most frequently used approach to transfer the desired DNA into a plant genome. Since the first report on tobacco transformation via *Agrobacterium* in 1985 (Horsch et al., 1985), this approach has been successfully used to transform many other plant species, such as tomato and potato. However, a number of agronomically important plant species, such as soybean, rice and wheat, are still resistant to *Agrobacterium*-mediated transformation but can be transformed with direct gene transfer techniques (Barcelo and Lazzeri, 1998). Most of the foreign protein-producing plants reported to date have been generated by stable transformation. The main advantages of stable transformation include the capacity to generate a large number of independent transgenic plants at a time, ease at maintaining transgenic plants expressing high levels of recombinant protein and the ability to sexually cross transgenic lines to obtain a single plant simultaneously expressing multiple proteins. Transient expression using genetically modified plant viruses as expression vectors allows higher levels of protein expression rapidly, but requires constant inoculation of large numbers of plants with the modified virus. For the delivery of oral vaccines or "tolerogens" in humans, plant viruses may be an option. To date there have been no reports linking plant virus transmission to any animals.

To develop transgenic plants expressing β cell autoantigens, we chose mouse GAD67 as an initial candidate. The justification for utilizing mouse

GAD67 was twofold. First, in rodents and in humans, two isomers of GAD namely GAD65 and GAD67, have been identified, and both are implicated in the pathogenesis of IDDM (Tisch et al., 1993; Kaufman et al., 1993; Elliott et al., 1994). However, there is a variation in their expression in the islets of humans and animals. Both human and rat islets predominantly express GAD65, whereas GAD67 is the major isoform in mouse islets (Kim et al., 1993). Secondly, as GAD67 represents a major isoform of GAD in mouse, and as it was shown that administration of purified GAD67 to young NOD mice specifically prevented development of diabetes (Elliott et al., 1994), the selection of mouse GAD67 for initial expression in plants facilitated the design of animal experiments for testing plant-derived GAD. To create transgenic plants expressing GAD, mouse GAD67 cDNA was first used to replace the β-glucuronidase gene in pTRL2-GUS composed of a CaMV 35S promoter with a double enhancer sequence linked to a 5' untranslated tobacco *etch* virus leader sequence, GUS and a nopaline synthase (NOS) terminator (Carrington and Freed, 1990). The resulting expression cassette was then inserted into the plant expression binary vector, pBin19 (Bevan, 1984). The mouse GAD67 cDNA was introduced into tobacco and potato plants by *Agrobacterium tumefaciens*-mediated leaf explant transformation method (Horsch et al., 1985). Integration of mouse GAD67 cDNA into the plant nuclear genome was confirmed by Southern blot, and the expression of full-length GAD67 mRNA transcript was confirmed by northern analyses. Western blots of protein homogenates from tobacco leaf and potato tuber tissues showed a single protein band of the correct size. The expression level of GAD was estimated from blot densitometry to be approximately 0.4% of total soluble protein. GAD67 expression was similar in tobacco leaf and potato tubers, and was not detected in untransformed control plants.

5. ORAL IMMUNOGENICITY OF THE PLANT DERIVED GAD PROTEIN

GAD is an enzyme that catalyzes the decaboxylation of glutamic acid to produce γ-aminobutyric acid (GABA), the major inhibitory neurotransmitter in the brain (Baekkeskov et al., 1990). However, the immunogenicity of GAD is not dependent on its enzymatic activity but rather its primary amino acid sequences which form immunodominant epitopes upon processing in antigen presenting cells *in vivo*. To assess whether the plant-derived recombinant GAD retained its ability to alter T-cell responses, transgenic tobacco leaf tissues or potato tubers were added to the diet of NOD mice. The amount of GAD delivered was estimated to be approximately 1 mg per day per mouse which represents a very large amount of antigen, and is much higher than would be delivered relatively to humans. After 4 weeks of supplementation, NOD mice were immunized in the hind footpads with highly purified *E. coli* -derived recombinant GAD67 emulsified in incomplete Freund's adjuvant (IFA). Ten days later mice were sacrificed, spleen and lymph node T cells were isolated and analyzed for their capacity to proliferate in response to GAD67 *in vitro*, in "recall assays". As anticipated, proliferation of T cells isolated from GAD67 plant-fed mice was markedly reduced, while T cells from control plant-fed mice retained capacity to proliferate in response to GAD67. Mice were also assessed for anti-GAD antibody responses. Although serum levels of total anti-GAD IgG antibodies had little change in mice fed control plants, there was a twofold increase in anti-GAD IgG antibody in mice fed GAD67 transgenic plants as compared to that in control mice. Further analysis of the isotypes of anti-GAD IgG antibody revealed that the increased levels of serum IgG anti-GAD antibody production in mice fed GAD plants were due to an increase in anti-GAD IgG1 antibody, consistent with skewing of responses to Th2. No changes in IgG2a (Th1) anti-GAD antibody were found between GAD plant and control plant-fed groups. Moreover, cytokine analyses of supernatants derived from T cells of treated mice showed that the concentration of IFN-γ was reduced with an increase of IL-4 and IL-10 in mice fed GAD plants as compared to that in control mice. Taken together, these results confirmed that transgenic plants were capable of synthesizing the diabetes-associated autoantigen, mouse GAD67, in an immunogenic form and that oral delivery of the recombinant GAD protein through transgenic plant tissues to NOD mice was effective in inducing antigen-specific immune responses.

To determine the effects of transgenic plant tissues producing GAD67 on the suppression of diabetes, young pre-diabetic female NOD mice were fed transgenic potato tuber or transgenic tobacco leaf tissues as a diet supplement to standard mouse chow for a period of seven months starting at 5 weeks of age. The amount of GAD delivered was approximately 1mg per mouse daily. Control mice received an equivalent amount of corresponding vector control transgenic tobacco or potato tissue. NOD mice develop diabetes spontaneously and at age of 6 to 7 months; typically 75 to 85% of the female mice will become diabetic (Makino et al., 1981). As shown in Figure 34, 10/12 NOD mice fed GAD-containing potato tuber (n=6) or GAD-containing tobacco leaf tissues (n=6) remained free of disease (p=0.007 from controls). In contrast, 8 / 12 control plant-fed mice (67%), divided between potato and tobacco supplementation, developed diabetes. The mice tolerated plant tissue well, and no differences between groups were observed in the appearance or weight gains of mice.

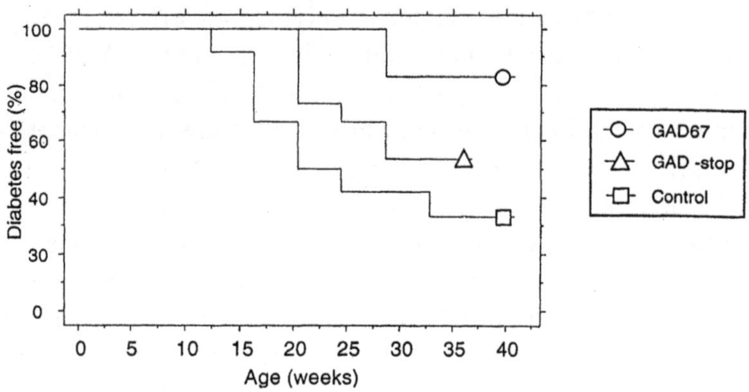

Figure 34. Oral administration of plant material expressing mouse GAD67 prevents development of diabetes in NOD mice. Female NOD mice (n=12/group) were fed plant GAD67 or control plant material from 5 weeks to 8 months of age. All mice were followed for the onset of hyperglycemia (defined as blood glucose > 16.7 mmol./l). 10/12 NOD mice fed either low-alkaloid tobacco (n=6) or potato GAD67 (n=6) remained free of disease (p=0.007 from controls) with equivalent protection using tobacco or potato. In contrast, 8/12 control plant-fed mice developed diabetes. In separate experiments, NOD mice were given GAD67 plant material for about 2 months (8 week duration) followed by standard chow, they showed a delay in disease-onset, but the incidence level of diabetes eventually returned to that of control groups (shown as "GAD-stop").

Although it is difficult to compare results between different laboratories and protocols, these results, are comparable to those previously reported with other methods of GAD immunization (Elliott et al., 1994; Kaufman et al., 1993; Tisch et al., 1993; Tian et al., 1996; Zhang et al., 1991), and suggest that oral plant GAD67 may offer an effective alternative approach to treatment of IDDM. There has been more work performed in the nature of protection. In separate experiments, NOD mice were given GAD 67 transgenic plant tissues for only two months starting at 5 weeks and followed by standard chow. While there was a delay in disease onset, the incidence level of diabetes eventually returned to that of control groups by the study completion, suggesting that continuous GAD feeding is necessary in order to maintain the tolerance status, at least when using "high-dose" GAD.

6. EXPRESSION OF CHOLERA TOXIN B SUBUNIT (CTB)-INSULIN FUSION PROTEINS

Insulin is a major constituent of β cells in the pancreas and is a true "auto" antigen in that it is produced by β-cells and not in surrounding cells. Like GAD, it appears to have an important role in the pathogenesis of diabetes. Injection of NOD mice with insulin or insulin peptides was shown to inhibit both insulitis (an early stage of diabetes involving infiltration of the islet cells with inflammatory lymphocytes) and the subsequent development of diabetes (Atkinson et al., 1990). Insulin therapy by injection has also been shown to delay IDDM in humans (Keller et al., 1993), and such observations have directed clinical trials in diabetes prevention with insulin. Oral insulin therapy has the advantages of easy administration and lack of toxicity. Recently, the results of the first multi-centred human clinical trial (Diabetes Prevention Trial Type 1) testing oral insulin in newly diagnosed diabetes patients were released (Chaillous et al., 2000). While no improvement was seen, continued testing of the effects of immune interventions in recent-onset type 1 diabetes was recommended as there were several possible explanations for a lack of effect. Doses and formulation of oral insulin for example may not have been adequate, or different modalities of oral administration of insulin are required. Recently, Arakawa et al. (1998) generated transgenic potato plants synthesizing a fusion protein consisting of human insulin and the CTB subunit. Transgenic potato tubers produced 0.1% of total soluble protein as the pentameric CTB-insulin fusion protein, which retained GM1-ganglioside binding affinity and native antigenicity of both CTB and insulin. NOD mice fed transgenic potato tuber tissues containing small (μg) amounts of the CTB-insulin fusion protein showed a reduction in insulitis, and a delay in the progression of clinical diabetes, while those receiving transgenic potato tissues producing insulin or CTB protein alone were not protected.

It appears therefore that insulin conjugated with a CTB subunit can suppress development of diabetes in NOD mice at lower doses than that required for insulin alone, which may be a useful strategy in cases when auto-antigen expression is limiting in transgenic plants. Again, a possible concern with the use of CTB as a mucosal adjuvant or carrier molecule for conjugated antigens in the induction of oral tolerance, is that prolonged feeding of CTB may induce neutralizing mucosal and systemic antibodies specific for CTB and prevent long lasting tolerance induction. In NOD mice fed transgenic potato tubers containing CTB-insulin fusion protein, serum and mucosal anti-CTB antibodies were detected (Arakawa et al., 1998).

7. EXPRESSION OF HUMAN GAD65

Recently the expression of human GAD65 in transgenic tobacco and carrot plants has been reported (Porceddu et al., 1999). Western analysis of tobacco leaf homogenates revealed a unique polypeptide which is comparable in size to rhGAD65 produced in the baculovirus-insect cell expression system. The expression level of human GAD65 in transgenic tobacco leaf reached 0.04% of the total soluble protein, while in carrot taproots it only accounted for 0.015% of the total protein. It appears that even in tobacco, the expression level of human GAD65 is lower than those we reported for mouse GAD67 (0.4%). The difference in expression may reflect structural differences between the two isoforms. Human GAD65 is membrane-anchored by signals located in the NH_2-terminal region (Namchuk et al., 1997; Shi et al., 1994), while GAD67 is a cytosolic isoform of GAD. The association of GAD65 with membranes in human cells is mediated by way of protein-protein interactions, with accessory effects contributed by palmitoylation (Namchuk et al., 1997). Interestingly, immunogold labeling and electron microscopy of transgenic tobacco tissue showed the selective targeting of human GAD65 to chloroplast thylakoids and mitochondria, supporting the existence of similar protein-protein interactions in membranes of plant cells. Based on studies using human autoantibodies directed against conformational epitopes of the autoantigen (Falorni et al., 1996) and the immunoenzymatic assay indicated that human GAD65 is folded correctly in plants. At present there are no data available on animal trials determining if oral administration of human GAD65 transgenic plants can induce protective oral tolerance in NOD mice, but it is clear it is feasible to use transgenic plants to produce a bioactive form of human GAD65. This is a primary focus of our efforts along with optimization of oral tolerance through novel immune augmentation strategies.

8. CONCLUSIONS AND FUTURE PROSPECTS

The use of transgenic plants and oral delivery offers a promising new strategy for tolerance induction for the treatment of autoimmune diseases such as type I diabetes. A plant-based approach for inducing oral tolerance has the advantage of being effective, simple, low-cost, safe from pathogens and non-invasive. This approach may hold promise for the prevention and/or treatment of other auto-immune diseases, allergy and organ transplant rejection, but these need to be determined by a case by case basis. The availability of relevant animal models with spontaneous rather than induced disease may be critical to antigen selection and testing. We have recently expressed murine major histocompatibility complex (MHC) class II molecules in transgenic plants which may allow us to use the plant-based approaches to induce oral tolerance induction for the prevention of organ transplant rejection (Ma and Jevnikar, 1999).

As oral tolerance induction is usually less efficient in previously sensitized hosts (Zhang et al., 1991), one of the obvious challenges is to identify those at risk and treat prior to disease onset. It is also important to create plants with high expression levels of recombinant mammalian proteins. However, a persistent problem in the expression of antigenic proteins in transgenic plants has been low levels of expression, but several strategies are available to overcome this problem. Other areas that need particular attention includes selection of appropriate plant species as expression hosts, the identification and isolation of novel, more robust plant promoters, strategies to address gene silencing, and strategies to maximize transgene protein accumulation in plant cells. Tobacco chloroplasts have been recently used in high-yield production systems for several therapeutic proteins. Human somatotropin can be expressed as a soluble, biologically active, disulfide-bonded form in tobacco chloroplasts, reaching 7% total soluble protein. This is more than 300-fold higher than a similar gene expressed using a nuclear transgenic approach (Staub et al., 2000). Similarly, tobacco chloroplast transformation with an unmodified CTB-coding sequence resulted in the accumulation of up to 4.1% of total soluble leaf protein as functional CTB oligomers, again which is approximatley 400-fold higher than with related LTB (the B subunits of enterotoxigenic *Escherichia coli*) expressed via the nuclear genome (Daniell et al., 2001). These are very promising results, as high-level expression of many antigenic proteins may now be possible in a system which also offers reasonable gene containment.

Oral tolerance has clear benefits in animals, but has not yet been conclusively shown in humans. One limitation of this strategy, namely cost and delivery, is well addressed by plant systems. It is likely that we will need to enhance the efficacy of oral tolerance in clinical autoimmune disease including the use of combined antigens, the use of adjuvants like CTB and other strategies to skew the response to Th2 when this is known to be beneficial. Plants are not only a competitive production system, but also a convenient and effective delivery system. We are very hopeful that effective immunotherapies for human autoimmune diseases will soon include therapy based on transgenic plants expressing autoantigens, given as a simple dietary supplement.

ACKNOWLEDGMENTS

The authors would like to thank Dr. Jim Brandle and Dr. Rima Menassa for significant contributions to our work in transgenic plants and oral tolerance, as well as many helpful discussions.

REFERENCES

Arakhong DKX, Hough J, Engen PC and Langridge WHR 1998. A plant-based cholera toxin B subunit-insulin fusion protein protects against the development of autoimmune diabetes. Nat Biotechnol, 16:934-938

Atkinson MA and Maclaren NK, 1994. The pathogenesis of insulin-dependent diabetes mellitus. N Engl J Med, 331:1428-1436

Atkinson MA, Maclaren NK and Luchetta R, 1990. Insulitis and diabetes in NOD mice reduced by prophylactic insulin therapy. Diabetes, 39:933-937

Baekkeskov S, Aanstoot H, Christgau S, Reetz A, Solimena M, Cascalho F, Folli F, Richter-Olesen H and DeCamilli P, 1990. Identification of the 64Kautoantigen in insulin-dependent diabetes as the GABA-synthesizing enzyme glutamic acid decarboxylase. Nature,347:151-156

Barcelo P and Lazzeri PA, 1998. Direct gene transfer: chemical, electrical and physical methods. In *Transgenic plant research*, Ed. KLindsey Harwood. Academic Publishers, Amsterdam, pp 35-55

Bergerot I, Fabien N, Maguer V and Thivolet C, 1994. Oral administration of human insulin to NOD mice generates $CD4^+$ T cells that suppress adoptive transfer of diabetes. J Autoimmun

7:655-663

Bergerot I, Ploix C and Petersen J, 1997. A cholera toxoid-insulin conjugate as an oral vaccine against spontaneous autoimmune diabetes. Proc Natl Acad Sci USA, 4:4610-4614

Bevan MW, 1984. Binary *Agrobacterium* vectors for plant transformation. Nuc Acids Res, 12:8711-8721

Bosi E and Botazzo GF, 1995. Autoimmunity in insulin-dependent diabetes mellitus. Clin Immunother, 3:125-135

Carrington JC and Freed DD, 1990. Cap-independent enhancement of translation by a plant potyvirus 5' nontranslated region. J Virol, 64:1590-1597

Chaillous L, Lefevre H and Thivolet C, 2000. Oral insulin administration and residual β-cell function in recent-onset type 1 diabetes: a multicentre randomized controlled trial. The Lancet, 356:545-549

Cramer CL, Weissenborn DL and Oishi KK, 1996. Bioproduction of human enzymes in transgenic tobacco. Ann NY Acad Sci, 792:62-71

Czerkinsky C, Sun JB and Lebens M, 1996. Cholera toxin B subunit as transmucosal carrier-delivery and immunomodulating system for induction of antiinfections and antipathological immunity. Ann NY Acad Sci, 778:185-193

Daniell H, Lee S-B, Panchal T and Wiebe PO, 2001. Expression of the native cholera toxin B subunit gene and assembly as functional oligomers in transgenic tobacco choloroplasts. J Mol Biol, 311, 1001-1009

Elliott JF, Qin HY and Bhatti S, 1994. Immunization with the larger isoform of mouse glutamic acid decarboxylase (GAD67) prevents autoimmune diabetes in NOD mice. Diabetes, 43:1493-1499

Falorni A, Ackefors M, Carlberg C, Daniels T, Persson B, Robertson J and Lernmark A, 1996. Diagnostic sensitivity of immunodominant epitopes of glutamic acid decarboxylase (GAD65) autoantibodies in childhood IDDM. Diabetologia 39:1091-1098

Goddijn OJM and Pen J, 1995. Plants as bioreactors. Trends Biotechnol 13:379-387

Gomez N, Wigdorovitz A, Castanon S, Gil F, Ordas R, Borca MV and Escribano JM, 2000. Oral immunogenicity of the plant derived spike protein from swine-transmissible gastroenteritis coronavirus. Arch Virol, 145:1725-1732

Hiatt AC, Cafferkey R and Bowdish K, 1989. Production of antibodies in transgenic plants. Nature, 342:76-78

Haq TA, Mason HS, Clements JD and Arntzen CJ, 1995. Oral immunization with a recombinant bacterial antigen produced in transgenic plants. Science, 268:714-716

Higgins PJ and Weiner HL, 1988. Suppression of experimental autoimmune encephalomyelitis by oral administration of myelin basic protein and its fragments. J Immunol, 140:440-445

Holmgren J, Lycke N and Czerkinsky C, 1993. Cholera toxin and cholera B subunit as oral-mucosal adjuvant and antigen vector systems. Vaccine, 11:1179-1184

Horsch RB, Fry JE, Hoffmann NL, Eichholtz D, Rogers SG and Fraley RT, 1985. A simple and general method for transferring genes into plants. Science, 227:1229-1231

Husby S, Mestecky J, Moldoveanu Z, Holland S and Elson CO, 1994. Oral tolerance in humans. T cell but not B cell tolerance after antigen feeding. J Immunol, 152:4663-4670

Kaufman DL, Clare-Salzler M and Tian J, 1993. Spontaneous loss of T cell tolerance to glutamic acid decarboxylase in murine insulin-dependent diabetes mellitus. Nature, 366:69-72

Keller RJ, Eisenbarth GS and Jackson RA, 1993. Insulin prophylaxis in individuals at high risk of type 1 diabetes. Lancet, 341, 927-928

Kim J, Richter W, Aanstoot HJ, Shi Y, Fu Q, Rajotte R, Warnock G and Baekkeskov S, 1993. Differential expression of GAD65 and GAD67 in pancreatic islets of man, rat, and mouse. Diabetes, 42:1799-1808

Lee JS, Choi SJ and Kang HS, 1997. Establishment of a transgenic tobacco cell suspension culture system for producing murine granulocyte-macrophage colony stimulating factor. Mol Cell, 7:783-787

Ma SW, Zhao D-L Yin, Z-Q and Mukherjee M, 1997. Transgenic plants expressing autoantigens fed to mice to induce oral immune tolerance. Nat Medicine, 3:793-796

Ma SW and Jevnikar AM, 1999. Autoantigens produced in plants for oral tolerance therapy of autoimmune diseases. In *New chemicals via higher plant bioengineering*. Eds. Shahidi F and Kolodziejczyk P. John Wiley and Sons, New York, pp 179-194

Ma JK-C, Hiatt A and Hein MB, 1995. Generation and assembly of secretory antibodies in plants. Science, 268:716-719

Makino, S Hunimoto, K Muraoka Y and Katagiri K, 1981. Effects of castration on the appearance of diabetes in NOD mice. Exp Anim, 3:137-140

Mason HS, Lam DMK and Arntzen CJ, 1992. Expression of hepatitis B surface antigen in transgenic plants. Proc Natl Acad Sci USA, 89:11745-11749

Mason HS, Ball JM, Shi JJ and Arntzen CJ, 1996. Expression of Norwalk virus capsid protein in transgenic tobacco and potato and its oral immunogenicity in mice. Proc Natl Acad Sci USA, 93:5335-5340

Mason H, Haq T, Clements J and Arntzen, C, 1998. Edible vaccine protects mice against *E. coli* heat-labile enterotoxin (LT): potatoes expressing a synthetic LT-B gene. Vaccine, 16:1336-1343

McGarvey PB, Hammond J and Dienelt MM, 1995. Expression of the rabies virus glycoprotein in transgenic tomatoes. Bio/Technology, 13:1484-1487

Mosmann TR and Coffman RL, 1989. Th1 and Th2 cells: different patterns of lymphokine secretion lead to different functional properties. Ann Rev Immunol, 7:145-173

Mowat AM, 1987. The regulation of immune response to dietary protein antigens. Immunol Today, 8:93-98

Namchuk M, Lindsay L, Turck CW, Kanaani J and Baekkeskov S, 1997. Phosphorylation of serine residues 3, 6, 10, and 13 distinguishes membrane anchored from soluble glutamic acid decarboxylase 65 and is restricted to glutamic acid decarboxylase 65 alpha. J Biol Chem, 272:1548-1557

Nussenblatt RB, Caspi RR and Mahdi R, 1990. Inhibition of S-antigen induced experimental autoimmune uveoretinitis by oral induction of tolerance with S-antigen. J Immunol, 144:1689-1695

Porceddu A and Falorni A, 1999. Transgenic plants expressing human glutamic acid decarboxylase (GAD65), a major autoantigen in insulin-dependent diabetes mellitus. Mol Breed, 5:553-560

Porta C and Lomonossoff GP, 1996. Use of viral replicons for the expression of genes in plants. Mol Biotechnol, 5:209-221

Richter LJ, Thanavala Y, Arntzen C and Mason S, 2000. Production of hepatitis B surface antigen in transgenic plants for oral immunization. Nat Biotechnol, 18:1167-1171

Rollwagen FM and Baqar S, 1996. Oral cytokine administration. Trends Immunol Today, 17:548-550

Sandhu JS,Krasnyanski SF and Domier L, 2000. Oral immunization of mice with transgenic tomato fruit expressing respiratory syncytial virus-F protein induces a systemic immune response. Transgenic Res, 9:127-135

Shi Y, Veit B and Baekkeskov S, 1994. Aminoacid residues 24-31 but not palmitoylation of cysteines 30 and 45 are required for membrane anchoring of glutamic acid decarboxylase, GAD65. J Cell Biol. 124:927-934

Staub JM, Garcia B, Graves J, Haidukiewicz PTJ, Hunter P, Nehra N, Paradkar V, Schlittler M, Carroll JA, Spatola L, Ward D, Ye G and Russell DA, 2000. High-yield production of a human therapeutic protein in tobacco chloroplasts. Nat Biotech, 18:333-338

Sun JB, Holmgren J and Czerkinsky C, 1994. Cholera toxin B subunit: an efficient transmucosal carrier-delivery system for induction of peripheral immunological tolerance. Proc Natl Acad Sci USA, 91:10795-10799

Sun JB, Rask C and Olsson T, 1996. Treatment of experimental autoimmune encephalomyelitis by feeding myelin basic protein conjugated to cholera toxin B subunit. Proc Natl Acad Sci USA 93:7196-7201

Tacket CO, Mason HS, Losonsky G, Clements JD, Levine MM and Arntzen CJ, 1998. Immunogenicity in humans of a recombinant antigen delivery in a transgenic potato. Nature Med, 4:607-609

Tacket CO, Mason HS, Losonsky G, Estes MK, Levine MM and Arntzen, CJ, 2000. Human immune responses to a novel Norwalk virus vaccine delivered in transgenic potatoes. J Infect Dis, 182:302-305

Thompson HS and Staines NA, 1986. Gastric administration of type II collagen delays the onset and severity of collagen-induced arthritis in rats. Clin Exp Immunol, 64:581-586

Tian J, Atkinson MA and Clare-Salzler M, 1996. Nasal administration of glutamate decarboxylase (GAD65) peptides induces Th2 responses and prevents murine insulin-dependent diabetes. J Exp Med, 183:1561-1567

Tisch R, Yang XD and Singer SM, 1993. Immune response to glutamic acid decarboxylase correlates with insulitis in non-obese diabetic mice. Nature, 366:72-75

Weiner HL, Friedman A and Miller A, 1994. Oral tolerance: Immunological mechanisms and treatment of animal and human organ-specific autoimmune diseases by oral administration of autoantigens. Annu Rev Immunol, 12:809-837

Weiner HL, 1997. Oral tolerance: immune mechanisms and treatment of autoimmune diseases. Immunol Today, 18:335-343

Wells HG, 1911. Studies on the chemistry of anaphylaxix. III. Experiments with isolated proteins, especially those of hens's egg. J Infect Dis, 9:147-151

Zhang ZJ, Davidson L, Eisenbarth G and Weiner HL, 1991. Suppression of diabetes in nonobese diabetic mice by oral administration of porcine insulin. Proc Natl Acad Sci USA, 88:10252-10256

Chapter 9

THE PRODUCTION AND DELIVERY OF THERAPEUTIC PEPTIDES IN PLANTS

L. Erickson, W-J. Yu, J. Zhang, C.F.M. deLange[1], B. McBride[1] and S. Du

Department of Plant Agriculture, [1]Department of Animal and Poultry Science, University of Guelph, Guelph, Ontario, N1G 2W1, Canada

Keywords: growth factors, antimicrobial peptides, fusion proteins, inducible promoters

Abstract There is a wide range of naturally-occurring peptides with immunophysiological function of particular relevance to the intestinal health and development of mammals. Our goal is to produce and deliver therapeutic peptides, such as growth factors and antimicrobial peptides, in feed and food plants. The chief advantage of this approach is the ability to produce large amounts of these peptides at low cost to augment endogenous levels of these factors in the intestine, especially in disease conditions when the levels of these factors may be low or otherwise inadequate, or to provide proteins with other novel immunophysiological properties not normally present in the gut. A major obstacle in this effort is achieving adequate levels of such peptides in plant tissue, due to problems with transcription, translation, degradation of the peptide or negative effects of the foreign protein on the growth and development of the plant. Our approach to this problem is to develop platform technologies involving fusion proteins and inducible expression systems, which can enhance the expression of small, biologically-active peptides in plant tissues for oral delivery.

1. PLANT SCIENCE IN THE CONTEXT OF VETERINARY AND MEDICAL HEALTH

There is a rapidly-growing interest on the part of the public as well as in the medical community in the role food plays in health and development, especially

L. Erickson et al. (eds.),
Molecular Farming of Plants and Animals for Human and Veterinary Medicine, 197–222.

in the immunophysiological impact of food over and above the role of basic nutrition. As a result, there is an expanding opportunity and need for those working in plant biology to conduct research in a broader scientific and social context than in the past. Such research requires a multidisciplinary approach and, despite the potential for discovery and wider relevance of such research, there are not many examples of plant scientists straying far beyond the boundaries of their disciplines to collaborate with colleagues, for example, in the medical community. Yet, the significance of plant-based food and feed products far exceeds that of providing the biochemical building blocks for humans and their animals. Many constituents of food have both a beneficial and harmful impact on health through effects on the immune, endocrine, nervous, circulatory and digestive systems. Our long-term goal is to combine our understanding of animal physiology, protein biochemistry and plant biology to form multidisciplinary teams dedicated to the discovery and engineering of feed/food proteins of veterinary and medical significance, with particular emphasis on intestinal disease.

With this focus, we are proposing a new approach to improving the intestinal health and development of human populations and their agricultural or companion animals, which is intermediate between traditional herbal medicines, on the one hand, and conventional pharmaceuticals, on the other. This approach entails the production and delivery of therapeutic peptides in feed and food plants. The chief advantage of this approach is the ability to produce large amounts of these peptides at low cost to augment endogenous levels of these factors in the intestine, especially in disease conditions when the levels of these factors may be low or otherwise inadequate.

2. SIGNIFICANCE OF INTESTINAL DISEASE

2.1 Human

Gastrointestinal disorders have a major impact on the health and productivity of the human populations worldwide, resulting in health costs and lost productivity valued at $100 billion annually in Canada alone. Peptic ulcers, non-ulcer dyspepsia, Crohn's disease, colitis and food allergies are examples of diseases commonly afflicting the intestine, along with many others resulting from infection by enteric pathogens, such as *E. coli*, *Salmonella*, *Lysteria*, *Cryptosporidium* and various viruses. In addition, scientific evidence is mounting for the role of anxiety or stress in intestinal dysfunction, mediated probably by the very extensive enteric nervous system (Soderholm and Perdue, 2001). Moreover, the burden imposed by these diseases is very likely to grow in the future, due to increased exposure to pollutants and pathogens, accelerating socioeconomic change with associated psychoneurophysiological stresses, and an aging population.

2.2 Animal

The mammalian gastrointestinal tract has evolved the capacity to digest and absorb a wide range of nutrients and accommodate numerous stresses. However, the nutrient delivery capacity of the small intestine of ruminant and non-ruminant species in modern livestock systems may be inadequate to allow full expression of the genetic potential for production (Kreikemeier et al. 1991). Such limitations may be particularly acute during nursing, weaning, and infections by intestinal pathogens (bacteria and viruses) and parasites. The increased use of antibiotics in animal feeds to control such pathogens has prompted fears of antibiotic resistance, especially in related human pathogens, as well concerns regarding drug residues in animal products. Other forms of stresses are imposed on intestinal function by the change and range of feeds utilized, the high grain diets in cattle feedlots leading to acidosis, and the administration of antibiotics and antihelminthics.

3. INTESTINAL DISEASE AND THE ROLE OF THE EPITHELIUM

The gastrointestinal tract is lined by a single layer of epithelial cells, joined together at their apical poles by tight junctions. The epithelium is the critical interface separating the internal milieu of the body from the external environment, including food and microbial antigens in the gut lumen. The permeability of the epithelial lining in the intestine is increased in a number of enteropathies, such as inflammatory bowel disease, celiac disease and food allergies (Bjarnason et al., 1995). Animal models and epithelial cell culture studies have demonstrated that many factors can increase epithelial permeability, including nutritional deficiencies (Hadfield et al., 1995), certain medications (Wallace, 1996), stress (Soderholm and Perdue, 2001), bacterial infections and/or toxins (Go et al., 1995; Philpott et al., 1996) and immune cell cytokines such as interleukin-4 (Colgan et al., 1994). Fortunately, a variety of mechanisms, such as mucins, secreted antibodies and antibiotic peptides have evolved to protect intestinal tissues from the formidable, daily onslaught of antigens, toxins and pathogens. In addition, epithelial cells have a rapid turnover rate, and an array of peptide growth factors continuously assist in maintaining or restoring the integrity of the epithelial barrier.

4. GROWTH FACTORS AND THE EPITHELIUM

4.1 Epidermal Growth Factor

Although there are many trophic peptides affecting gut tissues, we have focused on two major ones because of the wealth of data available regarding their structures and functions. One of the best characterized of all mammalian growth factors is epidermal growth factor (EGF), a single-chain polypeptide composed of 53 amino acids. It occurs in a variety of secreted fluids such as sweat, prostatic fluid and saliva (Fisher and Lakshmanan, 1990), promotes epithelial cell proliferation and plays a major role in gut growth, development, and regeneration (Marti et al., 1989). The EGF molecule is small, compact and highly stable due to 3 disulfide bonds. These features, as well as the fact that it is secreted into the gastrointestinal tract where it promotes the growth of epithelial cells, make it an attractive candidate for oral treatments of intestinal diseases. In addition, the small size of the final secreted peptide facilitates gene synthesis for experimentation with recombinant forms in heterologous hosts which may entail changes in codon usage. The 3-dimensional structure of all mammalian EGF molecules characterized to date appears to be highly conserved to the extent that EGF peptide of one species binds readily to the EGF receptor of another.

Human EGF, produced in *E. coli* and yeast, is commercially available and, although very expensive (>$1,000/mg), has been used in experiments with oral feeding. Oral administration of EGF to young rabbits (O'Laughlin et al., 1985), rats (Puccio and Lehy, 1988) and pigs (Jaeger et al., 1990) has produced multiple effects such as increases in stomach and pancreas weights, epithelial cell number in ileal mucosa, antral mucosal height, and jejunal lactase and sucrase activities. EGF seems to be especially important for gut development in newborn mammals; ingestion of EGF by newborn rodents has increased gastric and intestinal weight, stimulated calcium-binding protein levels and calcium transport, increased vitamin D receptors, augmented gastric acid secretion, and increased the activities of ornithine decarboxylase, trehalase, lactase, glucoamylase and alkaline phosphatase, as well as functional activities of the developing gut (Fisher and Lakshmanan, 1990). In addition, the administration of exogenous EGF has been shown to enhance uptake in existing cells of nutrients such as glucose, amino acids and electrolytes (Schwartz and Storozuk, 1988; Opleta-Madsen et al., 1991).

Of particular interest are the effects of ingested EGF under disease conditions. The intestines of rats damaged with methotrexate showed significant increases in mucosal disaccharidases and leucine aminopeptidases following oral treatment with EGF, compared with the non-EGF control (Petschow et al., 1993). Zijlstra et al.(1994) demonstrated a beneficial effect of oral administration of EGF in the recovery of baby pigs infected with rotavirus, but levels much higher than those observed physiologically were required. Young rabbits fed EGF along with enteropathogenic *E. coli* did not experience the diarrhea and reduced weight gain of the non-EGF control (Buret et al., 1998); associated with these effects was a reduced colonization of the small and large intestine by the *E. coli*, less microvillous injury and increased maltase and sucrase activties in the jejunum. EGF also performs other cytoprotective functions in the adult gastrointestinal tract, such as decreased gastric acid secretion, increased healing of ulcers, and increased crypt cell production rates after toxic injury (Marti et al. 1989; Fisher and Lakshmanan, 1990). These experiments suggest that there is an opportunity to develop treatments for intestinal disease based on oral delivery of EGF.

4.2 Glucagon-like Peptide-2

In 1971, an endocrine tumor secreting glucagon was shown to induce intestinal hyperplasia (Gleeson et al., 1971). In mammals, mRNA transcripts from a single proglucagon gene are translated and processed differently in various tissues to yield up to 6 different peptides (Drucker, 1998). Glucagon-like peptide-2 (glp2) is one such peptide which is secreted from enteroendocrine cells of the small and large intestine. To date, the predominant form of glp2 appears to be a 33-aa molecule, the coding region of which is highly conserved across species; there is only one aa difference between the human and mouse peptide (Drucker, 1998). Although not studied nearly to the extent as EGF, data are rapidly accumulating from rodent studies on the potent stimulatory effect of this peptide on proliferation of intestinal epithelial cells. Injection of glp2 has been shown to increase intestinal weight (Tsai et al., 1997; Litvak et al., 1998), disaccharidase levels and nutrient absorption (Brubaker et al., 1997), and reduce enterocyte apoptosis (Tsai et al., 1997). The potential therapeutic value of GLP2 was demonstrated in mice with induced severe colitis (Drucker et al., 1999); treatment by injection with an analogue of GLP2 significantly reversed weight loss, and increased colon length, crypt depth, mucosal area and integrity of the colon. Additional attractive features of this peptide as a potential therapeutic agent are its specificity to epithelial cells and evidence that the small bowel growth induced by exogenous GLP2 appears to be physiologically normal intestinal tissue (Drucker, 1998).

5. ANTIMICROBIAL PEPTIDES AND INTESTINAL HEALTH

Antimicrobial proteins are synthesized by a very wide range of organisms, including plants, and could be considered as components of an innate immunity system. Many of these are classified as cationic, and contain two broad structural categories: alpha helices and beta sheets (Hancock and Lehrer, 1998). Examples of the former are the magainins from amphibians (Simmaco et al., 1998), and cecropins of moths (Gudmundson et al. 1991), whereas examples of the latter are the defensins, protegrins and tachyplasins. Vertebrate defensins are 30-40 amino acids in length and contain 6 cysteines that form three disulphide bonds. These endogenous peptides have been found in epithelial cells of the gastrointestinal, respiratory, and genitourinary tracts of mammals where they are thought to provide resistance to invasion by bacteria, mycobacteria, fungi, and enveloped viruses (Lehrer and Ganz, 1996). A porcine beta-defensin has recently been cloned and the data suggest that the final secreted form consists of 38 amino acids (Zhang et al. 1998). In swine this gene is highly expressed in epithelial cells of the tongue, but is also expressed in tissues lining the respiratory and digestive tracts. It is interesting that, although there is no homology between defensins and epidermal growth factor, both proteins are small and contain three disulfide bonds which may confer stability in the GI tract.

As defensins are cationic and amphipathic, they have an affinity for the anionic-phospholipid-rich antimicrobial membranes, but not for the membranes typical of mammals, for example, which contain more neutral phospholipids and cholesterol. The mechanisms by which defensins interfere with microbial membranes are not well understood, but are characterized by an efflux of potassium ions and influx of calcium ions, probably through selective pores formed or induced by defensin (Kagan et al. 1990; Thevissen et al., 1997).

Plants also synthesize antimicrobial proteins which have been designated as defensins, since they are similarly small proteins and contain multiple cysteine residues involved in disulfide bonds much like the mammalian beta-defensins (de Samblanx et al., 1997). Thevissen et al. (1997) have suggested that plant defensins do not interact directly with phospholipids, but rather bind to specific receptors or sites on fungal membranes. Although there are several reports of expression of cecropins and derivatives thereof in plants for the purpose of enhancing resistance to plant pathogens (e.g. Osusky et al., 2000), there are no reports to date of synthesis of mammalian beta-defensins in plants . It is not known to what extent such peptides expressed in plants would confer resistance to plant pathogens. However, they constitute another category of proteins with considerable potential to enhance the health and productivity of farm animals, when produced and delivered as an additive derived from crop plants.

6. PLANTS ENGINEERED TO CONTAIN NOVEL THERAPEUTIC PEPTIDES

6.1 Medicinal Plants

The folklore and tradition of plant-based medicinal treatments pre-date recorded history, and is experiencing a renewed interest in many countries along with other forms of so-called "alternative medicine". Although most medicinal compounds from plants are non-proteinaceous in nature, plants produce a variety of proteins which have immunophysiological functions when ingested (Arai, 1996). Some examples are the hypotensive effects of a zein peptide from maize and the lowering of blood cholesterol by a glycinin peptide from soybean

(Arai, 1996). There is a burgeoning interest in North America in the utilization of "natural" products, from both food and non-food plants, referred to broadly as nutriceuticals and functional foods.

There are many advantages to the synthesis and delivery of various peptides with immunophysiological function in food plants and plant-derived products. The first is the ability to produce large amounts of protein at very low cost, which in many cases is likely to be necessary for oral administration. Another is the perception that plant foods promote health in general and the addition of a foreign protein to enhance their healthful properties seems a logical extension to this perception. For many, eating a "medicinal" apple or mixing a "therapeutic" soy powder into their fruit juice is more desirable than popping a pill containing a "synthetic" peptide produced perhaps in yeast or *E. coli.* Finally, plants are known to contain many compounds which modify the effects of associated elements in the diet, enhancing an immune response, for example. Depending on the peptide, its concentrations in the food product and the health concerns of the consumer, dosage of the peptide may not be closely controlled, i.e. an apple or two a day could not be harmful. In cases where dosage might be prescribed, pills or powders with plant extracts containing known amounts the peptide could be administered. In addition, in some current social environments plants and foods genetically-engineered for medicinal purposes would probably be more acceptable than those engineered to tolerate pesticides, for example.

The development of transformation techniques has now extended the potential of feed/food plants to produce a much broader range of proteins with medicinal/physiological effects than occur naturally.

6.2 Expression of Xenoproteins in Plants: Problems and Strategies

There are now many examples of "successful" expression of foreign genes in plants, depending on the purpose of expression. Several bacterial enzymes have been expressed at satisfactory levels with very little modification, other than regulatory sequences at the 5' and 3' ends. A well-known example is the gene for neomycin phosphotransferase, an antibiotic resistance gene used as a selectable marker. The protein levels needed for an enzyme, however, may be considerably less than what is necessary for non-enzymatic proteins. The early attempts to express in plants the Bt gene, from *Bacillus thuringiensis*, are probably more typical of this type of research. Initial constructs containing the naturally-occurring form of the gene or truncated versions thereof produced levels of the protein which could not be detected on Western blots, although some biological activity (binding to insect gut epithelia, mortality) was evident from feeding trials. It was only after extensive re-designing of the gene that adequate levels of the Bt protein were achieved in plant tissue for commercial purposes (Perlak et al., 1991). There are many other examples, as well, (and probably many unreported) of expression levels of non-plant proteins which are very low, i.e. 0.001% of total protein or less.

Several factors leading to low expression levels of foreign genes in plants have been postulated, including high A/T content, codons atypical of plants, premature poly-A signals, mRNA destabilizing sequences and fortuitous intron-like sequences. Some support for these assertions has been provided by increased expression of re-engineered versions of these genes, as in the Bt example. However, there are now probably numerous examples which demonstrate that addressing these factors may not be sufficient. Experience in our lab with a swine viral antigen has shown that the complete re-synthesis of genes with these factors in mind may not be rewarded with increased levels of the protein commensurate with the investment of time and money (see ch. 15, this volume). As synthetic gene constructs contain many alterations of the original gene aside from codon usage, the value of "optimizing" codon usage, i.e. adapting it to those most frequently utilized by the host plant, has yet to be conclusively demonstrated.

Several strategies have been adopted to enhance expression of xenoproteins in transgenic plants, such as targeting the transgene protein to the endoplasmic reticulum (ER), apoplast, or chloroplast, and translational fusions. A prominent example of the latter was expression of enkephalin, a 5-aa neuropeptide, as part of a seed storage protein, reaching levels of 0.1% of total extractable protein (van Dekerckhove et al., 1989). Another variation of this approach was the synthesis in plants of a malarial epitope as a part of the coat protein of the tobacco mosaic virus for use as a vaccine (Dalsgaard et al., 1997). Fusions to beta-glucuronidase, a bacterial reporter gene, and green fluorescent protein have been used to not only detect expression of the attached transgene protein, but also determine its location and distribution in tissues. Finally, attachment to an endogenous plant protein was effective for the synthesis in plants of two mammalian proteins with pharmaceutical potential: the human cytokine, granulocyte-macrophage colony-stimulating factor, fused to the signal peptide and first 8 aa of a rice glutelin protein (Sardana et al., 1998); the human anticoagulant peptide hirudin, produced as an appendage to the 3' end of the oleosin gene (Parmenter et al., 1996). In this latter case the fusion to oleosin localized the hirudin to the oil bodies of canola, as the oleosin protein is attached to the membranes of those bodies, which facilitates isolation of the protein by flotation of the oil bodies after grinding of the seed in aqueous buffer.

In our lab, we have been experimenting with fusing portions of the S gene, the major antigen of transmissible gastroenteritis virus (TGEV) of swine, to plant proteins to enhance expression (see Erickson et al., ch. 15, this volume). Attempts to express the full-length cDNA clone of this gene using various promoters, including the 35S CaMV promoter did not result in levels of the S protein detectable on Western blots, although antibodies to the viral protein were induced by injecting plant tissue extracts into pigs and mice. As an alternative, portions of the S gene encoding a linear epitope (D) responsible for generating neutralizing antibodies to the virus, were fused to two plant genes truncated at the 3' end. One of these plant proteins was a beta amylase (Tuboly et al., 2000), and the other was a pollen-specific protein from alfalfa (Qiu et al., 1997). Both fusion proteins were driven by a chimeric mannopine/octopine synthase promoter (Ni et al., 1995). In tobacco and alfalfa plants containing these fusion constructs, the levels of the fusion proteins were in the 0.1% range of total soluble protein. In contrast, plants containing a truncated, synthetic version of the S gene synthesized S protein at much lower levels of only 0.01% of total soluble protein.

It is likely that a major limitation imposed on the expression of foreign proteins in plants is the possible toxic effect of that protein on the host plant. It has been assumed perhaps by some that, as the foreign protein they intend to express in a transgenic plant has no significant homology to any plant protein, the transgenic protein will have no biological activity in the plant. However, it is impossible to predict what interactions a foreign protein may have with other plant proteins, such as enzymes and receptors, plant membranes, such as those of the endoplasmic reticulum, Golgi apparatus, vacuole and plasmalemma, or the host of other molecules critical to the growth and development of the plant. For any of these reasons, the expression of a foreign protein in a transgenic plant may be somewhat toxic or inhibitory to growth and development. Such effects may be even more pronounced in the case of peptides with potent biological activity in other non-plant organisms. As a result, the only plants which may regenerate from tissue in a transgenic culture are those that express the transgenic protein at a very low level. Hence, with constitutive promoters, there may be a selection for low expression levels of the foreign protein. Such negative interactions may be part of the reason for the low protein levels observed for many transgenes, even when such genes have been reconstructed to optimize expression in plants. Mason et al. (1998), for example, enhanced the levels of the B subunit of the *E. coli* heat labile toxin (LTB) in potato shoots and tubers using a synthetic form of the gene driven by the 35S promoter. However, they also noted that as LTB levels increased, rate of shoot growth and tuber yield dropped in plants grown in greenhouse conditions. When LTB synthesis was controlled by a tuber-specific promoter, shoot growth was normal, but tuber development was impaired. Similarly, Richter et al. (2000) noted poor shoot growth and low tuber yield in potato as they increased levels of hepatitis surface antigen B by adding plant terminator sequences to their construct driven by an enhanced 35S promoter. To overcome this barrier of toxic effects on the host plant, there is growing interest in the development of inducible expression systems (Zuo and Chua, 2000), and efforts are underway in our lab to devise such a system for commercial application with alfalfa grown and harvested in a conventional field setting (see Erickson et al., ch. 15, this volume).

6.3 Expression of Mammalian Proteins in Plants

There is a vast range of mammalian proteins that one may wish to express in plants for a variety of reasons. Antimicrobial peptides and derivatives thereof might provide plants with some degree of resistance to bacteria and fungi (During, 1996; Hancock and Lehrer, 1998). There is also a major interest in using plants as bioreactors to produce high-value proteins with pharmacological properties (e.g. Parmenter et al., 1996; Sardana et al., 1998; Khoudi et al. 1999). The principal obstacle to the use of plants for such purposes is not the production of the proteins, as plants are very cost-effective protein factories, but, rather the extraction and purification of proteins from plant tissues. A third approach is to orally consume plant tissue in some form to obtain the benefit of the new protein produced therein. Presumably such proteins would be active primarily within the digestive tract (e.g. phytase, antibiotic peptides), affect mucosal tissues and development (e.g. epidermal growth factor), or stimulate an immune response, particularly a mucosal immune response, which would result in the secretion of protective IgA molecules into the intestinal tract.

6.3.1 Synthesis of Epidermal Growth Factor in Plants

Given the potential therapeutic value of delivering large doses of EGF to intestinal tissues, it is not surprising that there have been attempts to produce it in edible plant tissue. Such tissue could provide a simple, low-cost, natural feed additive for farm animals such as cattle, swine, sheep, horses and goats. There is a report of the expression of human EGF in tobacco at levels <0.001% of total protein (determined by ELISA) (Higo et al., 1993), but an examination of the data is not convincing. The construct utilized by these researchers was designed for expression in *E. coli* and contained no signal peptide. This is probably a fatal design flaw as EGF requires the appropriate oxidative environment and enzymes for proper folding and formation of disulphide bonds, conditions which do not exist in the cytosol of plants.

In our lab, we have been experimenting with a synthetic porcine EGF gene in tobacco. The original gene was synthesized using overlapping oligos based on published DNA sequence data (Pascall et al., 1991), with codons modified to those most frequently utilized by dicotyledonous plants (Ikemura, 1993). A 25-aa signal peptide, MNFLKSFPFLOFGQYFVAVTHA, from a pathogenesis-related protein in tobacco was added to the amino terminus of the EGF coding region, as well as a 6 x histidine tail at the carboxyl terminus (Figure 35). This signal sequence has been utilized to direct the synthesis of foreign proteins to the endoplasmic reticulum where conditions exist to ensure disulphide bond formation and proper folding.

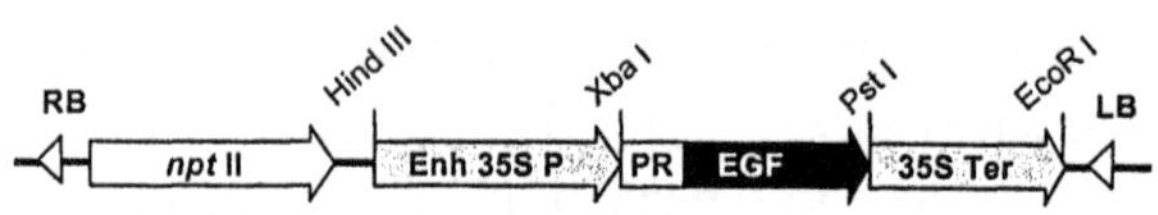

Figure 35. T-DNA region of the binary vector containing synthetic porcine EGF gene. RB: right border; LB, left border; npt II, neomycin phosphotransferase gene; Enh 35S P, enhanced 35S promoter; PR, signal peptide of plant pathogenesis-related protein; EGF; synthetic DNA coding for the mature protein region of porcine EGF; 35S Ter, 35S gene terminator.

In plants proteins synthesized in the endoplasmic reticulum (ER) are secreted from the cell if the coding region does not contain signals for retention in the ER or translocation to another intracellular locus such as the vacuole. This signal sequence and secretion strategy have been used for expression of a number of foreign proteins in plants (Sijmons et al., 1990; Firek et al., 1993;

Verwoerd et al., 1995). This gene construct was cloned into a pBIN19 derivative containing an enhanced 35S promoter and transferred to tobacco with *Agrobacterium*. Eight of ten transgenic plants had high levels of EGF mRNA of the expected size on Northern blots. However, we were unable to detect porcine EGF protein in any transgenic tobacco plants using a variety of methods including western blots, ELISA and HPLC. Attempts to translate mRNA extracts from these plants using a commercial wheat germ lysate kit also yielded no detectable EGF protein.

The presence of elevated levels of EGF may be problematic for plant cells. The EGF motif of 6 cysteines in a conserved pattern of amino acids, which gives rise to 3 disulfide bonds and a characteristic 3-dimensional structure, has been found in a broad range of organisms, and usually within extracellular or cell-surface proteins. EGF domains have been located in wall-associated kinases of plants (He et al., 1996); EGF-binding proteins have been detected in plant extracts (Komatsu et al., 1996) and recently membrane-associated proteins have been discovered in plants with homology to EGF receptor proteins (Ahmed et al. 1997).These observations raise the possibility that plant proteins with EGF-like domains may have some physiological function in plants and that exogenous EGF from transgenes may have toxic or deleterious effects. The apparent absence of EGF protein in our plants may, therefore, be due to rapid metabolic turnover of the free protein via specific receptors or simply due to non-specific protease activity, which has been responsible for degradation of other small foreign proteins in plants (Owens and Heutte, 1997). Fusion of EGF to a carrier protein, under the control of an inducible promoter, may result in more useful levels of the protein.

6.3.2 Synthesis of Antimicrobial Peptides in Plants

Genes for antimicrobial peptides, AMPs, constitute one of many groups of genes transferred to plants for the purpose of combatting pathogens (rev. by Datta et al., 1999). Included in these are genes coding for glucanases, chitinases, ribosome-inactivating proteins, thaumatin-like proteins and proteinase inhibitors. However, the objective in all cases was to enhance resistance to plant pathogens, and surprisingly little mention is made of possible effects of these proteins on microbial populations, beneficial or otherwise, resident in the gastrointestinal tract of livestock eating such plant material. Similarly, AMP-coding genes transferred to plants to date have not generally been intended for therapeutic use in animals (Table 5). However, there are valuable insights to be gleaned from these studies for those planning such applications.

AMP genes transferred to plants have originated generally from insects or plants. One group of genes most commonly harnessed for this purpose is known as the cecropins, named for their origin in the moth *Hyalophora cecropia*. The mature proteins coded for by the major cecropins, A, B, and C, are basic with a broad range of antibacterial activity, induced by infection and secreted into the hemolymph of pupae (Boman and Hultmark, 1987). Hightower et al. (1994) reported levels of cecropin A fused to a signal peptide from cecropin B at 0.0026% of total protein. Jaynes et al. (1993) transferred a gene for mature cecropin B and a derivative thereof to tobacco, the former under the control of the 35S promoter and the latter controlled by a wound-inducible promoter. No signal peptide was apparently fused to either coding region, but the protein was visible on a western blot. The authors estimated the levels reached for the synthetic derivative at "...no more than 0.1% of total protein...", which would be surprisingly high given the levels reported by others for these genes. Other constructs containing cecropin B fused to plant signal peptides have produced variable results (Table5), ranging from non-detectable on a western blot (Florack et al. 1995) to reasonably strong signals. Unfortunately, the levels have not been determined in several cases, and it is difficult to draw conclusions regarding expression as the vectors differ in more than one feature. However, a study by Sharma et al. (2000) is more instructive for comparing strategies for expressing small foreign peptides in plants. In this study a cecropin B gene was transferred to rice in two different fusion formats. In one construct, the coding region for the mature peptide was fused to the signal peptide of the rice chitinase gene. In the other construct, the coding region for the mature peptide was retained intact in the coding region for the naturally-occurring prepropeptide which contains an endogenous signal peptide. This latter strategy has been used successfully to express defensin in bacteria (Piers et al., 1993) Although cecropin protein was detected in rice plants containing either construct, the levels were significantly higher in plants containing the construct with the mature peptide fused to the signal peptide of rice chitinase.

Also relevant to this approach is the study by Okamoto et al. (1998), in which the gene for sarcotoxin was fused in various configurations to the GUS gene. Sarcotoxin is a 39-amino-acid mature peptide secreted into the hemolymph of the flesh-fly, *Sarcofaga peregrina*, in response to injury or infection. Initial experiments with fusions only to a tobacco PR signal peptide resulted in levels of sarcotoxin which were barely detectable. However, when fused to the GUS gene, sarcotoxin levels in transgenic tobacco were considerably increased, as determined by visual inspection of Western blots . This increase was irrespective of the location of the sarcotoxin gene at the amino or carboxyl terminus of the GUS gene, and whether a signal peptide was included; the signal peptides utilized were either endogenous to the sarcotoxin gene or the PR1a signal peptide from tobacco.

Table 5. Genetic engineering of plants with non-plant AMPs

AMPs	Natural host	Trans-genic host	Signal sequence	Promoter	Protein level[a]	Effect on pathogen	Reference
Tachyplesin I	horseshoe crabs	potato	barley α-hordothionin	CaMV 35S	0.003%	slightly increase resistance to *Erwinia* spp.	Allefs et al., 1996
Cecropin B analogue	giant silk moth	tobacco	no details	proteinase inhibitor II from potato	<0.1%	delayed wilt from *Pseudomonas solanacearum*	Jaynes et al. 1993
Cecropin A	giant silk moth	tobacco	cecropin B	CaMV 35S	0.0026%	no effect on *Pseudomonas syringae* pv. *tabaci*	Hightower et al., 1994
Cecropin	giant silk moth	tobacco	barley α-amylase	proteinase inhibitor II from potato	ND	increase disease resistance to *Pseudomonas syringae* pv. *tabaci*; reduce plant growth.	Huang et al., 1997
Cecropin B	giant silk moth	tobacco	barley leaf thionin	CaMV 35S	--	no protein detected	Florack et al., 1995
Cecropin-melittin	giant silk moth- bee venom	potato	No details	2xCaMV35S	ND	resistance to *Phytophthora, Fusarium, Erwinia spp.*	Osusky et al. 2000
Cecropin B	giant silk moth	rice	rice chitinase	E7WIn(Enh 35S)	ND	inhibited *Xanthomonas oryzae* pv. *oryzae*	Sharma et al., 2000
Sarcotoxin IA	flesh fly	tobacco	PR1a from tobacco and endogenous	E7WIn(Enh 35S)	ND	extract inhibited growth of *E. coli* in liquid culture	Okamoto et al. 1998
Sarcotoxin IA	flesh fly	tobacco	PR1a from tobacco	E12W(Enh 35S)	0.01%	enhanced resistance to *Pseudomonas syringae* pv. *tabaci* and *Erwinia carotovora*	Ohshima et al., 1999
Attacin E	giant silk moth	pear	attacin E	Ca2MV 35S	ND	increase resistance to *Erwinia amylovora*	Reynoird et al., 1999

a. percentage of soluble protein. ND = not determined.

No quantitative comparisons were made of the amounts of sarcotoxin peptide present in the plant extracts. There were two other interesting observations in this experiment. Protein extracts from plants containing sarcotoxin fused to GUS protein were less active against *E. coli* than those from plants containing the unfused peptide. It was also noted that the plants with the highest levels of sarcotoxin, i.e. those with the peptide fused to GUS protein, had abnormal morphology and pronounced electrolyte leakage, possibly due to membrane disruption by the toxin.

7. LEAF-BASED PRODUCTION SYSTEMS: ADVANTAGES AND DISADVANTAGES

To develop a commercially viable strategy for edible health products based on plants, it is useful to begin with the animal/problem complex and work back down through the crop species/tissue/processing options available. For example, if one wishes to address a problem in the intestine of cattle, a logical crop vehicle would be alfalfa, and the leaf then becomes the main tissue of interest. The synthesis of foreign proteins in leaves entails a number of advantages and disadvantages. An advantage frequently mentioned is the biomass produced in such cropping systems. For example, the average dry matter yield from a typical-farm stand of alfalfa harvested three times per growing season in the Great Lakes region of North America is 12-15 tonnes. In regions with longer growing seasons and under more intensive management regimes, the yield can be easily doubled. A feature of such cropping systems that is often overlooked, however, is that the large vegetative mass consists initially of at least 80% water. An advantage to using an existing crop species for such vegetative production is that simple, low-cost methods and conventional farm equipment are available to address the expense and difficulty of handling large volumes of such watery plant material. For forage crops such as alfalfa, it is possible to cut the plant material and leave it in a swath in the field to dry down to a moisture level (20-23%) where it can be baled in various formats, transported, handled and stored. At this moisture level, the plant material and its components are stable. All these operations are highly mechanized and result in the production of large volumes of high-protein plant material at low cost.

Another disadvantage of using leaves as a plant molecular factory is the low protein levels of such tissue. Not only is 80% of the leaf fresh weight attributable to water, but also the remaining dry matter generally contains only 2-9%, protein, depending on species, maturity and environmental conditions. Stevens et al. (2000) have demonstrated that soluble protein in the middle and lower leaves of tobacco drops to 2-5% of dry matter as the plant matures, and varies with light and temperature. This decline is typical of most crop plants as nitrogen is re-mobilized for flowering and seed-set. Even some of these estimates of protein may be inflated, as the methods used to measure protein are based on N determination, not measurement of soluble protein. In immature alfalfa, for example, 20-25% of measured N may be composed of non-protein N (Howarth, 1988). Hence, reports of foreign proteins in leaves at levels of 8-10 % of total soluble protein may not be as significant commercially as anticipated if the amount of soluble protein per tonne of harvested material is very low. The legume forage crops are an exception in this respect; the protein content of a typical alfalfa crop of reasonable quality is 20% of dry matter, and in the leaves alone it is in the 28-30% range (Sheaffer et al., 2000). The separation of leaves from the stem can be easily accomplished by simple physical processes, as the leaves readily fall off the stems when the moisture levels drop below 20-25%.

An advantage of leaves as a production vehicle is the opportunity to control expression of the transgene by inducible promoters (Zuo and Chua, 2000). Ideally, synthesis of the foreign protein could be induced by simple, low-cost application of a treatment to plants in the field, resulting in high levels of the protein within hours and continuing up to a day or so. Leaf tissue is the most obvious target for such rapid induction, as it would present the largest biomass available for induction and probably the most amenable to rapid metabolic change.

There are a number of advantages to such an approach. The phenomenon of transgene silencing may be less likely to occur when such transgenes are not constitutively expressed. The potential deleterious effects of expressing a foreign protein on the growth and development of a plant are avoided. The synthesis of the transgenic product can be timed to coincide with optimal conditions of growth and stage of development of the crop plant. Induction may be targeted to specific harvestable tissues, thereby avoiding synthesis in other tissues which entails additional metabolic costs and perhaps regulatory issues, as well. Finally, inducible transgene systems offer a method of biological containment, i.e. the foreign protein is not present in the crop until the application of the inducing treatment, at which time the crop is harvested.

8. DELIVERY OF THERAPEUTIC PEPTIDES TO THE INTESTINAL TRACT

A successful delivery system for therapeutic peptides produced in plant tissues must include strategies to avoid degradation in the gut. To reduce the risk of degradation one might select at the outset genes for proteins which can resist such degradation, either because they are adapted to that environment or because of their intrinsic properties. Examples of the former are proteins such as intestinally-secreted growth factors and antimicrobial peptides, and an example of the latter is the beta-conglycinins of soybean (Astwood et al., 1996), perhaps a candidate for a carrier protein. There are other strategies one can adopt, as well, to reduce degradation, such as localization in cell walls. Plant cell walls, depending on species, tissue and environment can produce a limitless array of physicochemical structures which can protect a protein in planta, as well as in the digestive tract of animals that ingest the plant tissue.

Dietary context is another very significant factor affecting both degradation and biological activity. It represents the entirety of biochemical components of the feed, many of which have varied effects on the immunophysiological functions of the intestinal mucosa, as well as on the digestion process itself.

The amount and type of processing can also affect the amount and condition of peptides in a plant-derived product that reach the target tissues of the gastrointestinal tract. Some seed-derived feed ingredients, such as soya meal, are heat-treated during processing at temperatures ranging from 60° to 80° C for periods ranging from a few to several minutes, depending on the process. It has been shown in many nutritional studies that such heating is necessary to inactivate enzyme inhibitors and other antinutritional factors to enable utilization of these components in animal diets. The potential for denaturation of a therapeutic protein using such a production and delivery system is obvious, but it not may be as extensive as anticipated since incorporation of the peptide within plant tissues may reduce the extent of denaturation compared to subjecting such proteins to heat as isolated molecules. There may be, in fact, a beneficial effect from heating plant-derived feedstuffs to enhance utilization of the protein component especially (Conrad and Klopfenstein, 1988). This process of rendering proteins less digestible is particularly attractive for ruminant utilization of the high protein content of alfalfa, much of which is rapidly degraded microbially in the rumen and lost as methane and urea. Controlled heating converts much of the readily soluble and digestible plant protein to "bypass protein" which escapes rapid degradation in the rumen, and is broken down more slowly in passage through the intestinal tract. This approach could be conceivably be utilized to reduce degradation of therapeutic peptides expressed in plant tissue for oral delivery to livestock animals.

8.0.1 Biological Activity

The levels of the foreign protein in the starting plant tissue, losses due to length and conditions of storage, and degradation in the gut will determine the amount of a therapeutic protein which will arrive at the physiological targets of interest in the animal. The biological activity of such protein, however, will depend on many factors. The amount and types of glycosylation resulting from synthesis of a protein in plants and the implications this has for biological activity have been well outlined by Lerouge et al. (see ch. 4, this volume). The amount and type of processing can also affect the function, as well as the amount of a peptides in a plant-derived product. Finally again, dietary context can have a very significant effect on the biological activity of peptides in the diet. There is a vast range of biochemical components in food and feed that can have many and varied effects on the immunophysiological functions of the intestinal mucosa, as well as on the digestion process itself. An edible delivery strategy must include considerations not only of the effects of the plant tissue containing the peptide, but also the effects of other components of the diet before, during and after ingestion.

9. REGULATORY CONSIDERATIONS

The utilization of crop plants to produce and deliver therapeutic peptides and oral vaccines to livestock or humans has received little comment to date from environmental or other public interest groups, but undoubtedly this will change as these products come closer to commercialization. Developments towards commercialization are probably most advanced in the United States, where a handful of companies are in advanced trials for growing, processing and testing crop plants such as corn, alfalfa and tobacco and tissues/extracts derived therefrom. One can only assume that such trials and developments would not have proceeded this far without prior consultation with government regulatory authorities such as the USDA, FDA and EPA, who must have signaled that such research could proceed and the prospective products would be considered for approval with the inevitable provisions. The nature of those provisions have not been spelled out in a public forum.

There are clear differences, however, between previous transgenic crops and these new crops with respect to how they will be grown, managed, marketed and utilized. The land area occupied by individual therapeutic crops will be much smaller generally and in some cases perhaps only a few hundred hectares. In addition, growth and harvesting of the transgenic crop will be under the tight control of a company directly or through a contractor. This allows for growing the crop in a wide range of environments, which can be selected for isolation from other crops, human populations and wild and domestic animals. Such flexibility provides the opportunity to maximize production, as well as containment until processing eliminates the possibility of contaminating other food and feed sources.

The employment of food and feed crops such as corn and soybeans for producing and delivering therapeutic proteins to livestock will undoubtedly face close scrutiny from the regulatory agencies for this possibility, especially following the problems with Starlink corn containing a form of the pesticidal Bt protein not registered for human consumption. The likelihood of this occurring would seem to be very remote given the small scale of cultivation and close control over the crop by the producer. Moreover, even if some medicinally-enhanced seed were accidentally diverted into conventional marketing channels, the concentration of the contaminating seed in the very large commercial stores typical of such commodity crops would be vanishingly small. In addition, it is highly unlikely that most therapeutic proteins would retain significant biological activity in food items following processing. A useful strategy in this connection might be to "label" such specialty seeds with highly visible genetic markers such as purple colouring for maize or black seed coat for soybean, which would enable rapid visual detection in a seed lot.

On the other hand, there are distinct advantages to using common field crops for production and delivery of therapeutic proteins, especially to livestock. The techniques for growing, harvesting and storing such crops are relatively simple and well-established, and the equipment for doing this is mass-produced and low-cost compared to a fermentation facility. Also significant is the common use and acceptance of these crops and derived foods/feeds for consumption by human and livestock populations. This is especially true of soybean, corn and alfalfa. This widespread acceptance is reflected in the government regulations of many countries regarding classification of allowable feedstuffs for consumption by livestock. Most conventional field crops and various by-products derived therefrom are acceptable according to most existing regulations in Europe and North America, tobacco being a notable exception.

ACKNOWLEDGEMENTS

This work was supported by grants from the Natural Sciences and Engineering Research Council of Canada program for Strategic Projects and the Ontario Ministry of Agriculture, Food and Rural Affairs.

REFERENCES

Ahmed SU, Bar-Peled M and Raikhel NV, 1997. Cloning and subcellular location of an Arabidopsis receptor-like protein that shares common features with protein-sorting receptors of eukaryotic cells. Plant Physiol, 114:325-336

Allefs S, De Jong E, Florack E, Hoogendoorn C and Stiekema W, 1996. Erwinia soft rot resistance of potato cultivars expressing antimicrobial peptide tachyplesin I. Molecular Breeding, 2:97-105

Arai S, 1996. Studies on functional foods in Japan – state of the art. Biosci Biotechnol Biochem, 60:9-15

Astwood JD, Leach JN and Fuchs RL, 1996. Stability of food allergens to digestion in vitro. Nat Biotechnol, 14:1269-1273

Bailey A, 2000. Expression in plants and immunogenicity of a fusion protein containing an epitope from the swine transmissible gastroenteritis virus S protein. M.Sc thesis, Department of Plant Agriculture, University of Guelph, Guelph, Ontario, Canada.

Bjarnason I, MacPherson A and Hollander D, 1995. Intestinal permeability: an overview. Gastroenterology, 108:1566-1581

Boman HG, Hultmark, 1987. Cell-free immunity in insects. Annu Rev Microbiol, 41:103-26

Brubaker PL, Izzo A, Hill M and Drucker DJ, 1997. Intestinal function in mice with small bowel growth induced by glucagon-like peptide-2. Am J Physiol, 272:E1050-E1058

Buret A, Olson ME, Gall DG and Hardin JA, 1998. Effects of orally administered epidermal growth factor on enteropathogenic *Escherichia coli* infection in rabbits. Infect Immun, 66:4917-4923

Colgan SP, Resnick MB, Parkos CA, Delp-Archer C, McGuirk D, Bacarra AE, Weller PF and Madara JL 1994. IL-4 directly modulates function of a model human intestinal epithelium. J Immunol, 153:2122-2129

Conrad HR and Klopfenstein TJ, 1988. Role in livestock feeding-greenchop, silage, hay, and dehy. In *Alfalfa and alfalfa improvement*. Eds. Hanson A, Barnes D and Hill R, Monograph 28, American Society of Agronomy, Madison, Wisc USA.pp 539-551

Dalsgaard K, Uttenthal A, Jones TD, Xu F, Merryweather A, Hamilton WDO, Langeveld JPM, Boshuizen RS, Kamstrup S, Lomonossoff GP, Porta C, Vela C, Casal JI, Meloen RH and Rodgers PB, 1997. Plant-derived vaccine protects target animals against a viral disease. Nat Biotechnol, 15:248-252

Datta K, Velazhahan R, Oliva N, Ona I, Mew T, Khush GS, Muthukrishnan S and Datta SK, 1999. Over-expression of the cloned rice thaumatin-like protein (PR-5) gene in transgenic rice plants enhances environmental friendly resistance to *Rhizoctonia solani* causing sheath blight disease. Theor Appl Genet, 98:1138-1145

de Samblanx GW, Goderis IJ, Thevissen K, Raemaekers R, Fant F, Borremans F, Acland DP, Osborn RW, Patel S and Broekaert WF, 1997. Mutational analysis of a plant defensin from radish (*Raphanus sativus* L.) reveals two adjacent sites important for antifungal activity. J Biol Chem, 272:1171-1179

Drucker DJ, 1998. Glucagon-like peptides. Diabetes,47:159-169

Drucker DJ, Yusta B, Boushey RP, DeForest L and Brubaker PL, 1999. Human [Gly2]GLP-2 reduces the severity of colonic injury in a murine model of experimental colitis. Am J Physiol, 276:G79-91

During K, 1996. Genetic engineering for resistance to bacteria in transgenic plants by introduction of foreign genes. Molecular Breeding, 2:297-305

Firek S, Ozcan S, Warner SA and Draper J, 1993. A wound-induced promoter driving npt-II expression limited to dedifferentiated cells at wound sites is sufficient to allow selection of transgenic shoots. Plant Mol Biol, 22:129-142

Fisher DA and Lakshmanan J, 1990. Metabolism and effects of epidermal growth factor and related growth factors in mammals. Endocr Rev, 11:418-442

Florack D, Allefs S, Bollen R, Bosch D, Visser B, and Stiekema W, 1995. Expression of giant silkmoth cecropin B genes in tobacco. Transgenic Res, 4:132-141

Gleeson MH, Bloom SR, Polak JM, Henry K and Dowling RH, 1971. Endocrine tumour in kidney affecting small bowel structure, motility, and absorptive function. Gut, 2:773-782

Go LL, Healey PJ, Watkins SC, Simmons RL and Rowe MI, 1995. The effect of endotoxin on intestinal mucosal permeability to bacteria in vitro. Arch Surg,130:53-58

Gudmundsson G, Dan-Anders L, Asling B, Gan R and Boman HG, 1991. The cecropin locus. J Biol Chem, 266:11510-11517

Hadfield RJ, Sinclair DG, Houldsworth PE and Evans TW, 1995. Effects of enteral and parenteral nutrition on gut mucosal permeability in the critically ill. Am J Respir Crit Care Med,152:1545-1548

Hancock RE and Lehrer R, 1998. Cationic peptides: a new source of antibiotics. Trends Biotechnol, 16:82-88

He ZH, Fujiki M and Kohorn B, 1996. A cell-wall associated, receptor-like protein kinase. J Biol Chem 271:19789-19793

Hightower R, Baden C, Penzes E and Dunsmuir P, 1994. The expression of cecropin peptide in transgenic tobacco does not confer resistance to *Pseudomonas syringae* pv. tabaci. Plant Cell Rep, 13:295-299

Higo K-I, Saito Y and Higo H, 1993. Expression of a chemically synthesized gene for human epidermal growth factor under the control of cauliflower mosaic virus 35S promoter in transgenic tobacco. Biosci Biotechnol Biochem, 57:1477-1481

Holtorf S, Apel K and Bohlmann H, 1995. Comparison of different constitutive and inducible promoters for the overexpression of transgenes in *Arabidopsis thaliana*. Plant Mol Biol, 29:637-646

Howarth RE, 1988. Antiquality factors and nonnutritive chemical components. In *Alfalfa and alfalfa improvement*. Eds. Hanson A, Barnes D and Hill R, Monograph 28, American Society of Agronomy, Madison, Wisc USA.pp. 493-514

Huang Y, Nordeen RO, Di M, Owens LD and McBeath JH, 1997. Expression of an engineered cecropin gene cassette in transgenic tobacco plants confers disease resistance to *Pseudomonas syringae* pv. tabaci. Phytopathology, 87:494-499

Ikemura T, 1993. Plant nucleic acids codon usage. In *Plant molecular biology labfax*. Ed. Croy RDD, BIOS Scientific Publishers Ltd, Oxford, pp 37-48

Jaynes JM, Nagpala P, Destefano-Beltran L, Huang JH, Kim J-H, Denny T and Cetiner S, 1993. Expression of a cecropin B lytic peptide analog in transgenic tobacco confers enhanced resistance to bacterial wilt caused by *Pseudomonas solanacearum*. Plant Sci 89:43-53

Jaeger LA, Lamar CH, Cline TR and Cardona CJ, 1990. Effect of orally administered epidermal growth factor on the jejunal mucosa of weaned pigs. Am J Vet Res, 51:471-474

Kagan BL, Selsted ME, Ganz T and Lehrer RI, 1990. Antimicrobial defensin peptides form voltage-dependent ion-permeable channels in planar lipid bilayer membranes. Proc Natl Acad Sci USA 87:210-214

Khoudi H, Laberge S, Ferullo J-M, Bazin R, Darveau A, Castonguay Y, Allard G, Lemieux R and Vezina L-P, 1999. Production of a diagnostic monoclonal antibody in perennial alfalfa plants. Biotech and Bioeng, 64:135-143

Komatsu S, Abe K and Karibe H, 1996. Epidermal growth factor binding proteins in rice (*Oryza sativa* L.) leaves. Biol Pharm Bull. 19:905-908.

Koziel MG, Beland GL, Bowman C, Carozzi NB, Crenshaw R, Crossland L, Dawson J, Desai N and Hill M, 1993. Field performance of elite transgenic maize plants expressing an insecticidal protein derived from *Bacillus thuringiensis*. Bio/Technology, 11:194-199

Kriekemeier KK, Harmon DL, Brandt RT Jr, Avery TB and Johnson DE, 1991. Small intestinal starch digestion in steers: effect of various levels of abomasal glucose, corn starch and corn dextrin infusion on small intestinal disappearance and net glucose absorption. J Anim Sci, 69:328-338

Lehrer RI and Ganz T, 1996. Endogenous vertebrate antibiotics. Defensins, protegrins, and other cysteine-rich antimicrobial peptides. Ann N Y Acad Sci 797:228-239

Litvak DA, Hellmich MR, Evers BM, Banker NA and Townsend CM Jr, 1998. Glucagon-like peptide 2 is a potent growth factor for small intestine and colon. J Gastrointest Surg, 2:146-150

Marti U, Burwen SJ and Jones AL, 1989. Biological effects of epidermal growth factor, with emphasis on the gastrointestinal tract and liver: an update. Hepatology, 9:126-138

Mason HS, 1998. Edible vaccine protects mice against *Escherichia coli* heat-labile enterotoxon (LT): potatoes expressing a synthetic LT-B gene. Vaccine, 16:1336-1343

Ni M, Cui D, Einstein J, Narasimhulu S, Vergara CE and Gelvin SB, 1995. Strength and tissue specificity of chimeric promoters derived from the octopine and mannopine synthase genes. Plant J, 7:661-676

Ohshima M, Mitsuhara I, Okamoto M, Sawano S, Nishiyama K, Kaku H, Natori S and Ohashi Y, 1999. Enhanced resistance to bacterial diseases of transgenic tobacco plants overexpressing sarcotoxin IA, a bactericidal peptide of insect. J Biochem (Tokyo), 125:431-435

Okamoto M, Mitsuhara I, Ohshima M, Natori S and Ohashi Y, 1998. Enhanced expression of an antimicrobial peptide sarcotoxin IA by GUS fusion in transgenic tobacco plants. Plant Cell Physiol, 39:57-63

O'Loughlin EV, Chung M, Hollenberg M, Hayden J, Zahavi I and Gall DG, 1985. Effect of epidermal growth factor on ontogeny of the gastrointestinal tract. Am J Physiol, 249:G674-678

Opleta-Madsen K, Meddings JB and Gall DG, 1991. Epidermal growth factor and postnatal development of intestinal transport and membrane structure. Pediatr Res 30:342-350.

Osusky M, Zhou G, Osuska L, Hancock RE, Kay WW and Misra S, 2000. Transgenic plants expressing cationic peptide chimeras exhibit broad-spectrum resistance to phytopathogens. Nat Biotechnol, 18:1162-1166

Owens LD and Heutte TM, 1997. A single amino acid substitution in the antimicrobial defense protein cecropin B is associated with diminished degradation by leaf intercellular fluid. Mol Plant Microbe Interactions, 10:525-528

Parmenter DL, Booth JG and Moloney MM, 1996. Production and purification of recombinant hirudin from plant seeds. In *Transgenic plants: a production system for industrial and pharmaceutical proteins*, Eds. Owen MRL and Pen J, Wiley, New York , pp 261-280

Perlak FJ, Fuchs RL, Dean DA, McPherson SL and Fischoff DA, 1991. Modification of the coding sequence enhances plant expression of insect control protein genes. Proc Natl Acad Sci USA, 88:3324-3328

Petschow BW, Carter DL and Hutton GD, 1993. Influence of orally administered epidermal growth factor on normal and damaged intestinal mucosa in rats. J Pediatr Gastroenterol Nutr,17:49-58

Philpott DJ, McKay DM, Sherman PM and Perdue MH, 1996. Infection of T84 cells with enteropathogenic Escherichia coli alters barrier and transport functions. Am J Physiol, 270:G634-645

Piers KL, Brown MH and Hancock REW, 1993. Recombinant DNA procedures for producing small antimicrobial cationic peptides in bacteria. Gene, 134:7-13

Puccio F and Lehy T, 1988. Oral administration of epidermal growth factor in suckling rats stimulates cell DNA synthesis in fundic and antral gastric mucosae as well as in intestinal mucosa and pancreas. Regul Pept, 20:53-64

Qiu X, Wu Y, Du S and Erickson L, 1997. A new arabinogalactan protein-like gene expressed in the pollen of alfalfa. Plant Sci, 124:41-47

Reynoird JP, Mourgues F, Norelli J, Aldwinckle HS, Briset MN, and Chevreau E, 1999. First evidence for improved resistance to fire blight in transgenic pear expression the attacin E gene from *Hyalophora cecropia*. Plant Sci, 149:23-31

Richter LJ, Thanavala Y, Arntzen CJ, and Mason HS, 2000. Production of hepatitis B surface antigen in transgenic plants for oral immunization. Nat Biotech 18:1167-1171

Sardana RK, Ganz PR, Dudani A, Tackaberry ES, Cheng X and Altosaar I, 1998. Synthesis of recombinant human cytokine GM-CSF in the seeds of transgenic tobacco plants. In *Recombinant proteins from plants-production and isolation of clinically useful compounds*, Eds. Cunningham C and Porter JR, Humana Press, Totowa, New Jersey pp.77-87

Schwartz MZ and Storozuk RB, 1988. Influence of epidermal growth factor on intestinal function in the rat: a comparison of systemic infusion versus luminal perfusion. Am J Surg 1988 Jan;155(1):18-22

Sharma A, Sharma R, Imamura M, Yamakawa M and Machii H, 2000. Transgenic expression of cecropin B, an antibacterial peptide from *Bombyx mori*, confers enhanced resistance to bacterial leaf blight in rice. FEBS Lett, 484:7-11

Sheaffer CC, Martin NP, Lamb JFS, Cuomo JR, Grimsbo JJ and Quering SR, 2000. Leaf and stem properties of alfalfa entries. Agron J, 92:733-739

Sijmons PC, Dekker BM, Schrammeijer B, Verwoerd TC, van den Elzen PJ and Hoekema A. 1990. Production of correctly processed human serum albumin in transgenic plants. Bio/Technology 8:217-221

Simmaco M, Mignogna G and Barra D, 1998. Antimicrobial peptides from amphibian skin: what do they tell us? Biopolymers, 47:435-450

Soderholm JD and Perdue MH, 2001. Stress and intestinal barrier function. Am J Physiol Gastrointest Liver Physiol, 280:G7-13

Stevens LH, Stoopen GM, Elbers I, Molthoff J, Bakker H, Lommen A, Bosch D and Jordi W, 2000. Effect of climate conditions and plant developmental stage on the stability of antibodies expressed in transgenic tobacco. Plant Physiol 124:173-182

Thevissen K, Ghazi A, De Samblanx G, Brownlee C,Osborn R and Broekaert W, 1996. Fungal membrane responses induced by plant defensins and thionins. J Biol Chem 271:15018-5025

Tsai CH, Hill M, Asa SL, Brubaker PL and Drucker DJ, 1997. Intestinal growth-promoting properties of glucagon-like peptide-2 in mice. Am J Physiol, 273:E77-E84

Van dekerckhove J,Van Damme J, Van Lijsebettens M, Botterman J, De Block M, Vandewiele M, De Clercq A, Leemans J,Van Montagu M and Krebbers E, 1989. Enkephalins produced in transgenic plants using modified 2S seed storage proteins. Bio/Technology 7:929-932

Verwoerd TC, van Paridon PA, van Ooyen A, van Lent J, Hoekema A and Pen J, 1995. Stable accumulation of *Aspergillus niger* phytase in transgenic tobacco leaves. Plant Physiol 109:1199-1205

Wallace JL, 1996. NSAID gastroenteropathy: past, present and future. Can J Gastroenterol, 10:451-459

Zhang G, Wu H, Shi J, Ganz T, Ross C and Blecha F, 1998 . Molecular cloning and expression of porcine beta-defensin-1. FEBS Lett 424:37-40

Zijlstra RT, Odle J, Hall WF, Petschow BW, Gelberg HB and Litov RE, 1994. Effect of orally administered epidermal growth factor on intestinal recovery of neonatal pigs infected with rotavirus. J Pediatr Gastroenterol Nutr,19:382-390

Zuo J and Chua NH, 2000. Chemical-inducible systems for regulated expression of plant genes. Curr Opin Biotechnol,11:146-151

Chapter 10

AN ORAL VACCINE IN MAIZE PROTECTS AGAINST TRANSMISSIBLE GASTROENTERITIS VIRUS IN SWINE

J. Jilka
Prodigene Inc., 1500 Research Parkway, College Station, Texas, TX 77845, USA

Keywords: antigen, mucosal immunity, transgenic plants, TGEV, enteric pathogen

Abstract Transmissible gastroenteritis (TGE) is a highly contagious enteric disease that is characterized by vomiting, severe diarrhea and high mortality in piglets less than two weeks of age. The development of edible vaccines offers the potential to aid in the control of enteric diseases such as TGE. Edible vaccines from plant material could be directly delivered in the feed and could be produced cheaply in large volumes thus avoiding many costs associated with the administration of conventional vaccines. Vaccines from plants are particularly suitable for stimulation of mucosal immunity, since edible plant products can be delivered orally to reach the gut mucosal tissue and elicit an immune response at mucosal surfaces. Recent advances in technology make it now possible to express vaccine antigens at high levels in plants. Corn expressing the S protein of TGEV was fed to 13-day-old piglets for ten days and subsequently challenged with a virulent Purdue strain of TGEV. This group of piglets was significantly protected from the disease as contrasted to the control group that was fed only corn. Results from a second trial duplicated these results demonstrating that the delivery of antigens delivered in an edible oral form are efficacious

1. INTRODUCTION

Swine transmissible gastroenteritis (TGE) (Saif et al., 1992*)* is recognized as one of the major causes of sickness and death in piglets particularly in areas with high concentrations of pigs. TGE is a highly contagious enteric disease that is characterized by vomiting, severe diarrhea and high mortality in piglets less

L. Erickson et al. (eds.),
Molecular Farming of Plants and Animals for Human and Veterinary Medicine, 223–236.

than two weeks of age. The causal agent of TGE is a pleomorphic, enveloped single-stranded RNA virus belonging to the genus *Coronavirus* of the family Coronaviridae. Replication of virus in the villous epithelial cells of the small intestine results in the destruction or alteration of function of these cells. These changes lead to a reduction in the activity of the small intestine that disrupts digestion and cellular transport of nutrients and electrolytes. In small piglets this can lead to a severe and fatal deprivation of nutrients and dehydration. Following infection, pigs that have survived the infection are immune to subsequent infections presumably due to local immunity in the intestinal mucosa. Thus, since active immunity towards TGEV involves local immunity in the intestinal mucosa, presumably through the activation and secretion of intestinal SIgA, vaccines that target activation of the intestinal mucosa immune system are particularly attractive in the control of this disease. In particular, the development of edible vaccines offers the potential to aid in the control of enteric diseases such as TGE. Edible vaccines from plant material could be directly delivered in the feed and could be produced cheaply in large volumes thus avoiding many costs associated with the administration of conventional vaccines. Vaccines from plants are particularly suitable for stimulation of mucosal immunity, since edible plant products can be delivered orally to reach the gut mucosal tissue and elicit an immune response at mucosal surfaces. Recent advances in technology make it now possible to express vaccine antigens at high levels in plants.

A number of different plant systems have recently been under investigation for use as an edible oral delivery systems (Gomez et al., 1998; Tuboly et al., 2000) in the development of an oral vaccine for TGEV. Of these, a system based on the use of transgenic maize seed appears to be the most realistic for a number of different reasons. Among these reasons include the ability to introduce a grain-based product directly into a producer's feed system, the ability to utilize the already existing infrastructure for the production, harvesting, transportation, storage, and processing of the grain, the ability to deliver a product (both monovalent and multivalent) at a cost competitive with contemporary vaccines due to a low cost of goods, and a plant system amenable to transformation with highly developed and characterized genetics.

TGEV virions contain three major structural proteins: a nucleocapsid protein (N), a small membrane-bound glycoprotein (M), and large spike or peplomer glycoprotein (S). In this study, we generated transgenic maize plants that express the spike protein at high levels. Corn expressing the S protein of TGEV was fed to 13-day-old piglets for ten days and subsequently challenged with a virulent Purdue strain of TGEV. This group of piglets was significantly protected from the disease as contrasted to the control group that was fed only corn. Results from a second trial duplicated these results demonstrating that the delivery of antigens delivered in an edible oral form are efficacious.

2. GENE CONSTRUCTS AND TRANSFORMATION OF MAIZE VIA AGROBACTERIUM

The amino acid sequences of the S, M, and N protein of an isolate of the Miller strain of TGEV were provided by Prem Paul, DVM, College of Veterinary Medicine, Iowa State University. The amino acid sequences of the various structural proteins of TGEV were back-translated using the Backtranslate program of the Wisconsin GCG Package against a codon table tabulated for highly expressed maize genes. The resulting DNA sequence was scanned for the presence of undesirable sequence, e.g. polyadenylation signals, 5' and 3' consensus splice sites, other mRNA destabilizing sequences, and undesirable endonuclease restriction enzyme sites. The DNA sequence was modified to eliminate these sites by choosing alternative codons. Alternative codons with a codon frequency of less than 10 percent for that amino acid were avoided. The resulting sequence was then constructed using a series of synthesized overlapping complementary oligonucleotides and the polymerase chain reaction (PCR) to amplify the resulting synthetic sequence. Convenient restriction sites were also engineered into the 5' and 3' ends of the optimized gene to facilitate cloning. The barley alpha-amylase signal sequence (Rogers J.C., 1985) was also synthesized using overlapping complementary nucleotides with maize-preferred codons. Inclusion of the translated leader sequence directs the protein of interest to the cell wall and allows high levels of accumulation.

Transgenic maize plants were generated using the method of Ishida et al. (1996). Essentially, maize corn ears were harvested at 9-12 days after pollination when embryos are approximately 1-2 mm in length. Whole ears were surface sterilized in 50% bleach (+teaspoon of Tween 20) for 30 min and given two rinses of sterile H_2O. Immature zygotic embryos (ZE) were sterilely isolated from the ears. Embryos were washed twice with co-cultivation medium and *Agrobacterium* was added directly by pouring bacterial solution into the ZE tube. Embryos with bacteria were vigorously vortexed for 30 seconds and allowed to incubate at room temperature for 5 minutes. Embryos were placed scutellum side up onto co-cultivation medium and incubated at 19°C in the dark for 3-5 days. Keeping scutellum side up, embryos were transferred to antibiotic-containing medium without selection for three days in the dark at 27-28°C. every subsequent 2 weeks, embryos and herbicide resistant calli were transferred to fresh selection medium. When sufficient callus from a single event had developed on selection medium (approximately two plates), the callus was transferred onto regeneration medium. Mature somatic embryos were placed in the light and allowed to germinate. Ten plants from each event were transplanted to soil in the greenhouse and allowed to flower and produce seed. The resulting seed (T1 seed) was screened by ELISA to determine the levels of the recombinant protein of interest.

3. PROTOCOLS FOR FEEDING AND CLINICAL TRIALS

3.1 Transgenic Grain Production

Highly expressing seeds were backcrossed into maize lines of commercial interest. For this study, pollen from T1 seed was crossed to commercial maize hybrids in order to bulk up the seed as fast as possible. The resulting grain (approximately 30 lbs.) was ground to cornmeal (600-micron particle size). The levels of S protein in this fraction were estimated to be 0.004% (w/w). Piglets were fed about 50 grams per day of transgenic corn. That roughly amounted to 2 mg of S protein per dose per day.

3.2 Swine Feeding Trials

All swine feeding trials were conducted at Ames, Iowa in collaboration with David Carter, D.V.M., Veterinary Resources, Inc. and Mark Welter, M.S. at Oragen Technologies. 10-day-old SPF TGEV sero-negative pigs from a low disease incidence herd were utilized in these trials.

3.3 Vaccination of Feed Test Groups

For the appropriate consecutive days, all piglets were withheld from feed overnight (including the MLV vaccinates) and all feed-test groups were vaccinated first thing in the morning. In groups receiving the TGEV-S corn, 50 grams of TGEV transgenic corn was needed per day per pig. The dry corn was mixed with a wooden stick to ensure distribution of the transgenic corn. Medicated milk replacer to a total of not less than 300 ml and not more than 600 ml was used as a base to which the ground corn was added and mixed so as to produce a thick oatmeal-type meal. The corn was stirred in with a clean wooden stick until thick with just a little milk settling to the top. This amounted to approximately 1000 grams of feed representing 100 grams per piglet feeding, containing 50 grams of transgenic corn per pig feeding. A line of vaccine meal was placed on a clean dry floor and the piglets allowed to consume the vaccine. Attempts were made to ensure each piglet received an adequate vaccine portion. After the vaccine was consumed, regular water and medicated weaning rations were replaced in the pen. Pigs in the treatment group receiving the modified live vaccine (MLV) were orally vaccinated with MLV TGEV according to label directions at day 0 and 7 days later.

3.4 Virus Challenge

In the case of TGEV-1, on day 12 (2 days after last feed vaccination and 5 days after MLV TGEV) all animals were orally challenged with a 2 ml oral dose of virulent TGEV challenge (Purdue strain, titre of $10^{7.6}$ $FAID_{50}$'s per dose). In the case of TGEV-2, on day 18 (2 days after last feed vaccination for the 16-day groups and 11 days after MLV TGEV) all animals were orally challenged with a 2 ml oral dose of virulent TGEV challenge (Purdue strain, titre of $10^{7.6}$ $FAID_{50}$'s per dose). Previous work has determined that this challenge strain and levels will produce a clinically typical TGEV watery diarrhea in 21 to 28 day old piglets that persists for 7 to 10 days. No mortality values have been observed with this challenge model in this age animal.

3.5 Data and Sample Collection

Persons performing daily observations were blinded as to treatment.

1) **Daily Observations:** Piglets were observed twice daily and were scored for any signs of diarrhea as below:

0 (Normal)

2 (Creamy, piles up in pen)
4 (Watery)

Additional clinical signs which were observed such as dehydration or depression, anorexia, vomitus and death were scored as below and the number added to fecal observation for a total clinical score as shown below. Any animal that died or appeared moribund was sacrificed and necropsied. A sample from the jejunum of the small intestine was collected and observed for villous atrophy and providing that the sample was not too necrotic it was assayed for TGEV.

1 (Dehydration & Depression)
1 (Anorexia)
3 (Vomitus)
10 (Moribund or Death)

Attempts were made to isolate TGEV from the feces of watery scouring animals so as to confirm the challenge. A fecal sample was collected and TGEV isolation was conducted by inoculating confluent ST cells and staining by specific immunofluorescence.

2) Weights: All animals were weighed on day 0, day 12 and day 24.

3) Blood Samples: Blood was collected on day 0, day 12 and day 26. Blood was allowed to clot and serum collected and stored at 20° C until assay. Sera was assayed for TGEV neutralizing titers and titre values calculated using a Spermen Karber table.

4) Fecal Samples: Fecals were collected from randomly selected animals within a group that showed watery diarrhea and fecals were checked for TGEV activity.

3.6 Data Analysis

The total clinical scores for all animals within their group were divided by the number of observations to give a group clinical score. Statistical differences between groups were compared. The clinical symptom data are presented as: Percent Morbidity Incidence (number of animals with clinical signs >2 divided by total number of animals); Percent Morbidity Incidence and Duration (total number of clinical observations ≥ 2 divided by total number of pig days) and;

Clinical Severity Index (total clinical score value divided by total number of pig days).

3.7 Swine Feeding Trial #1 (TGEV-1)

The study consisted of three treatment groups. Group A was fed transgenic corn expressing the spike protein (S) of TGEV, Group B was fed non-transgenic corn, and Group C was vaccinated with a commercial modified live TGEV vaccine (MLV TGEV). All animals were challenged on day 12. Table 6 shows a summary of the design of the study.

Table 6. Summary of study design (TGEV-1)

Group	Number of pigs	Vaccine description	Amount/day	Route	Timing	Day of challenge
A	10	TGEV	50 g	Oral	0 to 10 days	Day 12
B	10	Control corn	50 g	Oral	0 to 10 days	Day 12
C	10	MLV TGEV	N.A.	Oral	0 & 7 days	Day 12

3.8 Swine Feeding Trial #2 (TGEV-2)

This study consisted of five treatment groups. Groups A , B and C were fed transgenic corn expressing the spike protein (S) of TGEV for 4, 8 or 16 days. Group D was fed non-transgenic corn for 16 days, and Group E was vaccinated with a commercial modified live TGEV vaccine (MLV TGEV) on days 0 and 7. All animals were challenged on day 18. Table 7 shows a summary of the design of the study.

4. RESULTS FROM CLINICAL TRIALS

4.1 Observations of Clinical Symptoms for TGEV-1

The results from this experiment are summarized in Table 8. The data for morbidity incidence represent a compilation of the significant (>2) clinical scores for each treatment group. These data show that 100% of the pigs that were fed control corn developed TGEV clinical symptoms. Only 50% of those that received the TGEV corn exhibited symptoms and 78% of the pigs receiving the modified live vaccine developed symptoms.

Table 7. Summary of study design (TGEV-2)

Group	Number of pigs	Vaccine description	Amount/day	Route	Timing	Day of challenge
A	10	TGEV transgenic corn	50 g	Oral	0 to 4 days	Day 18
B	10	TGEV transgenic corn	50 g	Oral	0 to 8 days	Day 18
C	10	TGEV transgenic corn	50 g	Oral	0 to 16 days	Day 18
D	10	Control corn	50 g	Oral	0 to 16 days	Day 18
E	10	MLV TGEV	N.A.	Oral (0.5 ml)	0 & 7 days	Day 18

The data for the percent of morbidity incidence and duration are also a compilation of the significant (>2) clinical scores for each treatment group. As indicated by the data, 100% of the pigs that were fed control corn developed TGEV clinical symptoms. Only 50% of those that received the TGEV corn exhibited symptoms of similar duration, while 78% of the pigs receiving the modified live vaccine developed symptoms of similar duration.

Data for the clinical severity index for each treatment group are a compilation of the total clinical value divided by the total number of pig days for each treatment group. This data shows that the disease that developed in the pigs that were fed control corn was much higher than either the TGEV-S containing transgenic corn or the modified live vaccine treatment group.

4.2 Observations of Clinical Symptoms for TGEV-2

The results from this experiment are summarized in Table 9. As in TGEV-1, the data for morbidity incidence represent a compilation of the significant (>2) clinical scores for each treatment group. These data show that 36% of the pigs that were fed control non-transgenic corn developed TGEV clinical symptoms. Such symptoms were present in 50%, 0%, and 20% of the pigs that received 4 days, 8 days and 16 days of TGEV corn, respectively. 9% of the pigs receiving the modified live vaccine developed symptoms.

Table 8. Summary of clinical data for TGEV-1

Treatment group	Morbidity incidence	Morbidity incidence and duration	Clinical severity index
TGEV corn	50%	22%	0.96
Control corn	100%	38%	1.3
MLV TGEV	78%	18%	0.89

The data for percent morbidity incidence and duration in each of the treatment groups are also a compilation of the significant (>2) clinical scores for each treatment group. Pigs fed non-transgenic control corn scored 36% morbidity/duration, while pigs that received 4 days, 8 days and 16 days of TGEV corn scored 13%, 0% and 5% morbidity/duration. The corresponding score for pigs receiving the modified live vaccine was 2%.

Table 9. Summary of clinical data for TGEV-2

Treatment group	Dose duration	Percent morbidity incidence	Percent morbidity incidence and duration	Clinical severity index
TGEV corn	4 days	50	13	0.36
	8 days	0	0	0
	16 days	20	5	0.16
Control corn	16 days	36	5	0.15
MLV TGEV	NA	9	2	0.05

The data showing the clinical severity index are a compilation of the total clinical value divided by the total number of pig days for each treatment group. These data show that the disease that developed in the pigs that were fed control corn was much higher than either the TGEV-S containing transgenic corn or the modified live vaccine treatment group.

5. GENERAL DISCUSSION

Over the past decade, transgenic plants have been successfully used to express a variety of genes from bacterial and viral pathogens. Many of the resulting peptides induced an immunologic response in mice (Gomez et al., 1998; Mason et al., 1998; Wigdorovitz et al., 1999) and humans (Kapusta et al., 1999) comparable to that of the original pathogen. Characterization studies of these engineered immunogens have proven the ability of plants to express, fold and modify proteins in a manner that is consistent with the authentic source.

Numerous genes have been cloned into a variety of transgenic plants including many enzymes that have demonstrated the same enzymatic activity as their authentic counterparts (Hood et al., 1997; Moldoveanu et al., 1999; Trudel et al., 1992). Many additional genes have been expressed in plants solely for their immunogenic potential, including viral proteins (Gomez et al., 1998; Kapusta et al., 1999; Mason et al., 1996; McGarvey et al., 1995; Thanavala et al., Wigdorovitz et al., 1999) and subunits of bacterial toxins (Arakawa et al., 1997; Arakawa et al., 1999; Haq et al., 1995; Mason et al., 1998). Animal and human immunization studies have demonstrated the effectiveness of many plant-derived recombinant antigens in stimulating the immune system. The production of antigen-specific antibodies and protection against subsequent toxin or pathogen challenge demonstrates the feasibility of plant derived-antigens for immunologic use.

The utilization of transgenic plants for vaccine production has several potential benefits over traditional vaccines. First, transgenic plants are usually constructed to express only a small antigenic portion of the pathogen or toxin, eliminating the possibility of infection or innate toxicity and reducing the potential for adverse reactions. Second, since there are no known human or animal pathogens that are able to infect plants, concerns with viral or prion contamination are eliminated. Third, immunogen production in transgenic crops relies on the same established technologies to sow, harvest, store, transport, and process the plant as those commonly used for food crops, making transgenic plants a very economical means of large-scale vaccine production. Fourth, expression of immunogens in the natural protein-storage compartments of plants maximizes stability, minimizes the need for refrigeration and keeps transportation and storage costs low (Kusnadi et al., 1998). Fifth, formulation of multicomponent vaccines is possible by blending the seed of multiple transgenic corn lines into a single vaccine. Sixth, direct oral administration is possible when immunogens are expressed in commonly consumed food plants, such as grain, leading to the production of edible vaccines.

Some of the first attempts to make edible vaccines included transgenic potatoes expressing the *E. coli* heat-labile enterotoxin (LT-B) (Haq et al., 1995), and a Norwalk virus surface protein (Mason et al., 1996). In both cases, mice fed the antigenic tubers produced serum and secretory antibodies specific to the authentic antigen. Subsequently, many plant-expressed antigens, including those referenced above, have been shown to elicit an immune response when administered through an oral route. Several of these antigens have shown sufficient promise to warrant human clinical trials (Mason et al., 1998; Saif et al., 1994).

One of the most promising aspects of edible vaccines is the ability of orally administered immunogens to stimulate a mucosal immune response (Ruedl et al., 1995). Mucosal surfaces, the linings of the respiratory, gastrointestinal, and urogenital tracts, play an important physical and chemical role in protecting the body from invading pathogens and harmful molecules. The mucosal immune system is distinct and independent of the systemic, or humoral, immune system, and is not effectively stimulated by parenteral administration of immunogens (Czerkinsky et al., 1993). Rather, the mucosal immune system requires antigen presentation directly upon the mucosal surfaces. Since most invading pathogens first encounter one or more of the mucosal surfaces, stimulation of the mucosal immune system is often the best first defense against many transmissible diseases entering the body through oral, respiratory and urogenital routes (Holmgren et al., 1994). Transgenic plants could produce large quantities of immunologically active recombinant antigen very economically for vaccine production. Multicomponent vaccines could easily be formulated from the seed of multiple transgenic plant lines to generate an increased chance for successful virus neutralization in a stand-alone vaccination strategy, as a booster, or in combination with other vaccines and vaccination routes.

Previously the full spike (S) protein has been expressed in *Arabidopsis* (Gomez et al., 1998). In this case, expression of the S protein was not detectable yet a plant extract injected intramuscularly into mice resulted in the production of detectable anti-S serum. More recently, the S protein has been expressed in tobacco (Tuboly et al., 2000). The S protein was expressed at levels that could be detected by ELISA and of the expected size when analyzed by Western blotting. Leaf extracts from these plants were injected into 28-day old pigs. In contrast to pigs which were injected with nontransgenic plant extracts, the former produced measurable TGEV-specific antibodies, whereas the latter piglets did not.

We have extended these results by generating transgenic corn expressing the S protein which can be fed to pigs in a virulent TGEV challenge study. We report for the first time the protection of an economically important animal from a naturally occurring disease by an oral vaccination using an edible system. Moreover this system uses the conventional feed materials, e.g. corn, to deliver the antigen. One report (Modelska et al., 1998) has shown in the laboratory the amelioration of rabies symptoms in mice fed multiple doses of a chimeric plant virus expressing the rabies glycoprotein following challenge with an attenuated rabies strain. To our knowledge until our report no animals in conventional food animal husbandry have been vaccinated with edible vaccines and shown to be protected from the disease. The level of protection seen in this study includes general health and vigor, a decrease in clinical symptoms, lack of virus shedding and other observations known to be criteria for disease protection. The mode of protection is unknown but may be an active immune response by the animal, competitive inhibition of viral receptor sites leading to non-establishment of a viral infection, or interference with parts of the viral replicative process.

REFERENCES

Ahmad S, Lohman B, Marthas M, Giavedoni L, Amad Z, Haigwood NL, Scandella CJ, Gardner MB, Luciw PA and Yilma T, 1994. Reduced virus load in rhesus macaques immunized with recombinant gp160 and challenged with simian immunodeficiency virus. AIDS Res. Hum. Retroviruses, 10:195-204

Arakawa T, Chong DK, Merritt JL and Langridge WH, 1997. Expression of cholera toxin B subunit oligomers in transgenic potato plants. Transgenic Res, 6:403-413

Arakawa T, Yu J and Langridge WH, 1999. Food plant-delivered cholera toxin B subunit for vaccination and immunotolerization. Adv Exp Med Biol, 464:161-178

Bergmeier LA, Mitchell EA, Hall G, Cranage MP, Cook N, Dennis M and Lehner T, 1998. Antibody-secreting cells specific for simian immunodeficiency virus antigens in lymphoid and mucosal tissues of immunized macaques. AIDS, 12:1139-1147

Czerkinsky C, Svennerholm AM and Holmgren J, 1993. Induction and assessment of immunity at enteromucosal surfaces in humans: implications for vaccine development. Clin Infect Dis 16 Suppl, 2:S106-116

Garvey PB, Hammond J, Dienelt MM, Hooper DC, Fu ZF, Dietzschold B, Koprowski H and Michaels FH, 1995. Expression of the rabies virus glycoprotein in transgenic tomatoes. Biotechnology, 13:1484-1487

Gomez N, Carrillo C, Salinas J, Parra F, Borca MV and Escribano JM, 1998. Expression of immunogenic glycoprotein S polypeptides from transmissible gastroenteritis coronavirus in transgenic plants. Virology, 249:352-358

Guidry JJ, Cardenas L, Cheng E and Clements JD, 1997. Role of receptor binding in toxicity, immunogenicity, and adjuvanticity of *Eschericia coli* heat-labile enterotoxin. Immunology 65:4943-4950

Haq TA, Mason HS, Clements J and Arntzen CJ, 1995. Production of an orally immunogenic bacterial protein in transgenic plants: proof of concept of edible vaccines. Science, 268:714-716

Holmgren J, Czerkinsky C, Lycke N and Svennerholm AM, 1994. Strategies for the induction of immune responses at mucosal surfaces making use of cholera toxin B subunit as immunogen, carrier, and adjuvant. Am J Trop Med Hyg, 50:42-54

Hood EE, Withcher DR, Maddock S, Meyer T. Baszczynski CBM, Flynn P, Register J, Marshal L, Bond D, Kulisek E, Kusnadi A, Evangelista R, Nikolov Z, Wooge C, Mehigh RJ, Hernan R, Kappel WK, Ritland D, Li PC and Howard JA, 1997. Commercial production of avidin from transgenic maize: characterization of transformant, production, processing, extraction and purification. Mol Breeding, 3:291-306

Ishida Y, Saito H, Ohta S, Hiei Y, Komari T and Kumashiro T, 1996. High efficiency transformation of maize (*Zea mays* L.) mediated by *Agrobacterium tumefaciens*. Nat Biotechnol, 14:745-750

Kapusta J, Modelska A, Figlerowicz M, Pniewski T, Letellier M, Lisowa O, Yusibov V, Koprowski H, Plucienniczak A and Legocki AB, 1999. A plant-derived edible vaccine against hepatitis B virus. FASEB J, 13:1796-1799

Kusnadi AR, 1998. Production and purification of two recombinant proteins from transgenic corn. Biotechnol Prog, 14:149-155

Kusnadi A, Evangelista R, Hood EE, Howard JA and Nikolov Z, 1998. Processing of transgenic corn seed and its effect on the recovery of recombinant beta-glucuronidase. Biotechnol Bioeng, 60:44-52

Lu X, Kiyono H, Lu D, Kawabata S, Torten J, Srinivasan S, Dailey PJ, McGhee JR, Lehner T and Miller CJ, 1998. Targeted lymph-node immunization with whole inactivated simian immunodeficiency virus (SIV) or envelope and core subunit antigen vaccines does not reliably protect rhesus macaques from vaginal challenge with SIVmac251. AIDS, 12:1-10

Mason HS, Ball JM, Shi JJ, Jiang X, Estes MK and Arntzen CJ, 1996. Expression of Norwalk virus capsid protein in transgenic tobacco and potato and its oral immunogenicity in mice. Proc Natl Acad Sci USA, 93:5335-5340

Mason HS, Haq TA, Clements JD and Arntzen CJ, 1998. Edible vaccine protects mice against *Escherichia coli* heat-labile enterotoxin (LT): potatoes expressing a synthetic LT-B gene. Vaccine, 16:1336-1343

Modelska A, Dietzschold B, Sleysh N, Fu ZF, Steplewski K, Hooper DC, Koprowski H and Yusibov V, 1998. Immunization against rabies with plant-derived antigen. Proc Natl Acad Sci USA, 95:2481-2485

Moldoveanu Z, Vzorov AN, Huang WQ, Mestecky J and Compans RW, 1999. Induction of immune responses to SIV antigens by mucosally administered vaccines [In Process Citation]. AIDS Res Hum Retroviruses, 15:1469-1476

Pen J, Molendijk L, Quax WJ, Sijmons PC, Ooyen AJ van, Elzen PJ van den, Rietveld K, and Hoekema A, 1992. Production of active *Bacillus licheniformis* alpha-amylase in tobacco and its application in starch liquefaction. Bio/Technology, 10:292-296

Rogers JC, 1985. Two barley alpha-amylase gene families are regulated differently in aleurone cells. J Biol Chem, 260:3731-3738

Ruedl C and Wolf H, 1995. Features of oral immunization. Int Arch Allergy Immunol, 108:334-339

Saif LJ and Wesley RD, 1992. Transmissible gastroenteritis. In *Diseases of swine,* 7th Edition, Iowa State University Press 29, pp 362-386

Saif LJ, Cott JL van and Brim TA, 1994. Immunity to transmissible gastroenteritis virus and porcine respiratory coronavirus infections in swine. Vet Immunol Immunopathol, 43:89-97

Thanavala Y, Yang Y-F, Lyons P, Mason HS and Arntzen CJ, 1995. Immunogenicity of transgenic plant-derived hepatitis B surface antigen. Proc Natl Acad Sci USA, 92:3358-3361.

Trudel J, Potvin C and Asselin A, 1992. Expression of active hen egg white lysozyme in transgenic tobacco. Plant Sci, 87:55-67

Tuboly T, Yu W, Bailey A, Degrandis S, Erickson L, Du S and Nagy E, 2000. Immunogenicity of porcine transmissible gastroenteritis virus spike protein expressed in plants. Vaccine, 18:2023-2028

Wigdorovitz A, Carrillo C, Dus Santos MJ, Trono K, Peralta A, Gomez MC, Rios RD, Franzone PM, Sadir AM, Escribano JM and Borca MV, 1999. Induction of a protective antibody response to foot and mouth disease virus in mice following oral or parenteral immunization with alfalfa transgenic plants expressing the viral structural protein VP1. Virology, 255:347-353

Chapter 11

PRODUCTION OF ANTIBODIES IN ALFALFA (MEDICAGO SATIVA)

U. Busse, V. Levee, S Trepanier and L. Vezina
Medicago Inc., 2480 rue Hochelaga, Sainte Foy, Quebec, G1K 7P4

Keywords: alfalfa, molecular farming, biotechnology, antibody, transgenic plant, perennial plant, recombinant protein, expression, glycosylation, proteolysis, purification, large-scale production

Abstract Antibodies have long been recognized for their diagnostic and therapeutic potential. Advances in medicine and molecular biology, combined with the knowledge we will gather from the recent sequencing of the human genome, will pave the way to an even greater demand for antibody molecules. In order to satisfy this growing demand, alternative ways to current limiting production methods are being actively explored. Molecular farming of transgenic plants has gained a lot of attention in recent years as a promising large scale production and processing system. In this chapter, we will focus on a perennial legume crop, *Medicago sativa* (alfalfa), and its use in antibody production. Alfalfa benefits from several agronomic advantages such as nitrate fertilization-free cultivation, the existence of agricultural infrastructures and the possibility to rapidly produce clonal transgenic populations, making it a choice candidate for molecular farming purposes. Medicago's molecular farming protocol comprises five steps. First, the gene encoding the protein of interest is inserted into alfalfa-specific expression vectors; second, the resulting constructs are transferred into alfalfa cells, followed by the regeneration of transgenic alfalfa plants. The transgenic population is then grown and harvested at the vegetative state. Finally, recombinant proteins are extracted from harvested leaves and purified, followed by the disposal of residual biomass. A case study illustrates the production in alfalfa of an IgG antibody identical to a monoclonal antibody produced in hybridomas. Strategies to improve antibody recovery and quality are discussed.

L. Erickson et al. (eds.),
Molecular Farming of Plants and Animals for Human and Veterinary Medicine, 237–257.

1. INTRODUCTION

Antibodies are secreted proteins involved in the immune response. In humans, they fall into five major classes on the basis of the structure of their backbone (IgG, IgA, IgM, IgD and IgE), each harboring distinct features and functions (Haynes and Fauci, 1994). The basic antibody molecule is composed of two glycosylated heavy chains and two light peptide chains that are held together by interchain disulphide bonds. Co-expression of the two genes encoding these chains is one prerequisite to antibody assembly within the cell. The other prerequisite is correct intracellular post-translational processing that comprises translocation of the peptide chains into the endoplasmic reticulum via the N-terminal signal peptide, removal of the signal peptide, formation of disulphide bonds and domain folding. As the antibody progresses through the endoplasmic reticulum and the Golgi stacks, it is gradually glycosylated and finally secreted.

Specific antigen recognition is conferred by the N-terminal portions of both heavy and light chains that show considerable variations in their amino acid sequence. The remaining parts of the chains are relatively constant in sequence and structure and are required to mediate cellular and complement-mediated cytotoxicity (Haynes and Fauci, 1994). Genetic or enzymatic manipulations make it possible to eliminate most of the constant regions, generating Fab fragments or single chain variable fragments (scFv) that retain antigen binding capacity but loose effector functions.

1.1 Antibodies in Human and Veterinary Medicine

Monoclonal and recombinant antibodies are used for a wide variety of diagnostic and therapeutic applications. In the diagnostic field, *in vitro* use of antibodies identifies the presence of proteins related to disease status in clinical samples (serum or cells). Antibodies can also be coupled to radioactive molecules in order to localize specific cells *in vivo* by radioactive imaging. As our knowledge of the biology underlying human and animal diseases broadens, antibodies also gain a critical mass as a standard approach in therapy. More than one hundred clinical trials are currently testing the immunotherapeutic potential of monoclonal antibodies in the treatment of infectious and inflammatory diseases, central nervous system diseases and certain cancers. To date, ten antibodies have been approved worldwide for therapeutic use in humans and about forty are expected to be approved by 2010 (company data, Morgan Stanley Dean Witter). In parallel, antibodies are beginning to play an increasingly prominent role in the prevention of diseases. For example, antibodies against specific bacteria can be added to toothpaste or mouthwashes to prevent perodontic disease (Ma et al., 1998). Antibodies are also used in mammalian therapies. One such application is the minimization of newborn calf infection by *E. coli* K99 by the oral administration of Genecol® (Schering-Plough), a mouse monoclonal IgG antibody. Passive immunization to prevent gastrointestinal infections has also been investigated on other animals such as chicken (Wallach et al. 1990).

1.2 Antibody Production: from Fermentors to Molecular Farming

Many therapeutic antibodies require doses of hundreds of milligrams per treatment. For example, Herceptin® (Genentech), a humanized antibody against Her-2 that targets breast cancer cells, is administered at 2-4 mg/kg body weight/week. Antibodies added to toothpaste or mouthwashes will involve market requirements of more than 10,000 kg of antibody per year. The demand for monoclonal antibodies is constantly rising and high-volume manufacturing requirements may reach hundreds of kilograms in the near future. This trend has been challenging traditional antibody production methods for some years now and has paved the way to alternative, less expensive and more flexible protein production systems.

Antibodies have been recovered from hybridoma cell cultures since the discovery of techniques for the immortalization of antibody producing B cells (Koehler and Milstein, 1975). Hybridomas are still widely used in small and medium scale antibody production. Within the last two decades, advances in recombinant DNA technology have led to the recruitment of other cellular systems for the production of recombinant monoclonal antibodies. Mammalian cell lines such as the Chinese hamster ovary (CHO) cells are a viable alternative source for recombinant antibodies. This production system is less expensive than hybridomas but, as any other production system in cell culture, it requires sophisticated equipment and careful elimination of potential contaminants during purification. Production of recombinant antibodies in bacterial fermentors can solve problems of cost and quantity (Skerra, 1993). However, bacteria cannot produce full-size antibodies required for most immunotherapies nor perform most of the post-translational modifications inherent to animal cells. Bacteria often produce endotoxins that are difficult to remove during purification. In addition, expressed proteins frequently form inclusion bodies that make protein refolding procedures necessary during purification (Pen, 1996).

During the last decade, transgenic farm animals have been developed for the production of therapeutic proteins (Echelard, 1996). Among the advantages of such systems is the potential of expressing complex proteins in their proper three-dimensional configuration with intact post-transcriptional modifications such as glycosylation. Scale-up is possible through a program of animal breeding. However, biosafety concerns, especially for the transmission of viruses or prions, limit the use of transgenic animals since they are recognized host organisms for human pathogens. Political, ethical and regulatory constraints, added to long lag times (years), will further limit the use of this production system. Avian transgenics represent a viable alternative method for large scale antibody production (see chapter 13, this volume).

In parallel to transgenic animals, plants have been recruited as economical and efficient production systems for mammalian proteins (Fischer and Emans, 2000; Hood and Jilka, 1999). Plants possess higher eukaryote protein synthesis pathways similar to animal cells with only minor differences in protein glycosylation (Cabanes-Macheteau et al., 1999). High expression levels, accumulation to large quantities, storage capability and quality consistency of recombinant proteins have been routinely observed. Ease of genetic manipulation and rapid ramp up for field-scale production give plant production systems the potential to provide virtually unlimited quantities of recombinant molecules, in addition to satisfying safety concerns due to the absence of endogenous human or animal viral pathogens. Finally, plant production systems are characterized by higher flexibility and reduced capital and operating cost (Khoudi et al., 1999; Larrick et al., 1998).

Recombinant antibodies have been produced in plants for more than a decade now and their number is steadily increasing (Fischer et al., 1999). As first shown by Hiatt et al. (1989) for an IgG and Düring et al. (1990) for an IgM, light and heavy chains can be expressed in two separate plants that are then intercrossed to obtain one transgenic plant expressing the fully assembled antibody. Plants are also capable of synthesizing high levels of secretory IgA, a complex multi-subunit antibody suitable for passive mucosal immunotherapy

(Ma et al., 1995). These antibodies are functionally equivalent to those produced by hybridomas (Hiatt et al., 1989; Voss et al., 1995). Under some instances, plant-derived antibodies were used endogenously to increase resistance to plant pathogens or alter metabolic pathways (Conrad and Fiedler, 1998; De Wilde et al, 1998). In most cases, however, plants merely served as factories for recombinant therapeutic antibodies (Hood et al., 1999; Fischer and Emans, 2000). Some of these have already been tested in humans with promising results. For example, Guy's 13, a recombinant IgA antibody specific to *Streptococcus mutans*, can successfully prevent dental caries in humans (Ma et al., 1998).

Antibody production in plants has mostly been tested in annual plants such as tobacco and arabidopsis (Fischer and Emans, 2000). Despite many advantages, these model systems bear some disadvantages that could potentially limit their usefulness for large scale production in agricultural settings. One of these is the annual need for sexual reproduction through seeds and thus the segregation of genes in successive generations that invariably leads to large variations in antibody production. Seeds can be stored but not indefinitely, implying periodical reamplification of seed banks (Miele, 1997). Perennial plants, on the other hand, have the potential to live forever and thus provide a stable source of recombinant molecules. One such source that emerged as a very promising production system in recent years is *Medicago sativa* (alfalfa).

2. MOLECULAR FARMING USING ALFALFA

Alfalfa is a perennial forage crop cultivated worldwide. It is an autotetraploid plant (2n=16) which results in a multilevel of heterozygosity, providing a wide genetic variability and a good adaptability to soil and climate conditions as well as resistance to a broad spectrum of pathogens. In the field, alfalfa can be grown for 5 to 10 years. The plants become dormant in the fall and green up the following spring. The number of harvests during the growing season varies between 2 and 10, depending on the geographic location, soil quality and agricultural practices. As forage quality or yield greatly depends on cutting management system, so will be the utilization of forage for molecular farming since the content of recombinant proteins in the leaf depends on forage quality. Cutting management will have to be optimized to maximize yield of recombinant molecules per acre.

Alfalfa exhibits several agronomic qualities that underscore its usefulness in molecular farming. It is a legume plant and hence able to fix atmospheric nitrogen through symbiotic interaction with *Rhizobium*, obviating the need for nitrogen fertilization. Alfalfa cultures do not require pesticide applications, in contrast to other forage crops. Agricultural practices for alfalfa have already been developed and infrastructure and technology are in place, including large scale extraction procedures (wet fractionation) currently used for the production of animal food. As a perennial plant, alfalfa is easily propagated by stem cuttings and has a strong regenerative capacity. Growth of new stems arises from crown buds and axillary buds on the cut stem. Stem cutting allows ramp-up of clonal populations in a short time frame. Clonal populations offer stable expression of the transgene since every plant in the field is genetically identical to the characterized mother plant first selected in the laboratory based on accumulation, structure, and bioactivity of the recombinant protein. As no sexual crosses are required to propagate transgenic populations, seeds of the progeny do not need to be screened for the presence of the transgene.

Molecules are produced in leaves with harvests performed before flowering, thus diminishing the risk of gene flux into the environment due to pollination with related species. Alfalfa leaves contain very low levels of alkaloids and phenols (Jones et al., 1973; Goplen et al., 1980), and no human or animal pathogens. All these exceptional characteristics of alfalfa foliage render purification procedures easy and economical. In addition, recombinant proteins produced for oral veterinary applications can be administered without prior purification since alfalfa is already part of the diet of many farm animals.

3. PRODUCTION STRATEGIES IN ALFALFA

3.1 Genetic Programming

As illustrated (Figure 36), production of recombinant protein in alfalfa passes through a series of four steps that lead from the gene to the protein. A fifth step has been added in order to add value to the tonnes of alfalfa biomass produced during the process. In the genetic engineering step, the gene of interest is inserted into an expression vector under the control of a plant specific promoter and a sequence for transcriptional termination. To date, the cauliflower mosaic virus 35S promoter has been the most commonly used promoter for the production of recombinant proteins in plants (Odell et al., 1985; Jefferson et al., 1987). In alfalfa, the use of this promoter is limited since it confers weak levels of transcription (Narvàez-Vàsquez et al., 1992) and intellectual property issues limit the use of 35S for molecular farming purposes. At Medicago Inc. we have therefore developed a series of new strong promoters suitable for expression in alfalfa, based on promoter and terminator sequences endogenous to alfalfa. Concerns of biosafety and yield optimization have also driven the development of an inducible promoter. This promoter has been shown to be completely silent in the absence of the inducer and to confer levels of expression that are higher than those observed with the commercial 35S promoter upon induction (Figure 37).

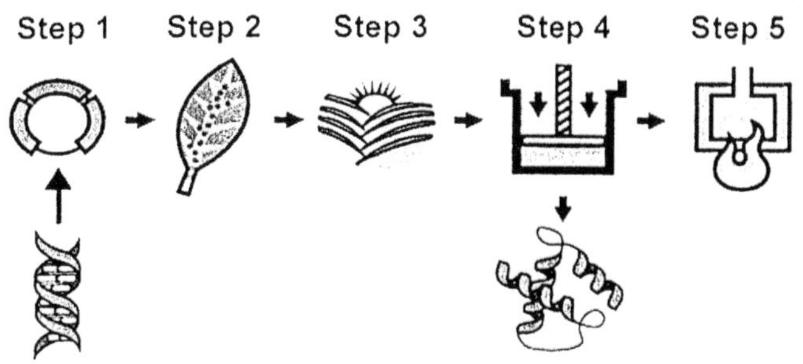

Figure 36. Production of recombinant proteins in alfalfa. The production schedule encompasses five steps: genetic programming (step 1), transformation (step 2), production of biomass (step 3), purification of recombinant protein (step 4) and disposal of biomass (step 5).

3.2 Gene Transfer

Once inserted into an expression vector, the gene of interest is transferred into alfalfa cells, followed by the regeneration of transgenic plants. Alfalfa transformation methods can be subdivided into two categories, indirect and direct methods, according to the need of an intermediate host to introduce DNA into the plant cell. Both methods have to be combined with a regeneration protocol in order to obtain a transgenic plant from the transfected cell. Regeneration is usually achieved through somatic embryogenesis which is a highly genotype dependent process in alfalfa (Brown and Atanassov, 1985;

Matheson et al., 1990). One alternative to somatic embryogenesis is pollination of flowers with transformed pollen grains.

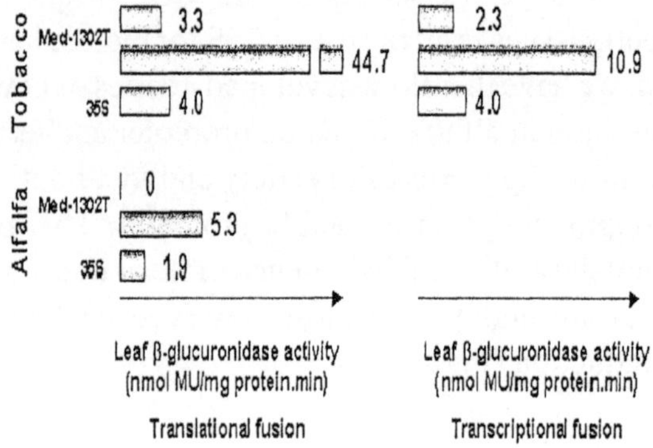

Figure 37. Inducibility of ß-glucuronidase (GUS) expression in alfalfa and tobacco leaves using Medicago's Med-1302T expression cassette. GUS was expressed either with the N-terminal portion of the protein naturally under the control of the inducible promoter (translational fusion) or without the N-terminal portion (transcriptional fusion).

The most currently used transformation method is based on infection by *Agrobacterium tumefaciens*. As a soil bacterium infecting plants, Agrobacterium is naturally able to transfer a small part of its genome into the genome of the plant cell. Genetic transformation of alfalfa using *Agrobacterium* was first reported in 1986 (Deak et al.,1986; Shahin et al., 1986). It relies on a host-pathogen interaction and its efficacy depends on the genotype of alfalfa and on the *Agrobacterium* strains used. Hence, efficient transformation protocols using *Agrobacterium* require genotypes of alfalfa that are both susceptible to *Agrobacterium* and capable to regenerate plants through somatic embryogenesis (Nowak et al., 1992; Desgagnés et al., 1995).

In order to bypass host specificity of *Agrobacterium*, direct transformation techniques allowing gene transfer without an intermediate host have been developed (Table 10). These techniques include physical as well as chemical methods to directly introduce foreign DNA into plant cells. Alfalfa has been transiently and stably transformed using direct methods. However, transgenic plants were only regenerated following transformation through particle bombardment.

The most currently used transformation methods, *Agrobacterium* and particle bombardment, are both highly protected by several patents. Hence, we are working on the development of our own transformation methods using electroporation of pollen grains. Germinated pollen grains exhibit a plasmic membrane that is susceptible to electroporation, allowing the entry of DNA. Electroporated pollen grains can then be used for pollination and seeds are harvested two months latter. Seedlings from electroporated pollen grains are then screened for the presence of the transgene. Transgenic plants generated by this transformation method are currently being screened to verify the presence of the transgene in the plant genome. Electroporation of pollen presents several advantages as a transformation method. It does not require sterile manipulations nor somatic embryogenic capacities to regenerate transgenic plants. Each seed results from a single pollen grain and each plant can be tested individually for the presence of the transgene. Hence, an antibiotic is not required to select transgenic plants. It is quite rapid since seeds are collected 2 months after electroporation and plants can be harvested for recombinant molecule purification 3 months after seed germination.

Table 10. Genetic transformation of *Medicago sativa* using direct DNA transformation methods.

Methods	Target cell	Transgene expression	Regeneration of transgenic plants
Particle bombardment	Petioles and calli from petioles	stable	Via somatic embryogenesis
	Immature embryos	stable	Via somatic embryogenesis
	Cell suspension culture	transient	no
	Pollen	stable	Via pollination
Microinjection	Protoplasts from cell suspension culture	stable	no
Electroporation	Protoplasts from cell suspension culture	transient	no
	Pollen	stable	Via pollination

3.3 Production of Biomass

Once transgenic plants have been obtained, they are multiplied either in greenhouses or open fields. Alfalfa plants can be propagated by a variety of methods, depending on the production scale needed (Table 11). Clonal propagation of transgenic alfalfa plants can be done by stem cutting or somatic embryogenesis and represents a fast and safe alternative to propagation through seeds. Stem cutting relies on the totipotency of meristematic structures within the stem. One transgenic alfalfa plant can generate a clonal population covering a 100 square metre greenhouse within 14 months, Typically, this population can supply recombinant protein in the order of 60 g per harvest (assuming an expression level of 0.5% of total protein). Stem cutting is also the method of choice when small quantities (milligrams) of recombinant protein are required within a limited time frame, as it is the case for preclinical studies, product testing and homologation.

Table 11. Propagation methods and field settings for alfalfa at different production scales of recombinant protein.

Harvest yield (4 weeks)	Lag time (months)	N° of plants/area	Propagation method	Site
10 mg	8	30 / m^2	Stem cutting	Greenhouse
10 g	14	10 000 / 100 m^2	Stem cutting	Greenhouse
2 kg	14	300 000 / 0.3 ha	Somatic embryogenesis	Greenhouse
100 kg	30	12 ha	Seeds	Field
1 tonne	30	120 ha	Seeds	Field

Somatic embryogenesis relies on alfalfa's endogenous potential to develop embryos from somatic tissues. This propagation method has the advantage of generating large-scale clonal populations from transgenic plants. Somatic embryogenesis is done from petioles that are sterilized and placed into liquid culture. After 2-3 weeks, totipotent cells liberated from the vascular layer of the petiole are layered onto a solid medium on which embryos develop within 3 weeks. Under ideal conditions, 3000-5000 embryos are produced from a 50 ml culture and the system is easily scaled up. Embryos can be dehydrated and frozen for long tem storage (Senaratna et al., 1989; McKersie et al., 1989; Senaratna et al., 1990). Under ideal conditions, one million plants can be produced from embryos generated from three 4-liter culture flasks.

Ultra-large scale production, providing tonnes of recombinant proteins, is currently only attainable by propagation through seeds. Seed production requires backcrosses to genetically broadbased populations, implying reselection of the transgenic phenotype and genotype.

As illustrated by the production schedule for a monoclonal antibody in transformed alfalfa plants (Figure 38), these propagation methods guarantee the production of sufficient quantities of recombinant molecules within the time frame of clinical studies.

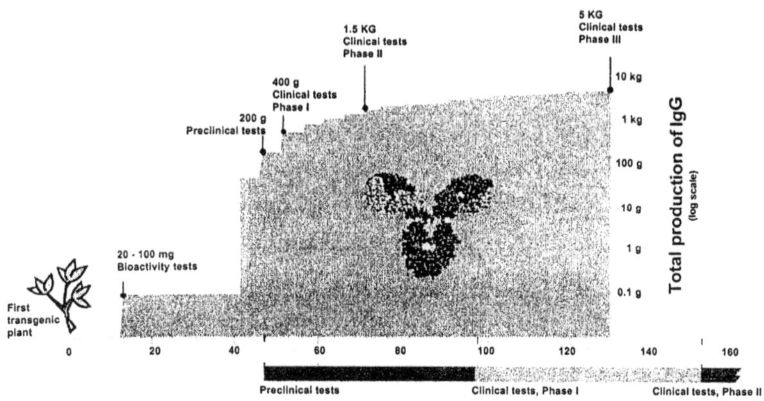

Figure 38. Production schedule for a monoclonal antibody in transformed alfalfa plants. Propagation is performed by stem cuttings; plants are grown in a 700 m^2 greenhouse; expression and purification yields (1% of total protein) are based on recent unpublished data.

3.4 Recovery of Recombinant Molecules

Initial processing of alfalfa leaves after harvest, also termed wet fractionation, involves the extraction of juices and handling of solid fractions. Since raw green alfalfa matter has been processed for years to nourish farm animals, industrial scale extraction processes already exist and adaptation to the extraction of alfalfa foliage containing recombinant protein is in progress. After harvest, alfalfa is chopped and passed through a press that separates the solid fiber containing fraction from the liquid colloidal fraction. This green juice contains most of the proteins and other ingredients of high nutritional value that can be isolated by heat-induced coagulation. Improvements towards the use in molecular farming of the actual industrial process have already been done in small scale trials, using an IgG antibody produced in transgenic alfalfa. The antibody retained its stability throughout the whole extraction process.

Purification of recombinant antibodies from alfalfa extracts can be performed following several strategies. Staphylococcal protein A is the most widely used ligand for a first affinity chromatography step to purify antibodies and has also be used in plants (Khoudi et al., 1999; Fischer et al., 1999). Recently, new strategies emerged to lower the cost of the purification process. Khoudi et al. (1999) were successful using the expanded bed technology to purify an IgG from alfalfa material. This latter technology allows the loading of partially clarified plant extracts containing colloidal material at a high-flow rate and guarantees a high recovery rate. Purification methods differ depending on the purity and the application of the compound isolated. In the case of oral administration of active molecules to animals, no purification process is required. For example, neutralizing antibodies against inflammatory gastro-intestinal diseases in cattle can simply be administered in the form of alfalfa pellets. The purification process is then cheaper and allows the production of an alfalfa product at a very low cost.

In general, alfalfa offers many advantages with regards to purification. Alfalfa contains few alkaloids and phenol compounds when compared to tobacco (Jones et al., 1973; Goplen et al., 1980) and no organic residues from pesticide treatments that have to be removed. Moreover, preliminary results at Medicago Inc. show that proteolytic activity in alfalfa extracts is weaker than in tobacco and an increased recovery is obtained (unpublished data). It is also crucial to remember that alfalfa is a safe expression system without risk of contamination by human or animal pathogens which allows us to skip a filtration step that is required to remove contaminants in other expression systems.

3.5 Disposal of Biomass

The production of recombinant molecules in alfalfa generates residues that are rich in water, fiber and proteins, and exempt of chemical additives, pesticides and fertilizers. For example, 1 kg of recombinant antibody will generate at least 2 tonnes of water, 300 kg of fibrous paste and 100 kg of different protein fractions. Large-scale and ultra-large-scale production of recombinant proteins requires infrastructures to eliminate or recycle these huge amounts of residual biomass. Residual alfalfa biomass can be transformed into valuable energy sources such as combustible pellets (for heating) or ethanol.

4. CASE STUDY

To illustrate the potential of alfalfa as a molecular factory for recombinant antibodies, we present here the production of a mouse monoclonal antibody, C5-1, in alfalfa (Khoudi et al., 1999). C5-1 is an IgG antibody directed against human immunoglobulins that is used in diagnostic agglutination tests to match blood donors and receivers.

cDNAs for both the heavy (H) and the light (L) chain of C5-1 antibody were cloned from the hybridoma cell line producing the antibody and inserted into two distinct plant expression vectors under the control of the 35S promoter. Transgenic plants were produced by Agrobacterium-mediated transformation methods. Heavy chain-expressing plants were then intercrossed with light chain-expressing plants and plants expressing both heavy and light chains were screened among the F1 progeny. Alfalfa C5-1 was purified by loading non clarified crude leaf extract onto an expanded bed affinity column (STREAMLINE rProteinA). This mainly yielded the tetrameric H2L2 form, the signal of the other intermediate forms representing less than 5% the signal obtained for the H2L2 form. Depending on the plants, the yield of the extracted C5-1 ranged from 0.13 to 1% of total soluble proteins.

Western blot analysis showed correctly assembled tetrameric H2L2 C5-1 antibodies that co-migrated with the fully assembled C5-1 antibody from hybridomas (Figure 39). These plants also contained other intermediate antibody forms such as HL, H2 and H2L. Migration in reducing conditions revealed that the size of alfalfa and hybridoma subunits were identical.

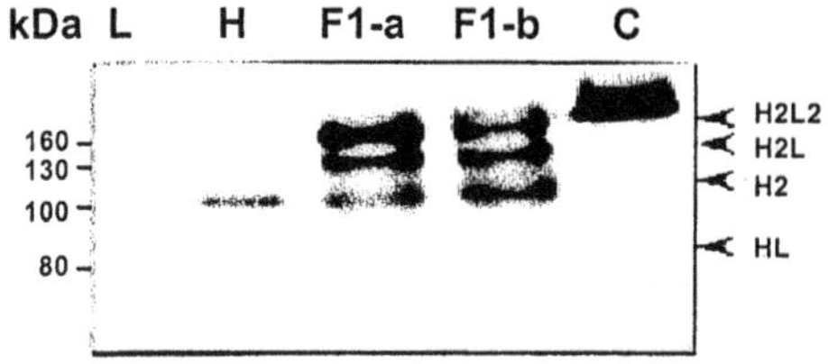

Figure 39. Protein blot analysis of C5-1 light and heavy chain assembly in alfalfa. Total protein extracts from plants producing the light chain (L), the heavy chain (H), and both chains (F1-a and F1-b) were analyzed and compared to purified C5-1 from hybridoma (C). From Khoudi et al., 1999.

To compare alfalfa and hybridoma derived C5-1 in terms of specific activity, several immunoaffinity and agglutination tests were performed. Results showed that both antibodies were comparable and that alfalfa C5-1 can be used as efficiently as a diagnostic reagent as hybridoma C5-1.

Comparative studies were also done to test for stability of antibody produced in alfalfa. While studies have showed proteolysis of recombinant IgGs in tobacco(Hiatt et al., 1989; Ma et al., 1995), our recombinant C5-1 antibody was very stable in alfalfa extracts stored in water as well as in a protective buffer (Figure 40)(Khoudi et al., 1999). This study was the first to show that perennial plants such as alfalfa offer a favourable cellular environment for the synthesis of complex recombinant proteins such as antibodies.

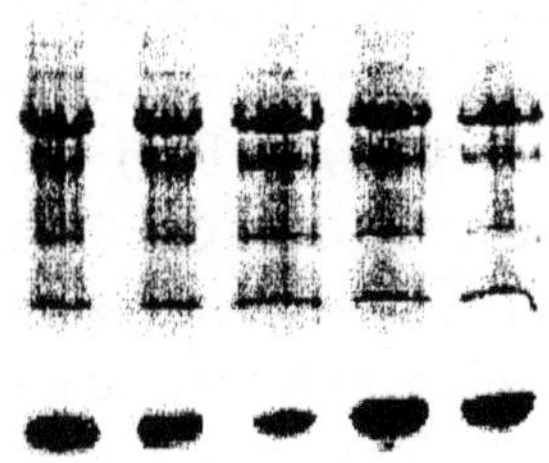

Figure 40. Stability of C5-1 in alfalfa. Transgenic alfalfa extracts were left up to 5 days at room temperature, total soluble proteins were extracted (upper part) and analyzed for the stability of C5-1 by protein gel blot (lower part). From Khoudi et al., 1999.

Since alfalfa is a perennial plant, the stability of C5-1 was also verified during clonal propagation. Accumulation of C5-1 remained stable in F1 lines harvested during 2.5 years. Protein yields from stem propagated clones and from the original F1 plants were identical. Furthermore, the immunoreactivity of C5-1 remained stable in dry hay within 12 weeks from harvest when stored in controlled conditions.

To test *in vivo* stabilities of alfalfa C5-1 and hybridoma C5-1, the antibodies were injected into the blood stream of mice. Degradation rates of both antibodies were very similar, with half lives of about 3 days (Figure 41).

In summary, the antibody produced in alfalfa was in all parts equivalent to its hybridoma derived monoclonal counterpart and very stable upon storage or *in vivo*. These characteristics, combined with the agronomic advantages of alfalfa, make the alfalfa bioreactor a very interesting system for the large scale production of antibodies.

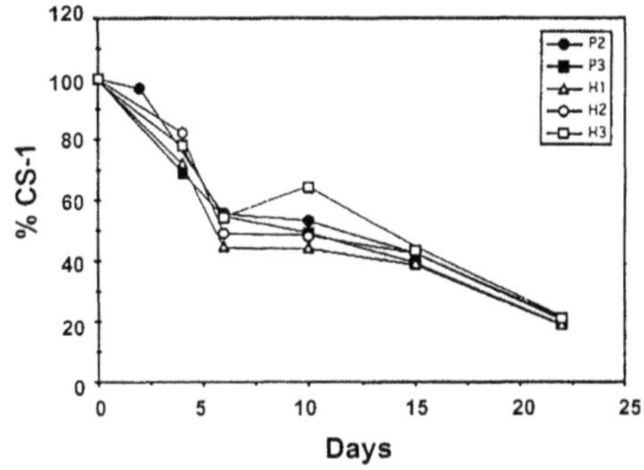

Figure 41. Degradation rate of alfalfa derived (P2, P3), and hybridoma derived (H1, H2, H3) C5-1 in the blood stream of mice. From Khoudi et al., 1999.

5. FUTURE DEVELOPMENTS

In this section, we would like to highlight some of the numerous challenges that production systems in general, and molecular farming in particular, will have to master in the near future. Current and future research projects will have to address product yield and quality as well as optimization of processing procedures. Product yield can be optimized by increasing expression levels of both RNA and protein, and by decreasing protein degradation before and during purification. Protein bioactivity can be maintained and quality consistency assured by optimizing post-translational processes such as correct protein folding and glycosylation as well as extraction procedures. As for all industrial processes, efforts will have to be made to render all steps more time and cost efficient. In addition, protocols developed at the laboratory scale for preliminary analyses need to be upgraded and adapted to large scale synthesis. The following paragraphs illustrate in more detail some of these challenges.

Messenger RNA levels of the recombinant gene in the plant cell are directly related to final yield of recombinant protein. One way to maximize expression is to avoid sequences such as ATTTA and polyadenylation sequences that might destabilize the mRNA. Another way is to modify gene sequences according to the codon usage of alfalfa. Codon usage in plants is different from that of animals or human and this can affect the level of expression of a recombinant protein. It has been shown that modifying the codon of a cDNA according to the codon usage of the host may increase the level of recombinant protein (Kusnadi et al., 1997).

Another solution to low yields of recombinant protein is to limit its degradation after synthesis. For proteins that are particularly sensitive to degradation by endogenous proteases, protein targeting to organelles and accumulation in sub-cellular compartments can dramatically increase yield (Michaud and Yelle, 2000). In several studies, a significant increase in intracellular accumulation was obtained when antibodies or their derivatives were targeted to the endoplasmatic reticulum of the plant's cells (Fischer et al., 2000). A study in 1992 demonstrated a 20-fold enrichment of vicilin (a pea seed storage protein) in alfalfa when targeted to, and retained within the endoplasmic reticulum using a carboxy-terminal KDEL sequence to the protein (Wandelt et al., 1992). Tabe et al. (1995) also showed a positive effect on protein accumulation in alfalfa when the KDEL retention signal was added. These two studies demonstrate that sub-cellular targeting can provide a solution to overcome rapid protein turnover by proteolysis in alfalfa, thus providing high levels of recombinant protein.

In the field of therapy, an important concern is to limit the body's immune response to the therapeutic agent upon administration. Since monoclonal antibodies are generally produced in mice, human anti-mouse immune responses during immunotherapy are a frequent drawback to therapy. Part of this response is directed against the constant regions of the antibody molecule and can be suppressed by "humanizing" the sequence of these regions by genetic engineering (Östberg and Queen, 1995). The other part of the immune response is elicited by N-linked glycans added to the heavy chains during post-translational processing through the endoplasmatic reticulum and the Golgi stacks. Glycan chains play a role in glycoprotein bioactivity and they may influence the *in vivo* half life of glycoproteins in circulation (Matsumoto et al., 1995; Boyd et al., 1995; Wright and Morrison, 1997). As mentioned earlier, plant glycosylation pathways bear some small but nonetheless important differences when compared to animal glycosylation, pathways (see chapter 4, this volume). Differences in glycosylation patterns with regard to their mammalian counterparts have indeed been observed for many plant derived antibodies (Faye and Chrispeels, 1988; Faye et al., 1993; Cabanes-Macheteau et al., 1999; Bakker et al., 2001, in press). To overcome this problem, proteins can be retained in the endoplasmic reticulum by a C-terminal KDEL peptide, avoiding the passage through the Golgi apparatus and thus eliminating fucose and xylose addition on the glycan chain (Pagny et al., 2000). Furthermore, proteins can be targeted to other organelles whenever glycosylation is not necessary. A better understanding of the plant glycosylation machinery will provide new strategies and allow the synthesis of human-like N-linked glycans in plants (see chapter 4, this volume). Glycosylation patterns in alfalfa are currently under investigation.

6. CONCLUSION

We have illustrated how alfalfa, a perennial crop, can be used for sustainable production of functional antibodies and other complex therapeutic molecules. Rapid scale-up procedures through stem cutting and somatic embryogenesis guarantee mid-scale production of recombinant protein within a year. The exceptional characteristics of alfalfa foliage and the lack of pesticides simplify purification procedures. In some applications, there is no need for purification since alfalfa is used as a basic diet for ruminants. All these advantages make alfalfa a very promising system for the large scale production of therapeutic proteins.

High-volume manufacturing capacity, competitive pricing and improved efficacy will be essential for the success of any commercial antibody drug in the future since the use of therapeutic antibodies for the treatment of disease will likely be a very competitive market. Regulatory issues will have to be solved and the adverse reaction of the population against genetically modified organisms will have to be tamed. Despite all these hurdles, it is likely that the market release of the first recombinant pharmaceutical from transgenic plants will occur within the next five years. Molecular farming will evolve through a new kind of collaboration between agriculture and pharmaceutical industries.

REFERENCES

Bakker H, Bardor M, Molhoff J, Gomord V, Elbers I, Stevens L, Jordi W, Lommen A, Faye L, Lerouge P and Bosch D, 2001. Galactose-extended glycans on antibodies produced by transgenic plants. Proc Natl Acad Sci USA, 27:2899-2904

Boyd PN, Lines AC and Patel AK, 1995. The effect of the removal of sialic acid, galactose and total carbohydrate on the functional activity of Campath-1H. Mol Immunol, 32:1311-1318

Brown DC and Atanassov A, 1985. Role of genetic background in somatic embryogenesis in *Medicago*. Plant Cell Tiss Org Cult, 4:111-122

Brown DCW, Tian L, Buckley DJ, Lefebvre M, McGrath A and Webb J, 1994. Development of a simple particle bombardment device for gene transfer into plant cells. Plant Cell Tiss Org Cult, 37:47-53

Cabanes-Macheteau M, Fitchette-Laine AC, Loutelier-Bourhis C, Lange C, Vine ND, Ma JK, Lerouge P and Faye L, 1999. N-glycosylation of a mouse IgG expressed in transgenic tobacco plants. Glycobiology, 9:365-372

Conrad U and Fiedler U, 1998. Compartment-specific accumulation of recombinant immunoglobulins in plant cells: An essential tool for antibody production and immunomodulation of physiological functions and pathogen activity. Plant Mol Biol, 38:101-109

Deak M, Kiss GB, Koncz C and Dudits D, 1986. Transformation of *Medicago* by *Agrobacterium* mediated gene transfer. Plant Cell Rep, 5:97-100

Desgagnés R, Laberge S, Allard G, Khoudi H, Castonguay Y, Lapointe J, Michaud R and Vézina LP, 1995. Genetic transformation of commercial breeding lines of alfalfa (*Medicago sativa*). Plant Cell Tiss Org Cult, 42:129-140

De Wilde C, De Rycke R, Beeckman T, De Neve M, Van Montagu M, Engler G and Depicker A, 1998. Accumulation pattern of IgG antibodies and Fab fragments in transgenic *Arabidopsis thaliana* plants. Plant Cell Physiol, 39:639-646

Düring K, Hippe S, Kreuzaler F and Schell J, 1990. Synthesis and self assembly of a functional monoclonal antibody in transgenic *Nicotiana tabacum*. Plant Mol Biol, 15:281-293

Echelard Y, 1996. Recombinant protein production in transgenic animals. Curr Opin Biotech, 7:536-540

Faye L and Chrispeels MJ, 1988. Common antigenic determinants in the glycoproteins of plants, molluscs and insects. Glycoconjugate, 5:245-256

Faye L, Gomord V, Fitchette-Lainé AC and Chrispeels MJ, 1993. Affinity purification of antibodies specific for Asn-linked glycans containing a1,3) –fucose or ß(1,2)-xylose. Anal Biochem, 209:104-108

Fischer R and Emans N, 2000. Molecular farming of pharmaceutical proteins. Transgenic Res, 9:279-299

Fischer R, Liao Y-C and Drossard J, 1999. Affinity-purification of a TMV-specific recombinant full-size antibody from a transgenic tobacco suspension culture. J Immunol Meth, 226:1-10

Goplen BP, Howarth RE, Sarkar SK and Lesins K,1980. A search for condensed tannins in annual and perennial species of *Medicago*, *Trigonella* and *Onobrychis*. Crop Sci, 20:801-804

Hiatt A, Cafferket R and Bowdish K, 1989. Production of antibodies in transgenic plants. Nature, 342:76-78

Hood, EE and Jilka JM, 1999. Plant-based production of xenogenic proteins. Curr Opinion Biotechnol, 10:382-386

Jefferson RA, Kavanagh TA and Bevan MW, 1987. GUS fusions: ß-glucuronidase as a sensitive and versatile gene fusion marker in higher plants. EMBO J, 6:3901-3907

Jones WT, Anderson LB and Ross MD, 1973. Bloat in cattle XXIX. Detection of protein precipitant (flavolans) in legumes. New Zeal J Agr Res, 16:441-446

Khoudi H, Laberge S, Ferullo J-M, Bazin R, Darveau A, Castonguay Y, Allard G, Lemieux R and Vezina L-P, 1999. Production of a diagnostic monoclonal antibody in perennial alfalfa plants. Biotechnol Bioeng, 64:135-143

Koehler G, and Milstein C, 1975. Continuous culture of fused cells secreting antibody of predefined specificity. Nature, 256:495-497

Kusnadi AR, Nikolov ZL and Howard JA, 1997. Production of recombinant proteins in transgenic plants: Practical considerations. Biotechnol Bioeng, 56:473-484

Larrick JW, Yu L, Chen, J, Jaiswal S and Wycoff K, 1998. Production of antibodies in transgenic plants. Res Immunol, 149:603-608

Leite A, Kemper E, da Silva M, Luchessi A, Siloto R, Bonaccorsi E, El-Dorry HF and Arruda P, 2000. Expression of correctly processed human growth hormone in seeds of transgenic tobacco plants. Mol Breeding 6:47-53

Ma JK, Hiatt A, Hein M, Vine ND, Wang F, Stabila P, van Dolleweerd C, Mostov K and Lehner T, 1995. Generation and assembly of secretory antibodies in plants. Science, 268:716-719

Ma JK, Hikmat BY, Wycoff K, Vine ND, Chargelegue D, Yu L, Hein MB and Lehner T, 1998. Characterization of a recombinant plant monoclonal secretory antibody and preventive immunotherapy in humans. Nat Med, 4:601-606

Matheson SL, Nowak J and Maclean NL, 1990. Selection of regenerative genotypes from highly productive cultivars of alfalfa. Euphytica, 45:105-112

Matsumoto S, Ikura K, Ueda M and Sasaki R, 1995. Characterization of a human glycoprotein (erythropoietin) produced in cultures of tobacco cells. Plant Mol Biol, 27:1163-1172

McKersie BD, Senaratna T, Bowley SR, Brown DCW, Krochko JE and Bewley JD, 1989. Application of artificial seed technology in the production of hybrid alfalfa (*Medicago sativa* L.). In Vitro Cell Dev Biol, 25:1183-1188

Michaud D and Yelle S, 2000. Production of useful protease inhibitors in plants. In *Recombinant protease inhibitors in plants*. Ed. Michaud D. Landes Bioscience, Georgetown, TX, pp191-201

Miele L, 1997. Plants as bioreactors for pharmaceuticals: regulatory considerations. Trends Biotechnol, 15:45-50

Nàrvaez-Vàsquez, J, Orozco-Càrdenas, ML and Ryan CA, 1992. Differential expression of a chimeric CaMV-tomato proteinase inhibitor I gene in leaves of transformed nightshade, tobacco and alfalfa plants. Plant Mol Biol, 20:1149-1157

Nowak J, Matheson S, Maclean NL and Havard P, 1992. Regenerative trait and cold hardiness in highly productive cultivars of alfalfa and red clover. Euphytica, 59:189-196

Odell JT, Nagy F and Chua NH, 1985. Identification of DNA sequences required for activity of the cauliflower mosaic virus 35S promoter. Nature, 313:810-812

Östberg L and Queen C, 1995. Human and humanized monoclonal antibodies: preclinical studies and clinical experience. Biochem Soc Trans, 23:1038-1043

Pagny S, Cabanes-Macheteau M, Gillikin JW, Leborgne-Castel N, Lerouge P, Boston RS, Faye L and Gomord V, 2000. Protein recycling from the Golgi to the endoplasmic reticulum is very active in plants but has a minor contribution to calreticulin retention. Plant Cell, 12:739-755

Pen J, 1996. Comparison of host systems for the production of recombinant proteins. In *Transgenic plants: a production system for industrial and pharmaceutical proteins.* Eds. Owen MRL and Pen J. John Wiley and Sons, London, 149-167

Senaratna T, McKersie BD and Bowley SR, 1989. Desiccation tolerance of alfalfa *(Medicago sativa* L.) somatic embryos: Influence of abscisic acid, stress pretreatments and drying rates. Plant Sci, 65:253-260

Senaratna T, McKersie BD and Bowley SR, 1990. Artificial seeds of alfalfa (*Medicago sativa* L.) induction of desiccation tolerance in somatic embryos. In Vitro Cell Dev Biol 26:85-90

Shahin EA, Spielmann A, Sukhapinda K, Simpson RB and Yashar M. 1986. Transformation of cultivated alfalfa using disarmed *Agrobacterium tumefaciens.* Crop Sci, 26:1935-1939

Skerra A, 1993. Bacterial expression of immunoglobulin fragments. Curr Opin Immunol, 5:256-262

Tabe LM, Wardley-Richardson T, Ceriotti A, Aryan A, McNabb WC, Moore A and Higgins TJV, 1995. A biotechnological approach for improving the nutritive value of alfalfa. J Plant Sci, 73:2752-2759

Voss A, Nierbach M, Hain R, Hirsch H, Liao Y, Kreuztaler F, Fischer R and Hain R,1995. Reduced virus infectivity in *N. tabacum* secreting a TMV-specific full size antibody. Mol Breeding 1:39-50

Wallach M, Pillemer G, Yarus S, Halabi A, Pugatsch T and Mencher D, 1990. Passive immunization of chickens against *Eimeria maxima* infection with a monoclonal antibody developed against a gametocyte antigen. Infect Immun, 58:557-562

Wandelt CI, Khan MRI, Craig S, Schoeder HE, Spencer D and Higgins TJV, 1992. Vicilin with carboxy-terminal KDEL is retained in the endoplasmic reticulum and accumulates to high levels in the leaves of transgenic plants. Plant J, 2:181-192

Wright A and Morrison SL, 1997. Effect of glycosylation on antibody function: implications for genetic engineering. Trends Biotechnol, 15:26-32

Zeitlin L, Olmsted S, Moench T, Martinell Co M, Paradkar V, Russell D, Queen C, Cone R and Whaley K, 1998. A humanized monoclonal antibody produced in transgenic plants for immunoprotection of the vagina against genital herpes. Nat Biotechnol, 16:1361-1364

Chapter 12

THE PRODUCTION OF RECOMBINANT ANTIBODIES IN PLANTS AND PLANT CELLS

R. Fischer[1,2], N. Emans[1], , and S. Schillberg[2]
[1] *Institut für Biologie VII (Molekulare Biotechnologie), RWTH Aachen, Worringerweg 1, 52074 Aachen, Germany.*[2] *Fraunhofer Dept. for Molecular Biotechnology, IUCT, Grfschaft, Auf dem Aberg 1, D-57392 Schmallenberg, Germany*

Keywords: molecular farming, recombinant protein, transgenic plant, purification, biotechnology.

Abstract The production of recombinant proteins in plants, known as molecular farming, has the potential to produce large quantities of medicinal polypeptides for the diagnosis and treatment of many diseases. Many of these polypeptides under consideration are antibodies. In this chapter, we describe the production of antibodies in transgenic plants, plants infected by engineered plant viruses and in plant cell culture. Some comparisons are also drawn between plant and animal based production.

1. INTRODUCTION

Plants have long been used as a source of medicines, and almost a quarter of the medicines in use today come from them. Molecular biotechnology has changed the nature of medicines that can be isolated from plants, through the development of the technology to produce recombinant proteins in crops. Many therapeutically active proteins have been identified in the last 40 years, and their development into useful medicines has driven a demand for a large scale expression system that is similarly useful and productive.

Molecular farming can satisfy the world demand for recombinant proteins. It is defined as the production of recombinant proteins on an agricultural scale in plants, and it unites genetic engineering with modern molecular medicine. It

L. Erickson et al. (eds.),
Molecular Farming of Plants and Animals for Human and Veterinary Medicine, 259–285.

has the potential to produce medicinal polypeptides and technical enzymes on a scale previously only possible for chemically synthesized pharmaceuticals.

Molecular farming will bring significant economic benefits through harnessing the scale of agricultural production. Molecular farming in crop plants is also inherently safer than many alternatives because human and animal pathogens will not contaminate the recombinant proteins. Importantly, there appear to be few limitations on the nature of recombinant proteins that can be synthesized in plants.

Of all the recombinant proteins currently under trial as medicinal polypeptides, one in four are antibodies (source www.phrma.org). Although there are concerns about the immunogenicity of mouse derived monoclonal antibodies, antibody engineering has developed to the point where antibodies can be engineered and expressed that are essentially identical to the human versions. Antibody engineering produces recombinant proteins designed to specifically interact with a target pathogenesis protein. For example, antibodies are unparalleled as diagnostic reagents for detecting and treating cancer. However, making antibodies outside the animal immune system is expensive and for antibodies to reach their potential as therapeutics, an efficient expression system is needed. In this review, we discuss the exploitation of transgenic plants and plant cells to produce recombinant antibodies by molecular farming. Through this technology, antibodies can be safely made on a scale that may one day make them as widely available as aspirin.

In recent history, gene cloning has determined the structure of many proteins that are therapeutically valuable, but for these proteins to be clinically useful an expression system is required to efficiently produce recombinant proteins. Bacteria were the expression system that was often used to attempt to produce recombinant proteins. However, the bacterial protein synthesis machinery is incapable of reproducing the complexity of eukaryotic proteins, which often require extensive folding, post-translational modification and assembly. Many mammalian proteins are produced as insoluble mis-folded aggregates or lack the features required for their activity.

Eukaryotic life shares a complex set of internal organelles that direct protein synthesis, cell function and protein secretion. This endomembrane system is well conserved from yeast to animals and higher plants and many protein sorting and trafficking signals are functional across these kingdoms. This can be exploited by biotechnologists and is a central reason that molecular farming is feasible.

Plant cells carry out many of the post-translational modifications required for optimal biological activity of mammalian proteins. Plants are an attractive alternative to bacteria as an expression system because they are genetically tractable, easy to cultivate, and can produce complex assemblies of functional proteins. For example, plant cells remain as (probably) the only practical and economic production system to produce the secretory IgA complex, comprising two full size antibodies coupled by a joining chain and secretory component (Ma et al., 1995). As a further example of the admirable flexibility of the plant expression system, recombinant antibodies produced in tobacco have the same specificity and affinity as monoclonal antibodies produced by the original hybridoma cell line (Voss et al., 1995). These features have lead to plants becoming a system for the mass production of pharmaceutical proteins through molecular farming. The long term goal is to harness plant cultivation to the needs of 21st century medicine and to harvest crops rich in proteins that are currently in scarce supply.

Molecular farming in plants has become possible because of the convergence of several molecular technologies. Plant transformation has advanced to the point where many commercially interesting crops can be genetically modified (Christou, 1993; Christou, 1995). In parallel, antibody engineering has developed to the point where recombinant antibodies can be generated against almost any molecule with produced to suit almost any clinical purpose (Winter et al., 1994; Winter and Milstein, 1991). Through the convergence of these technologies, it has become possible to produce chimeric mouse-human and therapeutic antibodies in plants in sufficient quantities for pre-clinical trials (Vaquero et al., 1999; Zeitlin et al., 1998).

In this chapter, we discuss the uses of plants to produce therapeutic and diagnostic recombinant antibodies in plants by transient or stable plant transformation. Some comparisons are drawn with animal based production and we describe the use of whole plants versus suspension cultured cells to produce recombinant antibodies. We present our argument to support the hypothesis that molecular farming will become one of the predominant methods for recombinant therapeutic and diagnostic protein production in the next decade.

2. TRANSGENIC PLANTS AS BIOREACTORS FOR RECOMBINANT PROTEIN PRODUCTION

Since the development of plant transformation in 1983 (Fraley et al., 1983; Willmitzer et al., 1983), many new transgenic plant varieties including monocot cereal crops (Christou, 1993; Christou, 1995) have been developed through the technology and have improved or desirable features. These include herbicide insensitivity, delayed fruit ripening, improved nutritional quality and pathogen resistance. Molecular farming applies plant transformation to develop crops that express therapeutically or commercially valuable proteins. These molecular farming crops can be thought of as hectare sized protein synthesis factories fed by sunlight and rainfall, producing recombinant polypeptides until they are harvested.

The feasibility of exploiting plants for the production of recombinant mammalian proteins was first demonstrated in 1989, with the description of the expression of functional full size antibodies in transgenic tobacco (Hiatt et al., 1989). The therapeutic applications of recombinant antibodies are well known and monoclonal antibodies (mAbs) (Koehler and Milstein, 1975) are essential therapeutic and diagnostic tools used in medicine, human and animal health care, the life sciences and biotechnology. The significance of producing antibodies in plants was that it created the opportunity to produce antibodies on a very large scale. Until this point, monoclonal antibody production in cultured hybridoma cells or ascites fluid was the only practical method to make a single antibody on a scale that could be considered useful for treating disease. The clear implications from Hiatt et al.'s technical innovation was that molecular farming could replace these production systems as a less expensive, large scale alternative.

Molecular farming technology has rapidly expanded and many therapeutically valuable proteins have been successfully synthesized in transgenic plants. These include recombinant antibodies (rAbs) (Ma et al., 1995; Ma et al., 1994), enzymes (Hogue et al., 1990; Verwoerd et al., 1995), hormones (Leite et al., 2000; Staub et al., 2000), cytokines, interleukins (Magnuson et al., 1998), plasma proteins (Sijmons et al., 1990), human alpha-1-antitrypsin (Terashima et al., 1999) and edible vaccines (Mason and Arntzen, 1995). Thus, plant cells are capable of expressing a large variety of recombinant proteins and protein complexes. This is exemplified by the advances in recombinant antibody expression in plants.

2.1 Antibody Engineering

The polypeptides made by molecular farming are the product of both genetic and protein engineering. The development of cloning and molecular biology techniques has had a large impact on the kinds of molecules available for expression in plants. This is particularly true for recombinant antibodies, where the development of phage display and *in vitro* methods for antibody isolation have revolutionized the field (Burton, 1995). Antibody engineering has broadened the range of possible applications for recombinant antibodies beyond what was possible when monoclonal antibodies were first developed. Phage display technology has been central to this change, and is the display of libraries of cloned recombinant antibodies on the surface of bacteriophage particles. In these systems, a single antibody fragment is physically coupled to its cognate nucleic acid. This means that when the particles are selected via a particular property of the antibody, such as specificity to a given antigen, the gene encoding the antibody is co-selected and can be amplified (Winter et al., 1994). Thus, antibody properties can be isolated and fine-tuned *in vitro*, and these optimized proteins then produced in plants.

2.2 Antibody Production in Transgenic Plants

The development of recombinant antibody expression in plants has identified many of the salient issues that need to be addressed to make molecular farming practical (Hiatt, 1990; Hiatt et al., 1989; Hiatt and Ma, 1993; Hiatt and Mostov, 1992; Hiatt et al., 1992; Hiatt, 1991). Many forms of recombinant antibodies can be produced in plants and the variety ranges from single chain antibody fragments to full size antibodies and antibody complexes, as well as including antibody fusion proteins (Figure 42). The original studies focussed on full size IgGs (Baum et al., 1996; De Wilde et al., 1996; Düring et al., 1990; Hiatt et al., 1989; Ma et al., 1994; Voss et al., 1995). Since then, Fab fragments (De Neve et al., 1993) and single chain antibody fragments (scFvs) (Artsaenko et al., 1995; Fecker et al., 1996; Fiedler and Conrad, 1995; Firek et al., 1993; Owen et al., 1992; Schouten et al., 1996; Tavladoraki et al., 1993) have been functionally expressed in leaves and seeds of plants.

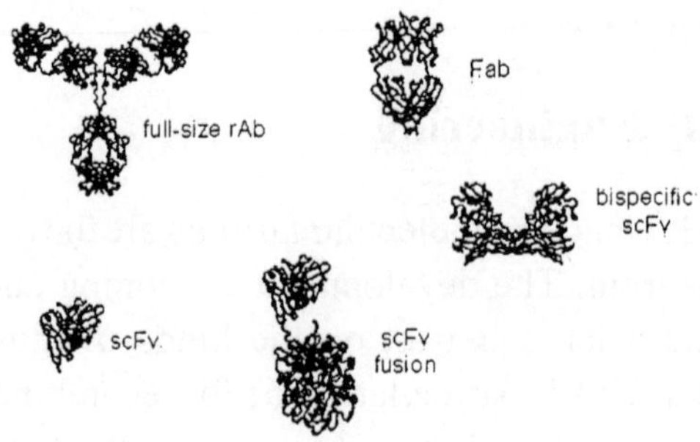

Figure 42. Forms of recombinant antibodies produced in plants. RAb: recombinant antibody; Fab: fragment antigen binding; scFv: single chain antibody fragment; dAb: single domain antibody.

As mentioned earlier, sIgA molecules can be more easily produced in transgenic plants than any other expression system (Ma et al., 1995). The flexibility of exploiting the plant protein synthesis pathway is further demonstrated by the production of bispecific single chain antibody fragments (Fischer et al., 1999), fusion proteins (Spiegel et al., 1999) and membrane anchored antibodies (Schillberg et al., 2000).

Remarkably, the specificity and affinity of recombinant antibodies is maintained after expression in plants. The recent studies have defined many criteria that are important for expression, which range from transcriptional control to codon optimization and sub-cellular protein targeting. The pattern of codon usage in plants is different to that of animals and altering the composition of the heterologous cDNA to meet the plant pattern increases the rate of translation for some recombinant proteins (Kusnadi et al., 1997).

Eukaryotic cells synthesize proteins in the context of their complex protein trafficking pathways. Expression analysis has shown that precise control of the sub-cellular destination of a recombinant antibody is desirable and important for the highest expression level. These observations seem applicable to all plant based expression systems, whether transient or stable and whether recombinant proteins are expressed in suspension cells, whole plants, or plant organs. Recombinant antibodies are generally unstable in the cytosol and their yield in plants increases when they are targeted to the secretory pathway using a signal peptide (Conrad and Fiedler, 1998) (Schillberg et al., 1999). Proteins that are co-translationally imported into the endoplasmic reticulum (ER) are often secreted in plants in the absence of any other targeting signals. While antibody delivery to the apoplast can be desirable for protein recovery, yields improved upon when the protein is retained in the endoplasmic reticulum. This is often achieved by the addition of a peptide signal (KDEL) to the protein's C-terminus that is recognized by a cellular quality control system that retrieves ER resident proteins when they escape the compartment. ER retained antibodies accumulate to at least 100 fold higher levels than their cytosolically directed counterparts, that are often produced at such low levels that they are beneath the limits of detection (Conrad and Fiedler, 1998) (Schouten et al., 1996). Increasing knowledge on how signals within proteins target them to their subcellular destinations has been exploited to direct recombinant proteins to many intracellular and extracellular destinations. These include the intercellular space (Voss et al., 1995), ER and chloroplasts (Staub et al., 2000).

As noted, the cytosol is an inappropriate location for full size antibody accumulation. Intracellular expression of rAbs in the cytoplasm has only been achieved using single chain Fv fragments (Owen et al., 1992; Schillberg et al., 1999; Schouten et al., 1997; Zimmermann et al., 1998), presumably because they only require little post-translational processing. Even with scFvs, measured expression levels in the cytosol are still meager (Owen et al., 1992; Schouten et al., 1996). There is a report where cytosolic scFvs have reached levels of up to 1.0% of total soluble protein (De Jaeger et al., 1998) and the reasons behind these differences need precise definition. Cytosolic expression is useful where antibodies are used as tools to mediate pathogen resistance (Schillberg et al., 2001) or to modulate metabolic pathways (Artsaenko et al., 1995; Phillips et al., 1997) found there, but as a tool for molecular farming it is of little interest. Importantly, the most attractive system for molecular farming is the expression of antibodies under the control of a leader peptide that directs them to and retains them within the ER. This statement is justified because the yields are highest yet observed.

A long-term goal of molecular farming is the production of proteins in fields of transgenic plants. Thus, a key issue is the total amount of a given protein that can be made in genetically modified crops, and this is tied to the expression level in individual plants. Expression levels of different antibodies in stably transformed plants vary, with expression of full size IgG under the control of the 35S promoter ranging from 0.35% (van Engelen et al., 1994) to 1.3% of the total soluble protein (TSP) in tobacco leaves (Hiatt et al., 1989). The expression level of a recombinant protein is related to the intrinsic properties of the antibody and improving expression is a focus of intense research. Transgenic plants have been identified with expression levels of scFvs in leaves reaching 6.8% of the TSP (Fiedler et al., 1997) and levels of secretory IgA up to 500 μg per gram leaf material (Ma et al., 1995). The basis of this variability is not clear, but our working hypothesis is that the selection method used to isolate the antibody generally influences its stability. We strongly support the concept that phage display selection selects for more stable antibody scaffolds. So far, all reported scFvs that were expressed in the cytosol at very low levels were directly converted from hybridoma cell lines, which means that the selection was based on the properties of the secreted full-size antibody. The only exception (De Jaeger et al., 1998) showing high scFv accumulation in plant cytosol was obtained by using phage display technology for the selection of the specific scFvs. Phage display biopanning procedure favours the selection of scFvs that are robust and tolerate the cytosolic reducing conditions because the selection procedure is usually performed for several rounds, and robust scFvs which tolerate the reducing environment of the bacterial cytosol and show antigen-binding activities could be preferentially enriched. scFvs that accumulate to the high levels in transgenic plants may represent a suitable framework for engineering recombinant antibodies produced in plants, and therefore, it may be used for establishment of "designer" libraries. Given that some proteins express at very high levels in plants, where they reach more than a quarter of the total soluble leaf protein (Ziegler et al., 2000) antibody expression levels may be significantly improved as soon as features influencing antibody stability in plants are identified.

One important aspect of protein production in plants is that the recombinant proteins can be stored in tissues and seed. Leaves from transgenic plants expressing ER retained scFvs can be dried and stored for more than three weeks without losses of scFv specificity or antigen binding activity (Fiedler et al., 1997). Seeds are storage organs that are often rich in protein and can be stored almost indefinitely (Conrad et al., 1998; Fiedler and Conrad, 1995). Seeds can be thought of as storage containers for recombinant proteins, since single chain antibodies can reach up to 4.0% of the total soluble seed protein (Phillips et al., 1997). Just as has been observed in plant leaves for recombinant antibodies, ER retention leads to an increase in scFv accumulation over the secreted counterpart. Potato tubers have also been used as storage containers with expression levels reaching 2% TSP. In tubers, the antibody was sufficiently stable that during cold storage for 18 months there was only a 50% loss of functional protein (Artsaenko et al., 1998).

A further attractive aspect of production via molecular farming in plants is that production levels can be rapidly increased to cope with a changing marketplace. In contrast to transgenic animals, the production system (plant, seed) and the protein itself (expressed in a plant or seed) can be stored for future uses and shelved when demand reduces. This is a key advantage over transgenic animal herds, where the herds have to continuously maintained.

In summary, molecular farming in transgenic plants has many desirable features for protein production. A promising approach for protein expression and in field production is to target the protein to the ER on a cellular level and, if long-term storage is required, to target the protein for seed specific expression. However, it is still unclear which crop is preferable for protein production. This issue has been raised in several reviews of molecular farming, but there is no consensus. The discussion of which crop to choose is broad, but the first molecular farming product on the market, recombinant avidin, is made in corn (Hood et al., 1997; Hood et al., 1999).

3. PROTEIN EXPRESSION IN PLANTS

Techniques have been developed for the expression of genes in stably transformed plants, where the gene is incorporated into the plant genome, or the transient production of the protein in plant tissues. Transient gene expression is rapid compared to stable plant transformation and gives results in days. However, transient gene expression is limited in scale and is generally used to test the constructs used for protein expression before stable transformation is performed. We feel that a transient expression system, such as agroinfiltration (Kapila et al., 1996), may be suitable for producing gram amounts of protein on a case by case basis. Transient expression is already carried out on a field scale in tobacco using viral vectors (http://www.lsbc.com).

Three systems are used to transiently express genes in cells, these comprise: bombardment of the cells with projectiles coated in naked DNA, vacuum infiltration with Agrobacteria (agroinfiltration) and viral vector infection. The number of cells transiently transformed varies between these systems. Particle bombardment usually reaches only a few cells and for transcription the naked DNA has to reach the nucleus of the cell (Christou, 1993; Christou, 1995). Agroinfiltration targets many more cells in a leaf and the T-DNA harboring the gene of interest is actively transferred into the nucleus with the aid of bacterial proteins (Kapila et al., 1996). A virus will systemically infect most cells in a plant after inoculation with the recombinant vector.

The most robust and useful technique that we have found in practice is agroinfiltration, because it is technically straightforward and uses the same constructs as needed for stable plant transformation.

3.1 Agroinfiltration

Agroinfiltration is the transient transformation of adult leaf cells and is achieved by the vacuum infiltration of bacteria into the intercellular spaces (Kapila et al., 1996). Agrobacterial proteins catalyze the transfer of the gene of interest into the host cell nucleus and the gene is carried in a conventional binary vector, such as are used for generating transgenic plants. Transient gene expression can be detected in isolated leaves after 2-3 days and our experience is that kilograms of leaves can be transformed without the need for sophisticated laboratory equipment. It is generally believed that transient transformation works because the transferred T-DNA, which may not integrate into the host chromosome, is transiently transcribed in the nucleus (Kapila et al., 1996).

Using agroinfiltration, we have transiently expressed scFvs, individual heavy and light chains as well as full size chimeric α-carcinoembryonic antigen (CEA) antibodies in plant leaves (Vaquero et al., 1999). For full size chimeric antibody expression, the mouse-human chimeric heavy and light chain genes were integrated into two vectors in two separate recombinant Agrobacterium strains and these two strains were simultaneously infiltrated into leaves. Functional full size chimeric antibodies were assembled *in vivo* by simultaneous expression of both genes in the host cells. This demonstrates the feasibility of using this system to express and assemble multi-component protein complexes. The partners in such complexes can be easily exchanged through combinations of recombinant Agrobacteria strains. In the transgenic plant system, this can only be achieved by time consuming crossing experiments with individual transgenic plants or by co-transfection of two independent expression cassettes.

3.2 Viral Vectors

Viral vectors (Scholthof et al., 1996) share some of the advantages of agroinfiltration Here, the gene of interest is cloned into the genome of a viral plant pathogen under the control of a strong subgenomic promoter. Infectious recombinant viral transcripts are used to infect plants and produce the target protein. Target genes are expressed at high levels because of the high level of gene multiplication during virus replication (Porta and Lomonossoff, 1996). Some plant viruses have a wide host range, are easily transmissible by mechanical inoculation and can spread from plant to plant, making it possible to rapidly infect large numbers of plants with recombinant viruses.

Plant virus genomes can be composed of DNA or RNA but the more than 80% of plant viruses that carry an RNA-genome are the focus of this review section. DNA viruses, such as the geminivirus family, have been used as extrachromosomal replicons for heterologous protein expression (Timmermans et al., 1994), but RNA viruses multiply to higher titers in infected plants, which makes them ideal vectors for protein expression. For vector construction, viral RNA genomes are reverse-transcribed *in vitro* and cloned as full-length cDNAs into *in vitro* or *in vivo* transcription vectors. There are different strategies for the insertion of foreign genes into plant viral genomes: i) gene replacement – where nonessential viral genes, like coat protein, are replaced by the gene of interest; ii) gene insertion – where the gene of interest is placed under the control of an additional strong subgenomic promoter and iii) gene fusion – where the gene of interest is a translational fusion with a viral gene. Some viruses have genomic size constraints and gene replacement is then used for the generation of the recombinant viral vector. Gene insertion is desirable when large coding sequences have to be expressed. Gene fusions have been mostly created with coat protein genes and are a very efficient method for the presentation of foreign peptide sequences on the surface of viral particles. For the inoculation of plants and protein expression, recombinant viral vectors are usually transcribed *in vitro* and the synthesized RNA is then mechanically inoculated onto plants by rubbing leaves with an abrasive. Extracts from infected plants can then be directly used for the inoculation of very large numbers of plants.

We transiently expressed an scFv in plants, using a tobacco mosaic virus (TMV) based vector (Verch et al., 1999). The scFv coding region was inserted into the viral genome under the control of the strong subgenomic coat protein promoter. This promoter is duplicated and drives the transcription of the viral coat protein gene. The coat protein is essential for long distance, systemic viral infection and the scFv was expressed throughout the entire plant. This approach has been adapted to express the heavy and light chain of a full size antibody from two different viral vectors (Verch et al., 1998). Improvement of this technique may involve the increase of inoculation efficiency by combining the cloned recombinant viral DNAs with particle bombardment or Agrobacterium based techniques, like agroinfection.

3.3 Expression in Stably Transformed Plants

Stable plant transformation is defined as the genomic integration of a transgene. Both the nuclear (Horsch et al., 1985) and plastid (Staub et al., 2000) genomes can be transformed through a variety of transgenesis methods. The principle transformation technologies currently used in plant biotechnology are Agrobacterium mediated gene transfer to dicots, such as tobacco and pea (Horsch et al., 1985), or biolistic delivery of genes to monocots, such as wheat, rice and corn (Christou, 1993). Agrobacterium based transformation does not succeed with certain monocot species but rice can be transformed (Hiei et al., 1997). Ease of transformation has been a critical bottleneck in plant biotechnology and still influences the choice of crops used in molecular farming. The basis of stable transformation is well understood but is uniformly time consuming. Three to nine months are needed in our hands, depending on the plant variety, to generate stably transformed plants available for testing the function and characteristics of the expressed protein. The transient expression assays described earlier now ensure that most errors and technical problems with gene expression can be identified and resolved before making stable transformants (Kapila et al., 1996).

3.4 The Ideal Crop for Production of Recombinant Antibodies

Crops currently being used to produce recombinant antibodies include tobacco, potato, arabidopsis and cereals, such as maize, wheat and rice. Most research into recombinant protein accumulation in plants has used tobacco to evaluate what factors are successful for production. However, tobacco and other plant species produce noxious chemicals such as alkaloids. Cereals do not produce such toxic compounds and can be regarded as an alternative production system for recombinant antibodies. A comparison of accumulation levels in different crop species has been performed with an scFv antibody fragment that binds to the carcinoembryonic antigen (CEA), one of the best characterized tumor-associated antigens. Consequently, anti-CEA antibodies are extensively used for tumor diagnosis and therapy and cost-effective systems for their production are needed. It has been demonstrated that the anti-CEA scFvT84.66

antibody fragment accumulates to the highest levels in leaves and seeds of rice followed by tobacco leaves, pea seeds and wheat (Perrin et al., 2000; Stöger et al., 2000; Stöger et al., submitted). Moreover, accumulation of the scFv in the ER significantly improved protein levels when compared to secretion into the apoplast (Figure 43). An advantage of rAb production in cereal seeds is that rAbs remain active even after prolonged seed storage (Stöger et al., 2000). However, it has to considered that production of pharmaceutical proteins in crops needs intensive care to avoid that these products enter the food and feed chain, which makes non food and feed crops such as tobacco an attractive alternative plant production system.

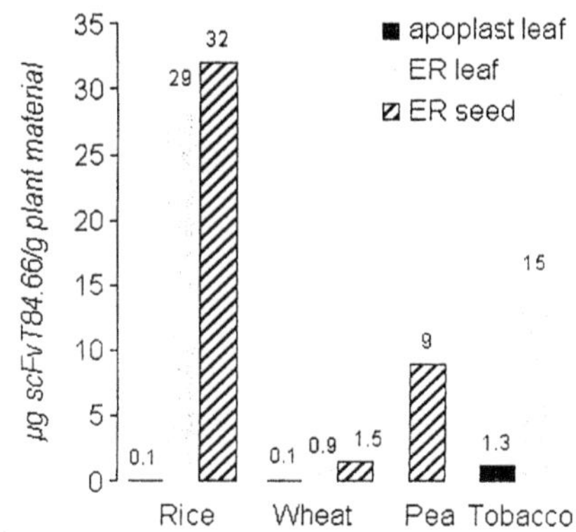

Figure 43. Maximum scFvT84.66 antibody production levels (µg/g plant material) in wheat and rice leaves and seeds, pea seeds and tobacco leaves. Accumulation of functional scFvT84.66 in the ER and apoplast of plant cells was analyzed by ELISA (Perrin et al., 2000; Stoger et al., 2000; Stoger et al., submitted).

The compromise between production costs and profit is likely to be a key in selecting the crops used because most pharmaceuticals will be produced by industry. We predict that these costs will dictate what crop is generally accepted as the best for rAb production.

4. PROTEIN PRODUCTION IN PLANT SUSPENSION CULTURED CELLS

It is beneficial to have the possibility to produce proteins in sealed bioreactors using plant cells. It aids in conforming to regulatory guidelines and permits recombinant protein production where public acceptance of transgenic plant cultivation in fields is poor, as in most of Europe. Plants demonstrate their versatility as a production system here because plant cells can be cultured as single cell suspensions. These cells can be used as an *in vitro* production system for recombinant proteins under controlled, certified conditions (Bisaria and Panda, 1991; Hooker et al., 1990; Nagata et al., 1992). Tobacco BY-2 cells are an example of a fast growing cell line that can be cultivated in shaker flasks or fermenters to produce recombinant proteins after transformation (Fischer et al., 1999). BY-2 cells can be stably and transiently transformed with Agrobacterium and there is no need for any additional construct engineering than is required for plant transformation.

Plant cells can be cultivated using different formats, which include hairy roots (Hilton and Rhodes, 1990), immobilized cells (Archambault, 1991) and free cell suspensions (Kieran et al., 1997). Suspension cell culture is regarded as the most suitable for large-scale biotechnology applications (Schlatmann et al., 1996; Wen, 1995). The diversity of plant species that can be cultivated as suspension cultures is broad, ranging from model systems like arabidopsis (Desikan et al., 1996) to *Catharanthus* (Van Der Heijden et al., 1989), *Taxus* (Seki et al., 1997), and important crop plants like rice (Chen et al., 1994), soybean (Hoehl et al., 1988), alfalfa (Daniell and Edwards, 1995) and tobacco (Nagata et al., 1992).

The number of applications of suspension cells for protein production is relatively small (Kieran et al., 1997), but there have been significant advances in the expression levels of certain proteins in plant cells, such that selected single chain antibodies can reach 25 milligrams per litre when they are secreted from suspension cells (Russell and Fuller, 2000). These advances and the application of functional genomics to suspension cell biology will contribute to increased yields of recombinant proteins and so to increased use of this production system. This is because plant suspension cells can be cultivated in sealed reactors and therefore are totally physically and biologically contained.

Plant cell suspensions can be generated from calli that are transferred to liquid medium in shaker flasks. If a homogenous culture can be maintained, the suspension cells can be cultivated by large scale fermentation. Plant cells can be grown using standard fermentation equipment through batch, fed-batch, perfusion and continuous fermentation (Hooker et al., 1990; Schlatmann et al., 1996; ten Hoopen et al., 1992).

4.1 Suspension Cell Transformation

Plant cell transformation and protein expression follows similar criteria to those described for transgenic plant generation. Transformation can be achieved using Agrobacterium (Horsch et al., 1985; Koncz and Schell, 1986), particle bombardment (Christou, 1993), protoplast electroporation (Lindsey and Jones, 1987) or viral vectors (Porta and Lomonossoff, 1996). BY-2 cells can be directly transformed by co-cultivation with Agrobacterium (An, 1985). Using this approach, gene expression can be detected within 3 days and transgenic cell lines can be isolated within 8 weeks, after selection on a restrictive medium.

In addition to the intracellular compartments where proteins are retained on the basis of trafficking signals, expressed proteins can be secreted from suspension cells into the medium. Protein secretion is dependent on the presence of signal peptides that direct the protein to the endomembrane system and then on the intrinsic permeability of the plant cell wall to protein (Carpita et al., 1979). Selected proteins smaller than ~60 kDa pass through the BY-2 suspension cell wall and are found in the culture medium while the exit of larger proteins is retarded in a generally size dependent manner. Protein secretion has advantages for downstream purification and the use of signals to retain recombinant polypeptides in certain plant cell compartments (ER, vacuole, plastids) (Kusnadi et al., 1997; Moloney and Holbrook, 1997) makes cell disruption a pre-requisite for protein isolation. This creates some complications through the release of phenolic compounds and proteases that can reduce protein yield. While intracellular retention may increase protein yield, our preferred approach is to target proteins for secretion and capture them from the culture supernatant or release them from the apoplast by mild enzymatic cell wall digestion (Fischer et al., 1999).

Plant suspension cells are as versatile as plants as an expression system for recombinant antibodies. We have expressed a range of recombinant antibodies in tobacco suspension cells, including full-size IgGs, Fab fragments, scFvs, bispecific scFvs and rAb-fusion proteins. We have demonstrated that the technology used for antibody production in tobacco cells is applicable to other cell lines derived from food crops, such as rice (Torres et al., 1999).

5. DOWNSTREAM PROCESSING OF RECOMBINANT ANTIBODIES FROM TRANSGENIC PLANT CELLS

Close to all the costs of producing a recombinant protein by molecular farming come from extracting the protein from the plant and purifying it (Evangelista et al., 1998). Thus, optimized, efficient schemes for protein isolation are as, if not more, important than the expression levels of recombinant proteins. Importantly, certified protein isolation and production schemes are essential for the isolation of pharmaceutical proteins if approval through the regulatory authorities is planned (Baker and Harkonen, 1990; Mariani and Tarditi, 1992; Miele, 1997; Murano, 1997). Therefore, the choice of a protein isolation strategy should be carefully thought out and planned from the beginning of molecular farming projects.

Compared to other expression systems, plants require different handling procedures early on in protein isolation. If the recombinant polypeptide is secreted, it can be recovered after removal of cell material by filtration and clarification of the media before purification begins. If the protein is retained within the cells, a simple, efficient method has to be developed to disrupt the cells and release the protein. Mechanical cell disruption devices like bead-mills are efficient but introduce complications related to heat generation, lysis of subcellular organelles, liberation of noxious chemicals (alkaloids, phenolics), and generation of fine cell debris, which can be difficult to remove.

Protocols for the recovery of recombinant antibodies from animal cells, serum and microbial cells are refined and well described. However, there are few reports of purification methods for recombinant antibodies from plant suspension cells, leaves or seeds (Moloney and Holbrook, 1997).

We have investigated the criteria important for the recovery of full size antibodies from suspension cells (Fischer et al., 1999) and developed an affinity purification protocol exploiting protein-A based matrices. We used partial enzymatic lysis to release full size antibodies from the intercellular space of plant cells and this was the superior method for isolation of functional antibodies. As the initial step, Protein-A affinity chromatography gave an efficient removal of contaminants and a 100-fold concentration of the recombinant protein. Gel filtration was applied as a polishing step for the removal of rAb-dimers and for exchange of the rAbs into a suitable storage buffer. The protocol is efficient for the more than 80% of the expressed full size IgG which can be recovered from suspension cultured plant cells (Fischer et al., 1999).

6. APPLICATIONS OF RECOMBINANT ANTIBODIES EXPRESSED IN PLANTS

Molecular farming is a practical approach to the production of therapeutic antibodies however, the applications of antibodies in plant biotechnology are wider than exploiting plant cells solely as a production system. The exquisite affinity of antibodies can be used to increase resistance to pathogens, alter metabolic or developmental pathways, increase the nutritional value of crops and potentially remove environmental pollutants (Conrad and Fiedler, 1998; Fischer et al., 1998; Fischer et al., 1998; Longstaff et al., 1998; Schillberg et al., 2001; Whitelam and Garry, 1996).

6.1 The Safety of Plant Produced Recombinant Antibodies

A question posed about plants as an expression system has been: Are the proteins produced dramatically different from the original proteins and are they safe? The general eukaryotic protein synthesis pathway is so well conserved between organisms that murine proteins are functional when expressed in tobacco plants. Many of the steps of post-translational modifications of secreted proteins are replicated. However, there are some differences in the post-translational glycosylation machinery in plants compared to the native synthetic pathway of the protein. Most recombinant proteins are targeted to the secretory pathway in plants where protein folding and glycosylation take place, and there were concerns that the plant pattern of glycosylation could be highly immunogenic (Ma and Hein, 1995).

Studies of the immune response of mice to a systemically administered recombinant IgG (Guys13) isolated from plants showed that, though there were some differences in the glycan present on the recombinant monoclonal, the antibody and the glycans were not immunogenic in mice (Chargelegue et al., 2000). Clearly, the difference in protein glycosylation in plants compared to animals is unlikely to provoke an immune response or to prevent plant produced proteins being used in humans. Further, plants are being successfully modified to express human galactosyltransferases so that the glycosylation pattern more closely matches the original. Expression of the human β1,4-galactosyltransferase has no gross effects on plant development and when this plant line was crossed with one expressing the heavy and light chain of a mouse antibody, the purified antibody had very similar pattern of *N*-glycosylation to that isolated from hybridoma cell cultivation (Bakker et al., 2001). It seems likely that plants will be developed where expressed proteins have glycosylation patterns that are indistinguishable from the original human version of the protein. Therefore, most proteins can be safely, functionally expressed in plants with the perspective of their use as human therapeutic molecular medicines.

6.2 USING RECOMBINANT PLANT-DERIVED ANTIBODIES IN HUMAN HEALTH CARE

The first clinical trial of plant-based immunotherapy was reported by Planet Biotechnology, Inc. (Mountain View, CA). The novel drug CaroRx™ is based on sIgA antibodies produced in transgenic tobacco plants and is designed to prevent the oral bacterial infection that contributes to dental carries (Ma et al., 1998). Planet Biotechnology has demonstrated that CaroRx™ can effectively eliminate *Streptococcus mutans*, the bacteria that causes tooth decay in humans and clinical trials are underway (Larrick et al., 1998). Planet Biotechnology is also engaged in the design and development of novel sIgA-based therapeutics to treat infectious diseases and toxic conditions affecting oral, respiratory, gastrointestinal, genital and urinary mucosal surfaces and skin.

Agracetus in Middleton, Wisconsin, has created a corn line producing human antibodies at yields of 1.5 kilograms of pharmaceutical-quality protein per acre of corn. A pharmaceutical partner of Agracetus plans to begin injecting cancer patients with doses of up to 250 milligrams of the antibody-based cancer drug purified from corn seeds. The company is also cultivating transgenic soybeans that produce humanized antibodies against herpes simplex virus 2 (HSV-2). These antibodies were shown to be efficient in preventing of vaginal HSV-2 transmission in mice. The *ex vivo* stability and *in vivo* efficacy of the plant and mammalian cell-culture produced antibodies were similar (Zeitlin et al., 1998). Plant-produced antibodies are likely to allow development of an inexpensive method for mucosal immuno-protection against sexually transmitted diseases.

A collaborative research group at Stanford University has developed a technology to produce a tumor-specific vaccine for the treatment of malignancies using a plant virus based transient expression system. The researchers created a modified tobacco mosaic virus vector that encodes the idiotype-specific scFv of the immunoglobulin from the 38C13 mouse B cell lymphoma. Infected *Nicotiana benthamiana* plants secreted high levels of secreted scFv protein to the apoplast. This antibody fragment reacted with an anti-idiotype antibody, suggesting that the plant-produced 38C13 scFv protein is properly folded. Mice vaccinated with the affinity-purified 38C13 scFv generated >10 μg/ml anti-idiotype immunoglobulins. These mice were protected from challenge by a lethal dose of the 38C13 tumor, similar to mice immunized with the native 38C13 IgM-keyhole limpet hemocyanin conjugate vaccine (McCormick et al., 1999). This rapid production system for generating tumor-specific protein vaccines may provide a viable strategy for the treatment of non-Hodgkin's lymphoma. The goal of the therapy is to create antibodies customized for each patient that will recognize unique markers on the surface of the malignant B-cells and target the cells for destruction.

7. PERSPECTIVES

Molecular farming is the application of molecular biotechnology to pharmaceutical production. It is specifically envisioned to serve as a synthesis machinery for the complex 'biotechnological drugs' that are under development. The convention on biological diversity defined biotechnology as 'any technological application that uses biological systems, living organisms, or derivatives thereof, to make or modify products or processes for specific use.' Biotechnological drugs can be further defined as therapeutic products produced through the use of living organisms.

In general the biotechnological drugs of interest for production by molecular farming are polypeptides, such as antibodies and plasma proteins, and there is a long history of the usefulness of these proteins in medicine. For example, immunological assays have unparalleled power to identify pathogens and reveal the nature of disease. We speculate that molecular farming will have its first impact as a production system for recombinant antibodies. More than a quarter of the biotechnological drugs in development and clinical trials are antibodies (see www.phrma.org). Their exquisite specificity and the ability to manipulate the properties of antibodies supports the contention that these polypeptides will remain as important in the near to medium term. Given the expense and difficulty of producing these antibodies conventionally, it seems that molecular farming is one of the most promising options for large-scale antibody production. The amounts of recombinant protein that are required to manage disease is difficult to estimate but is certainly on the kilogram scale for most antibodies and the tonne scale for plasma proteins (human serum albumin), when one considers the number of patients that may benefit. These demands - for large scale protein production - will be met through molecular farming using eukaryotic cells in the near future. This means that fields of transgenic plants in the field, transgenic animal herds or suspension cell culture in sealed fermenters will begin to complement pharmaceutical synthetic chemistry in the near future. The advantage of exploiting plants is that production is inexpensive and safe from the risks of pathogen contamination.

The twentieth century can be viewed as the hundred years when synthetic chemistry greatly benefitted medicine and offered many therapeutic chemicals for the management of disease. While synthetic chemistry based medicine will have the greatest share of the therapeutic market for the foreseeable future, molecular medicines will emerge as a large niche market in the near future. As molecular farming is applied to produce these molecular medicines on a larger scale, we will see molecular medicines emerge as a powerful tool for treating human disease, and one that is synthesized in plants.

REFERENCES

An G, 1985. High efficiency transformation of cultured tobacco cells. Plant Physiol, 79: 568-570

Archambault J, 1991. Large-scale (20 l) culture of surface-immobilized *Catharanthus roseus* cells. Enzyme Microb Technol, 13:882-892

Artsaenko O, Kettig B, Fiedler U, Conrad U and Düring K, 1998. Potato tubers as a biofactory for recombinant antibodies. Mol Breed, 4:313-319

Artsaenko O, Peisker M, zur Nieden U, Fiedler U, Weiler EW, Müntz K and Conrad U, 1995. Expression of a single-chain Fv antibody against abscisic acid creates a wilty phenotype in transgenic tobacco. Plant J, 8:745-750

Baker D and Harkonen W, 1990. Regulatory agency concerns in the manufacturing and testing of monoclonal antibodies for therapeutic use. Targeted Diagn Ther, 3, 75-98

Bakker H, Bardor M, Molthoff J, Gormord V, Elbers I, Stevens L, Jordi W, Lommen A, Faye L, Lerouge P and Bosch D, 2001. Galactose-extended glycans of antibodies produced by transgenic plants. Proc Natl Acad Sci USA, 98:2899-2904

Baum TJ, Hiatt A, Parrott WA, Pratt LH and Hussey, RS, 1996. Expression in tobacco of a functional monoclonal antibody specific to stylet secretions of the root-knot nematode. Mol Plant-Microbe Interact, 9:382-387

BisariaV and Panda A, 1991. Large-scale plant cell culture: methods, applicatons and products. Curr Opin Biotechnol, 2:370-374

Burton D, 1995. Phage display. Immunotechnology, 1:87-94

Carpita N, Sabularse D, Montezinos D and Delmer DP, 1979. Determination of the pore size of cell walls of living plant cells. Science, 205:1144-1147

Chargelegue D, Vine N, van Dolleweerd C, PMW, D and Ma J, 2000. A murine monoclonal antibody produced in transgenic plants with plant-specific glycans is not immunogenic in mice. Transgenic Res, 9:187-194

Chen MH, Liu LF, Chen YR, Wu HK and Yu SM, 1994. Expression of alpha-amylases, carbohydrate metabolism and autophagy in cultured rice cells is coordinately regulated by sugar nutrient. Plant J, 6:625-636

Christou P, 1993. Particle gun-mediated transformation. Curr Opin Biotechnol, 4:135-141

Christou P, 1995. Strategies for variety-independent genetic transformation of important cereals, legumes and woody species utilizing particle bombardment. Euphytica, 85:13-27

Conrad U and Fiedler, U 1998. Compartment-specific accumulation of recombinant immunoglobulins in plant cells: an essential tool for antibody production and immunomodulation of physiological functions and pathogen activity. Plant Mol Biol, 38:101-109

Conrad U, Fiedler U, Artsaenko O and Phillips J, 1998. High level and stable accumulation of single chain Fv antibodies in plant storage organs. J Plant Physiol, 152:708-711

Daniell T and Edwards R, 1995. Changes in protein methylation associated with the elicitation response in cell cultures of alfalfa *(Medicago sativa)* L. FEBS Lett, 360:57-61

De Jaeger G, Buys E, Eeckhout D, De Wilde C, Jacobs A, Kapila J, Angenon G, Van Montagu M, Gerats T and Depicker A, 1998. High level accumulation of single-chain variable fragments in the cytosol of transgenic *Petunia hybrida*. Eur J Biochem, 259:1-10

De Neve M, De Loose M, Jacobs A, Van Houdt H, Kaluza B, Weidle U, Van Montagu M and Depicker A, 1993. Assembly of an antibody and its derived antibody fragment in *Nicotiana* and *Arabidopsis*. Transgenic Res, 2:227-237

De Wilde C, De Neve M, De Rycke R, Bruyns AM, De Jaeger G, Van Montagu M, Depicker A and Engler G, 1996. Intact antigen-binding MAK33 antibody and Fab fragment accumulate in intercellular spaces of *Arabidopsis thaliana*. Plant Science, 114:233-241

Desikan R, Hancock JT, Neill SJ, Coffey MJ and Jones OT, 1996. Elicitor-induced generation of active oxygen in suspension cultures of *Arabidopsis thaliana*. Biochem Soc Trans, 24:199S

Düring K, Hippe S, Kreuzaler F and Schell J, 1990. Synthesis and self-assembly of a functional monoclonal antibody in transgenic *Nicotiana tabacum*. Plant Mol Biol, 15:281-293

Evangelista RL, Kusnadi AR, Howard JA and Nikolov ZL, 1998. Process and economic evaluation of the extraction and purification of recombinant beta-glucuronidase from transgenic corn. Biotechnol Prog, 14:607-614

Fecker LF, Kaufmann A, Commandeur U, Commandeur J, Koenig R and Burgermeister W, 1996. Expression of single-chain antibody fragments (scFv) specific for beet necrotic yellow vein virus coat protein or 25kDa protein in *Escherichia coli* and *Nicotiana benthamiana*. Plant Mol Biol, 32:979-986

Fiedler U and Conrad U, 1995. High-level production and long-term storage of engineered antibodies in transgenic tobacco seeds. Bio/Technology, 13:1090-1093

Fiedler U, Philips J Artsaenko, O and Conrad U, 1997. Optimisation of scFv antibody production in transgenic plants. Immunotechnology, 3:205-216

Firek S, Draper J, Owen MR, Gandecha A, Cockburn, B and Whitelam GC, 1993. Secretion of a functional single-chain Fv protein in transgenic tobacco plants and cell suspension cultures [published erratum appears in Plant Mol Biol 1994 Mar;24(5):833]. Plant Mol Biol, 23:861-870

Fischer R, Drossard J, Hellwig S, Emans N and Schillberg S, 1998. Transgenic plants as bioreactors for the expression of recombinant antibodies. In *Proceedings of the Xth international congress of immunology*, Ed. Talwar N. Monduzzi Editore Publishers, Bologna, pp 307-313

Fischer R, Drossard J, Liao YC and Schillberg S, 1998. Characterisation and applications of plant-derived recombinant antibodies. In *Methods in biotechnology Vol 3: Recombinant proteins in plants: Production and isolation of clinically useful compounds*. Eds. Cunningham C and Porter AJR. Humana Press, Totowa, NJ, pp 129-142

Fischer R, Emans N, Schuster F, Hellwig S and Drossard J, 1999. Toward molecular farming in the future: Using plant cell suspension cultures as bioreactors. Biotechnol Appl Biochem, 30:113-116

Fischer R, Liao Y-C and Drossard J, 1999. Affinity-purification of a TMV-specific recombinant full-size antibody from a transgenic tobacco suspension culture. J Immunol Meth, 226:1-10

Fischer R, Schumann D, Zimmermann S, Drossard J, Sack M and Schillberg S, 1999. Expression and characterization of bispecific single chain Fv fragments produced in transgenic plants. Eur J Biochem, 262:810-816

Fraley RT, Rogers, SG Horsch RB, Sanders PR, Flick JS, Adams SP, Bittner ML, Brand LA, Fink CL, Fry JS, Galluppi GR, Goldberg SB, Hoffmann NL and Woo SC, 1983. Expression of bacterial genes in plant cells. Proc Natl Acad Sci USA, 80:4803-4807

Hiatt A, 1990. Antibodies produced in plants. Nature, 344:469-470

Hiatt A, Cafferkey R and Bowdish K, 1989. Production of antibodies in transgenic plants. Nature, 342:76-78

Hiatt A and Ma JK, 1993. Characterization and applications of antibodies produced in plants. Int Rev Immunol, 10:139-152

Hiatt A and Mostov K, 1992. Assembly of multimeric proteins in plant cells: characteristics and uses of plant-derived antibodies. In *Transgenic plants: fundamentals and applications*, Ed. Hiatt A. Marcel Dekker Inc, New York, pp 221-236

Hiatt A, Tang Y, Weiser W and Hein MB, 1992. Assembly of antibodies and mutagenized variants in transgenic plants and plant cell cultures. Genet Eng, 14:49-64

Hiatt AC, 1991. Production of a monoclonal antibody in plants. Transplant Proc, 23:147-151

Hilton MG and Rhodes MJ, 1990. Growth and hyoscyamine production of 'hairy root' cultures of *Datura stramonium* in a modified stirred tank reactor. Appl Microbiol Biotechnol, 33:132-138

Hoehl U, Upmeier B and Barz W, 1988. Growth and nicotinate biotransformation in batch cultured and airlift fermenter grown soybean cell suspension cultures. Appl Microbiol Biotechnol, 28:319-323

Hogue RS, Lee JM and An G, 1990. Production of a foreign protein product with genetically modified plant cells. Enzyme Microb Technol, 12:533-538

Hood E, Witcher D, Maddock S, Meyer T, Baszczynski C, Bailey M, Flynn P, Register J, Marshall L, Bond D, Kulisek E, Kusnadi A, Evangelista R, Nikolov Z, Wooge C, Mehigh R, Hernan R, Kappel W, Ritland D, Li C-P and Howard J, 1997. Commercial production of avidin from transgenic maize: characterization of transformant, production, processing, extraction and purification. Mol Breed, 3:291-306

Hood EE, Kusnadi A, Nikolov Z and Howard JA, 1999. Molecular farming of industrial proteins from transgenic maize. Adv Exp Med Biol, 464:127-147

Hooker BS, Lee, JM and An G, 1990. Cultivation of plant cells in a stirred tank reactor. Biotechnol Bioeng, 35:296-304

Horsch R, Fry JE, Hoffman N, Eicholtz D, Rogers S and Fraley R, 1985. A simple and general method for transferring genes into plants. Science, 227:1229-1231

Kapila J, De Rycke R, van Montagu M and Angenon G, 1996. An *Agrobacterium* mediated transient gene expression system for intact leaves. Plant Sci, 122:101-108

Kieran PM, MacLoughlin PF and Malone DM, 1997. Plant cell suspension cultures: some engineering considerations. J Biotechnol, 59:39-52

Koehler G and Milstein C, 1975. Continuous cultures of fused cells secreting antibody of predefined specificity. Nature, 256:495-497

Koncz C and Schell J, 1986. The promoter of T_L-DNA gene 5 controls the tissue specific expression of chimeric genes carried by a novel type of *Agrobacterium* binary vector. Mol Gen Genet, 204:383-396

Kusnadi AR, Nikolov ZL and Howard JA, 1997. Production of recombinant proteins in transgenic plants: Practical considerations. Biotechnol Bioeng, 56:473-484

Larrick JW, Yu L, Chen J, Jaiswal S and Wycoff K, 1998. Production of antibodies in transgenic plants. Res Immunol, 149:603-608

Leite A, Kemper E, da Silva M, Luchessi A, Siloto R, Bonaccorsi E, El-Dorry H and Arruda P 2000. Expression of correctly processed human growth hormone in seeds of transgenic tobacco plants. Mol Breed, 6:47-53

Lindsey K and Jones MGK, 1987. Transient gene expression in electroporated protoplasts and intact cells of sugar beet. Plant Mol Biol, 10:43-52

Longstaff M, Newell CA, Boonstra B, Strachan G, Learmonth D, Harris WJ, Porter AJ and Hamilton W D, 1998. Expression and characterisation of single-chain antibody fragments produced in transgenic plants against the organic herbicides atrazine and paraquat. Biochim Biophys Acta, 1381:147-160

Ma J and Hein M, 1995. Plant antibodies for immunotherapy. Plant Physiol, 109:341-346

Ma J K, Hiatt A, Hein M, Vine N D, Wang F, Stabila P, van Dolleweerd C, Mostov K and Lehner T, 1995. Generation and assembly of secretory antibodies in plants. Science, 268:716-719

Ma JK, Hikmat BY, Wycoff K, Vine ND, Chargelegue D, Yu L, Hein MB and Lehner T, 1998. Characterization of a recombinant plant monoclonal secretory antibody and preventive immunotherapy in humans. Nat Med, 4:601-606

Ma JKC, Lehner T, Stabila P, Fux CI and Hiatt A, 1994. Assembly of monoclonal antibodies with IgG1 and IgA heavy chain domains in transgenic tobacco plants. Eur J Immunol, 24:131-138

Magnuson NS, Linzmaier PM, Reeves R, An G, HayGlass K and Lee JM, 1998. Secretion of biologically active human interleukin-2 and interleukin-4 from genetically modified tobacco cells in suspension culture. Protein Expr Purif, 13:45-52

Mariani M and Tarditi L, 1992. Validating the preparation of clinical monoclonal antibodies. Biotechnology, 10:394-396

Mason HS and Arntzen C J, 1995. Transgenic plants as vaccine production systems. Trends Biotechnol, 13:388-392

McCormick AA, Kumagai MH, Hanley K, Turpen TH, Hakim I, Grill LK, Tusé D, Levy S and Levy R, 1999. Rapid production of specific vaccines for lymphoma by expression of the tumor-derived single-chain Fv epitopes in tobacco plants. Proc Natl Acad Sci USA, 96:703-708

Miele L, 1997. Plants as bioreactors for pharmaceuticals: regulatory considerations. Trends Biotechnol, 15:45-50

Moloney MM and Holbrook LA, 1997. Subcellular targeting and purification of recombinant proteins in plant production systems. Biotechnol Genet Eng Rev, 14:321-336

Murano G, 1997. FDA perspective on specifications for biotechnology products--from IND to PLA. Dev Biol Stand, 91:3-13

Nagata T, Nemoto Y and Seiichiro H, 1992. Tobacco BY-2 cell line as the "HeLa" cell in the cell biology of higher plants. Int Rev Cytol, 132:1-30

Owen M, Gandecha A, Cockburn B and Whitelam G, 1992. Synthesis of a functional anti-phytochrome single-chain Fv protein in transgenic tobacco. Biotechnology (N Y), 10:790-794

Phillips J, Artsaencko O, Fiedler U, Horstmann C, Mock HP, Müntz K and Conrad U, 1997. Seed-specific immunomodulation of abscisic acid activity induces a developmental switch. EMBO, 16:4489-4496

Porta C and Lomonossoff GP, 1996. Use of viral replicons for the expression of genes in plants. Mol Biotech, 5:209-221

Russell D and Fuller J, 2000. Method for producing antibodies in plant cells (USA: Monsanto Company, St. Louis, Mo)

Schillberg S, Zimmermann S, Voss A and Fischer R, 1999. Apoplastic and cytosolic expression of full-size antibodies and antibody fragments in *Nicotiana tabacum*. Transgenic Res, 8:255-263

Schillberg S, Zimmermann S, Zhang, MY and Fischer R, 2001. Antibody-based resistance to plant pathogens. Transgenic Res, 10:1-12

Schlatmann JE, ten Hoopen HJ G and Heijnen JJ, 1996. Large-scale production of secondary metabolites by plant cell cultures. In *Plant cell culture secondary metabolism: Toward industrial application*. Eds. DiCosmo F and Misawa M. CRC Press, Boca Raton , pp. 25-41

Scholthof H, Scholthof K and Jackson A, 1996. Plant virus gene vectors for transient expression of foreign proteins in plants. Annu Rev Phytopathol, 34:299-323

Schouten A, Roosien J, de Boer JM, Wilmink A, Rosso MN, Bosch D, Stiekema WJ, Gommers FJ, Bakker J and Schots A, 1997. Improving scFv antibody expression levels in the plant cytosol. FEBS Lett, 415:235-241

Schouten A, Roosien J, van Engelen FA, de Jong GAM, Borst-Vrenssen AWM, Zilverentant JF, Bosch D, Stiekema WJ, Gommers FJ, Schots A and Bakker J, 1996. The C-terminal KDEL sequence increases the expression level of a single-chain antibody designed to be targeted to both cytosol and the secretory pathway in transgenic tobacco. Plant Mol Biol, 30:781-793

Seki M, Ohzora C, Takeda M and Furusaki S, 1997. Taxol (Paclitaxel) production using free and immobilized cells of taxus cuspidata. Biotechnol Bioeng, 53:214-219

Sijmons PC, Dekker BMM, Schrammeijer B, Verwoerd TC, van den Elzen PJM and Hoekema A, 1990. Production of correctly processed human serum albumin in transgenic plants. Bio/Technology, 8:217-221

Spiegel H, Schillberg S, Sack M, Holzem A, Nähring J, Monecke M, Liao Y-C and Fischer R, 1999. Expression of antibody fusion proteins in the cytoplasm and ER of plant cells. Plant Sci, 149:63-71

Staub J, Garcia B, Graves J, Hajdukiewicz P, Hunter P, Nehra N, Paradkar V, Schlittler M, Carroll J, Spatola L, Ward D, Ye G and Russell D, 2000. High-yield production of a human therapeutic protein in tobacco chloroplasts. Nat Biotechnol, 18:333-338

Tavladoraki P, Benvenuto E, Trinca S, De Martinis D, Cattaneo A and Galeffi P, 1993. Transgenic plants expressing a functional single-chain Fv antibody are specifically protected from virus attack. Nature, 366:469-472

ten Hoopen HJG, van Gulik WM and Heijnen JJ, 1992. Continuous culture of suspended plant cells. In Vitro Cell Dev Biol, 28:115-120

Terashima M, Murai Y, Kawamura M, Nakanishi S, Stoltz T, Chen L, Drohan W, Rodriguez R L and Katoh S, 1999. Production of functional human alpha 1-antitrypsin by plant cell culture. Appl Microbiol Biotechnol, 52:516-523

Timmermans MCP, Prem Das O and Messing J, 1994. Geminiviruses and their uses as extrachromosomal replicons. Annu Rev Plant Physiol Plant Mol Biol, 45:79-112

Torres E, Vaquero C, Nicholson L, Sack M, Stöger E, Drossard J, Christou P and Fischer R, 1999. Rice cell culture as an alternative production system for functional diagnostic and therapeutic antibodies. Transgenic Res, 8:441-449

Van Der Heijden R, Verpoorte R and Ten Hoopen HJG, 1989. Cell and tissue cultures of *Catharanthus-Roseus* L., a literature survey. Plant Cell Tissue Organ Cult, 18:231-280

van Engelen FA, Schouten A, Molthoff JW, Roosien J, Salinas J, Dirkse WG, Schots A, Bakker J, Gommers FJ and Jongsma MA, 1994. Coordinate expression of antibody subunit genes yields high levels of functional antibodies in roots of transgenic tobacco. Plant Mol Biol, 26:1701-1710

Vaquero C, Sack M, Chandler J, Drossard J, Schuster F, Schillberg S and Fischer R, 1999. Transient expression of a tumor-specific single chain fragment and a chimeric antibody in tobacco leaves. Proc Natl Acad Sci, USA, 96:11128-11133

Verch T, Lewandowski D and Fischer R, 1999. Expression of a single chain antibody in plants using a viral vector. *submitted*

Verch T, Yusibov V and Koprowski H, 1998. Expression and assembly of a full-length monoclonal antibody in plants using a plant virus vector. J Immunol Methods, 220:69-75

Verwoerd TC, van Paridon PA, van Ooyen AJJ, van Lent JWM, Hoekema A and Pen J, 1995. Stable accumulation of *Aspergillus niger* phytase in transgenic tobacco leaves. Plant Physiol, 109:1199-1205

Voss A, Niersbach M, Hain R, Hirsch H, Liao Y, Kreuzaler F and Fischer R, 1995. Reduced virus infectivity in *N. tabacum* secreting a TMV-specific full size antibody. Mol Breed ,1:39-50

Wen W S, 1995. Bioprocessing technology for plant cell suspension cultures. Appl Biochem Biotechnol, 50, 189-216

Whitelam G and Garry W, 1996. Antibody expression in transgenic plants. Trends Plant Sci, 1, 268-271

Willmitzer L, Depicker A, Dhaese P, De Greve H, Hernalsteens JP, Holsters M, Leemans J, Otten L, Schroder J, Schroder G, Zambryski P, van Montagu M and Schell J, 1983. The use of Ti-plasmids as plant-directed gene vectors. Folia Biol, 29:106-114

Winter G, Griffiths AD, Hawkins RE and Hoogenboom HR, 1994. Making antibodies by phage display technology. Annual Rev Immunol, 12:433-455

Winter G and Milstein C, 1991. Man-made antibodies. Nature, 349:293-299

Zeitlin L, Olmsted SS, Moench TR, Co MS, Martinell BJ, Paradkar VM, Russell DR, Queen C, Cone RA and Whaley KJ, 1998. A humanized monoclonal antibody produced in transgenic plants for immunoprotection of the vagina against genital herpes. Nat Biotechnol, 16:1361-1364

Ziegler M, Thomas S and Danna K, 2000. Accumulation of a thermostable endo-1,4-b-D-glucanase in the apoplast of *Arabidopsis thaliana* leaves. Mol Breed, 6:37-46

Zimmermann S, Schillberg S, Liao YC and Fischer R, 1998. Intracellular expression of TMV-specific single-chain Fv fragments leads to improved virus resistance in *Nicotiana tabacum*. Mol Breed, 4: 369-379

Chapter 13

IMMUNOTHERAPEUTIC POTENTIAL OF ANTIBODIES PRODUCED IN CHICKEN EGGS

Y, Mine and J. Kovacs-Nolan
Department of Food Science, University of Guelph, Guelph, Ontario, N1G 2W1, Canada

Keywords: chicken, egg yolk, IgY, passive immunization, prophylaxis, rotavirus, *E. coli*, *Salmonella*, infectious gastrointestinal diseases

Abstract Oral administration of specific antibodies is an attractive approach to establish passive immunity against gastrointestinal pathogens in human and animals. Recently, laying hens have attracted considerable attention as an alternative source of antibody for peroral immunotherapy. After immunization of the chicken, the specific antibodies (called IgY) are transported to the egg yolk from which the IgY then can be separated without sacrificing chickens. Egg yolk contains 70-250 mg of IgY per yolk, suggesting that more than 30g of IgY can be obtained per year. Eggs are also normal dietary components and mass production systems in chicken farms are already established. Thus, IgY is considered to be more hygienic and economical source of antibodies than antibodies from mammalian sera. Oral administration of IgY has proven to be successful for treatment of a variety of gastrointestinal infections. The passive immunization of IgY has been applied to the prevention of human rotavirus infections. A successful trial of passive immunization of rainbow trout against infection with *Y. ruckeri* has been achieved with specific IgY. Enterotoxigenic *Escherichia coli* (ETEC) is the major cause of diarrhea and death in neonatal calves and piglets. Anti-ETEC IgY was found to be effective in controlling diarrhea of new-born calves and piglets. *Salmonella* spp are also often pathogenic to human and animals. Anti-Salmonella IgY can provide a prophylactic function to prevent Salmonellosis in animals. Other applications for immunotherapeutic potential of IgY also have been documented in this chapter. IgY technology may provide great opportunities for designing prophylactic strategies against infectious gastrointestinal diseases in humans and animals.

L. Erickson et al. (eds.),
Molecular Farming of Plants and Animals for Human and Veterinary Medicine, 287–317.

1. INTRODUCTION

Antibodies are immunoglobulin molecules of the immune system developed by higher organisms to combat the invasion of foreign substances (antigens). Five classes of immunoglobulins (IgG, IgA, IgM, IgD and IgE) are known which are distinguishable in structure and immunological function. The major immunoglobulin is the IgG class which makes up about 75% of immunoglobulin in blood (Janeway and Travers, 1996). Due to their high specificity, antibodies have been investigated for: 1) targeting agents in cancer diagnosis and therapy; 2) inactivating toxic substances including drugs; and 3) passive immunotherapy for neoplastic or infectious disease (Reilly et al., 1997). Traditionally, commercially available polyclonal antibodies have usually been produced in mammals such as mouse, rat, rabbit, sheep, goat and horse. Recently, there has been increasing interest in the oral administration of antibodies for localized treatment of infectious or other conditions in the gastrointestinal tract (Reilly et al., 1997). The antibodies are generally obtained from sera after immunization of these animals. These antibodies can not be prepared on an industrial scale because of the difficulty in obtaining large amounts of blood, and also from an animal welfare point of view. Since hybridoma technology has been widely adopted as a method of choice for the preparation of monoclonal antibodies, various hybridomas have been cloned and grown in large quantity for indefinite periods of time. However, successful commercialization of any of the therapeutic monoclonal antibodies is still not feasible due to its expensive cost (Wang and Imanaka, 1995). Some of the real or potential immuno therapeutic applications of antibodies for animal or human infectious diseases will require kilogram quantities of highly purified antibody. Oral administration of specific antibodies is an attractive approach to establish immunity against gastrointestinal pathogens in humans and animals. The increase of antibiotic resistant bacterial infections emphasizes the need to find alternatives to antibiotics. Immunotherapy can also be used against pathogens that are difficult to treat with antibiotics such as viral disease. Several approaches have been initiated including use of bovine colostrum and colostral antibodies or monoclonal antibodies (Crabb, 1998; Yolken et al., 1990; Chmel, 1990). However, their quantity has limitations. Therefore, cost-efficient methods for producing large quantities of antibodies are desired.

Recently, chickens have attracted considerable attention as an alternative source of antibody for prophylaxis and therapy of infectious intestinal diseases (Carlander et al., 2000; Carlander et al., 1999; Sim and Nakai, 1994; Sim et al., 1999; Hatta et al., 1997; Larrson and Sjoquist, 1990; Losch et al., 1986; Kuklmann et al., 1988). The serum immunoglobulin of hens, referred to as immunoglobulin (Ig) Y, is deposited in the egg yolk, in an effort to give acquired immunity to the developing embryo (Janson et al., 1995). The use of chicken immunoglobulins provides several advantages over other methods of antibody production. In contrast to mammalian serum, egg yolk contains only the single class of antibody, IgY, which can be easily purified from the yolk by simple precipitation techniques. The phylogenetic distance between chickens and mammals renders possible the production of antibodies, in chickens, against highly conserved mammalian proteins, and much less antigen is required to produce an efficient immune response (Gassmann et al., 1990). As well, the method of producing antibodies in hens is much less invasive and therefore less stressful on the animal (Schade et al., 1991) and sustained high titres in chickens reduce the need for frequent injections (Gassmann, et al., 1990). Hens also provide a more hygienic, cost efficient, convenient and plentiful source of specific antibodies, as compared to the traditional method of obtaining immunoglobulins from mammalian serum (Losch et al., 1986). Gottstein and Hemmeler (Gottstein and Hemmeler, 1985) estimated that the productivity of antibodies in hens is nearly 18 times greater than that by rabbits based on the weight of antibody produced per head. Since passive immunization techniques generally require the continuous administration of antibodies (Hammarstrom, 1999), such an efficient method of generating large volumes of neutralizing antibodies would be desirable. The potential application of orally administered IgY for the prevention of gastrointestinal infections caused by enteric pathogens such as rotavirus, *Escherichia coli*, *Yersinia*, or *Salmonella* has been studied by several research groups.

This chapter focuses on the issue of the potential immunotherapy with chicken antibody for the prevention and treatment of enteric infections and speculates on the future possibilities for IgY technology.

2. PHYSIOLOGY OF CHICKEN EGG FORMATION

Under modern husbandry conditions, a chicken can lay an average of 250-280 eggs per year. The egg is the largest biological cell which originates from one cell division and is composed of various important chemical substances for the next generation of birds. An egg is composed of three main parts, shell, albumen and yolk, The yolk is surrounded by an albumen layer and compartmentalized by an eggshell. The formation of an egg involves the conversion of the feed into egg constituents through a number of intricate and highly coordinated steps as a storehouse of nutrients. The hen normally starts laying from 16-26 weeks of age. The reproductive system of the hen is shown in Figure 44 consisting of the ovary and oviduct (Romanoff and Romanoff, 1949). The ovary, which is the site of assembly of the yolk, is a small organ. When the chicken becomes mature (about 150 days old), the ovary has grown to about 7g, and rapidly increases to about 40g (around 170 days old) (Epple and Steson, 1980). A mature ovary contains many oocytes and at least 600-700 of the oocytes will become mature yolk. Each oocyte becomes a follicle after being covered with a granular layer. The follicles in the ovary are surrounded by the hens veins (Burley and Vadehra, 1989). Yolk constituents are synthesized in the liver and they are transported to the follicular walls by the blood. The follicle proceeds a rapid development during which most of the yolk is deposited 6-10 days prior to ovulation, when the yolk has accumulated sufficiently. The follicle in the ovary is ovulated into the oviduct where the yolk is enveloped in albumen and the shell. It takes 24-27 hrs for this development. In laying hens, the oviduct is 40-80cm long with an average weight of 40g and consists of five regions, infundibulum, magnum, isthmus, uterus and vagina (Burley and Vadehra, 1989). The infundibulum is the top portion of the oviduct with a broad funnel shaped anterior end (8-9 cm) and a narrow posterior end to receive the ovulated follicles. The ovulated follicle is held for 15-30 cm and the yolk probably acquires the outer layer of the vitelline membranes and the chalaza layer of the albumen (Burley and Vadehra, 1989). The albumen-secreting region is the largest part of the oviduct, about 30 cm long and the follicle is held here for 2-3 hrs while the egg albumen is secreted to cover the yolk. The isthmus is about 11 cm long and the shell membranes are synthesized here. The egg yolk enveloped with albumen is immediately wrapped by the membrane. The complete synthetic process of the shell formation takes place in the uterus (shell gland) for about 20 hrs, while calcium from the blood is deposited in the shell by assembling a crystalline-like calcium structure on the shell membranes. However, its mechanism still is not well understood. The vagina is the last portion of the oviduct and the egg of the vagina connects with the cloaca. It takes only 5 minutes to pass through this portion.

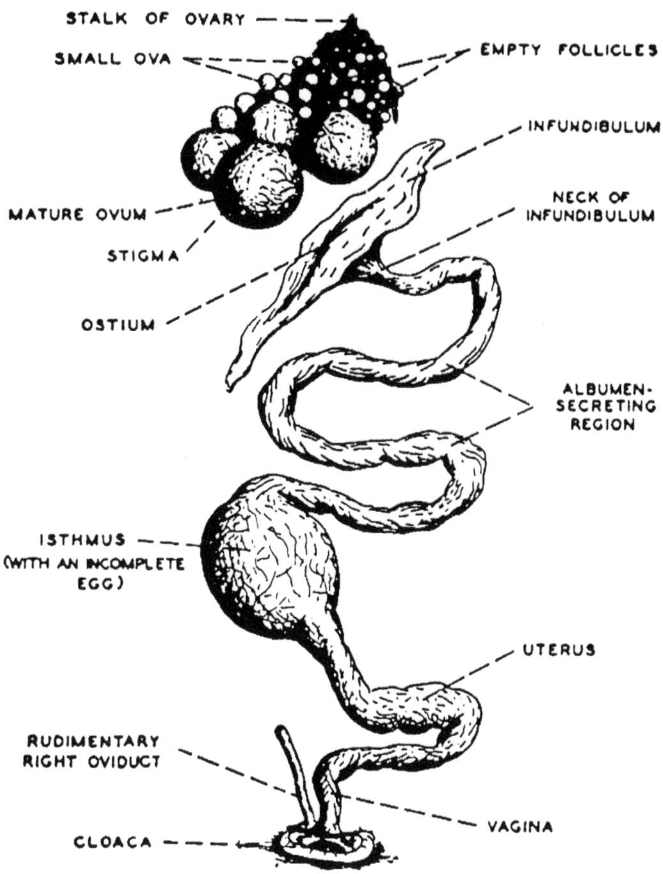

Figure 44. The reproductive system of the hen, ovary and oviduct. (From Romanoff and Romanoff (1949), reproduced by permission of John Wiley & Sons, Inc., New York).

3. CHICKEN EGGS AS AN ALTERNATIVE SOURCE OF ANTIBODIES

Antibodies are immunoglobulin molecules of the immune system developed by higher organisms to combat the invasion of foreign substances (antigens). Each antibody has specific activity for an antigen. Chickens have a unique immune system. The protection of the chickens against pathogens occurs through transmission of antibodies from the mother via the egg because of the immune incompetence of newly hatched chicks. The first report of antibodies present in the yolk of an egg was made by Klemperer (Klemperer, 1983). Currently, it is well known that the IgG found in the blood sera of chickens is deposited to the yolk of eggs laid by the chicken to give acquired immunity. In 1969, Leslie and Clem proposed that the antibody in egg yolk should be called IgY rather than IgG (Leslie and Clem, 1969). In egg, the egg white contains IgA and IgM at concentrations of about 0.15 mg and 0.7 mg/ml, respectively. On the other hand, yolk contains a considerably high concentration of IgG (IgY) at 25 mg/ml (Rose et al., 1974). Chicken IgA and IgM are secreted together with other proteins of egg white at the oviduct, while the serum IgG is specifically transferred through the membrane into the yolk during its maturation. A receptor specific to IgG translocation is known to exist on the surface of the yolk membrane (Loeken and Roth, 1983).

Traditionally, the species chosen for antibody production has been mammals, most frequently rabbit, sheep, goat and horse. The antibodies are generally obtained from sera after immunization. Recently, chicken egg has attracted considerable interests as a source of antibodies, especially for practical application of antibodies for immunotherapeutic approaches (Carlander et al., 2000; Carlander et al., 1999; Kuklmann et al., 1988). There are several advantages to using chicken antibodies instead of other animals: 1) With the conventional method, animals are inevitably sacrificed for the specific IgG in their circulating blood, whereas only the eggs from the immunized hens are used; 2) Egg yolk contains only IgY, making the isolation of IgY from the yolk easier than IgG from animal blood sera; 3) Large scale feeding of hens for egg production is already being carried out, which provides a convenient and practical means for collecting the source of a specific antibody; 4) The immunization of hens has been in practice for a long while and thus is much more systematic than in other animals; 5) Egg yolk as a source of IgY is more hygienic than mammal's sera from which IgG is separated and finally 6) Due to phylogenetic differences, the hen has the ability of producing specific antibodies whose formation is difficult or impossible in mammals (Hatta et al., 1997).

3.1 STRUCTURAL CHARACTERISTICS OF CHICKEN IGY

Although IgY is classified as an IgG class immunoglobulin, the structure of IgY is considerably different from mammalian IgG. IgY has two heavy (H) and two light (L) chains and has a molecular mass of 180 kDa, larger than that of rabbit IgG (-159 kDa). IgY possess a larger molecular weight H chain (68 kDa) versus the mammalian one (50 kDa). The H chain of IgG consists of four domains: the variable domain (V_H) and three constant domains (Cγ1, Cγ2 and Cγ3). The Cγ1 domain is separated from Cγ2 by a hinge region, which gives considerable flexibility to the Fab fragments. In contrast, the H chain of IgY does not have a hinge region, and possesses four constant domains (Cv1-Cv4) in addition to the variable domain (Figure 45). Sequence comparisons between IgG and IgY have shown that the Cv2 and Cv3 domains of IgG are closely related to the Cv3 and Cv4 domains, respectively of IgY, while the equivalent of the Cv2 domain is absent in the IgG chain, having been replaced by the hinge region (Warr et al., 1995). The isoelectric point of IgY is lower than that of IgG (Polson et al., 1980). IgY does not associate with mammalian complements and the binding of IgY with Fc receptor on cell surface is less as compared to IgG (Gardner and Kaya, 1982). IgY does not bind to *Staphylococcus* protein A (Kronvall et al., 1974) or rheumatoid factor in blood unlike IgG (Larsson and Sjoquist, 1988). Recently, cloning and sequencing of genes encoding the H and L chains of IgY were done and the primary structure of these polypeptides has been determined (Reynaud et al., 1983; Parvari et al., 1988). Sequence data support a phylogenetic tree in which IgY gave rise to both IgG and IgE. Sequence comparisons of the H, L and E chains show that they are more closely related to each other than to either μ (IgM) or α (IgA). In fact, IgY is more closely related to IgE than it is to IgG. The close similarity of IgY to IgE is apparent from the number and organization of the intradomain and intrachain disulfide bonds. Unlike rabbit IgG, IgY has two additional Cys residues, Cys331 and Cys338, in the Cv2-Cv3 juncton. These Cys residues are likely to participate in the inter-v chain disulfide linkages. (Warr et al., 1995).

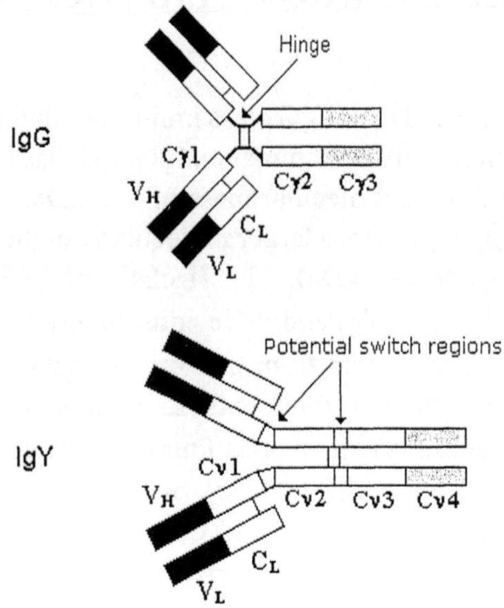

Figure 45. Structure of IgG and IgY (Adapted from Warr et al., 1995).

Both mammalian and chicken IgG are known to contain Asn/linked oligosaccharides. However, the structure of oligosaccharides in chicken IgY differ from those of any mammalian IgG. Chicken IgY contains unusual monoglucosylated oligomannose-type oligosaccharides with $Glc_1Man_{7\text{-}9}$ $GlcNAc_2$ structure (Ohta et al., 1991; Matsuura et al., 1993). Shimizu et al studied the molecular stability of chicken IgY (Shimizu et al., 1992). Although the stability of both immunoglobulins was similar in alkaline conditions, IgY showed much less stability than that of rabbit IgG to acid denaturation. IgY was found to be fairly resistant to trypsin or chymotrypsin digestion, but more sensitive to peptic digestion (Otani et al., 1991). It was reported that IgY was significantly more sensitive than rabbit IgG to temperatures above 70°C (Shimizu et al., 1992). The structural factors that affect those differences in the stability of the two immunoglobulins are unknown, since immunoglobulins are large complicated molecules with heterogeneous polypeptides. Shimizu and co-workers predicted that the lower content of β-structure in IgY may indicates that the conformation of the IgY domain is more disordered and less stable that that of rabbit IgG domain (Shimizu et al., 1992). The flexibility of the hinge region of IgY could be another factor affecting the molecular stability. The lower flexibility of Cν1 and Cν2 domains of IgY may cause the rapid inactivation of the antibody by various treatments, because the flexibility of the hinge regions is considered to influence the overall properties of the IgG molecules (Janeway and Travers, 1996).

3.2 Production and Isolation of Chicken IgY

Chicken eggs can be an alternative antibody source because IgG in blood is transported and accumulated in egg yolk. An attractive and important application of IgY is for oral passive immunization in which the IgY is specific to antigens of certain infectious pathogens. The feasibility of cost-effective, large-scale production of antibodies has favored the use of IgY. An industrial scale production of IgY from eggs is easier as compared to that of IgG from sera, since a large number of chicken farms and automatic egg breaking/processing facilities are already available. Collecting eggs from laying hens does not require the bleeding or sacrifice of animals for antibody production which is suitable from an animal welfare point of view. Antigens (mainly proteins) need to be administered with an adjuvant to assure a high quality/quantity, memory-enhanced antibody response by animals. An important consideration which affects IgY production is adjuvant. Freund's complete adjuvant (FCA) still continues to be the most commonly used for antibody production. The level of IgY was higher when FCA or Freund's incomplete adjuvant (FIA) was administered than without adjuvant (Shimizu et al., 1989). FCA did not influence egg production as much as the antigen itself (Schade et al., 1994). Egg yolk weight and the percentage hen day production were considered to be other important factors for the efficient production of IgY (Li et al., 1998). Recently, Bollen and Hau reported that FCA significantly reduces the egg laying frequency in comparison to the use of FIA and Hunter's titre Max (Bollen and Hau, 1999). Therefore, development of a suitable vaccination method is critical for the induction of high IgY titers. Table 12 summarizes the comparison of mammalian IgG and IgY. Hens usually lay about 280 eggs in a year. Yolk contains a considerably higher concentration of 100-150mg of IgY (Rose et al., 1974). Therefore, an immunized hen yields more than 40g of IgY through eggs a year, equivalent to that from 40 rabbits (Schade et al., 1991). Jensenius et al. (1981) reported that IgY corresponding to almost half a litre of serum may be recovered from a chicken in one month. This is 5-10 times higher than that from the blood of a rabbit. Over a period of slightly less than 6 weeks, 298mg of specific IgY against *Echino coccusgranulosus* was obtained from eggs compared with only 16.6mg from the rabbit's blood, 18 times more from yolk (Gottstein and Hemmeler, 1985).

Table 12. Comparison of mammalian IgG and chicken IgY

Animals	Rabbit (IgG)	Chicken (IgY)
Source of antibody	Blood serum	Egg yolk
Kind of antibody	Polyclonal	Polyclonal
Antibody sampling	Bleeding	Collection of eggs
Antibody amount	200 mg/bleed (40 ml blood)	100 - 150 mg/egg
Quantity of antibody (per year)	1,400 mg	40,000 mg
Amount of specific antibody	~5 %	2-10 %
Protein A/G binding	yes	no
Interaction with mammalian IgG	yes	no
Interaction with rheumatoid factors	yes	no
Activation of mammalian complements	yes	no

Based on Gottstein and Hemmeler (1985), and Schade et al. (1991).

Separation of IgY from egg yolk has been extensively studied and an excellent review was documented by Nakai et al. (1994). Egg yolk is a fluid emulsion with a continuous phase of lipoprotein particles. Egg yolk lipids therefore exist as lipoproteins (Burley and Cook, 1961). The major problem in isolating IgY from egg yolk is separating lipoproteins from egg yolk prior to purification of the IgY from the water-soluble fraction which contains the IgY fraction. Based on this strategy, many purification methods of IgY have been reported. These include: lipoprotein separation by ultracentrifugation (McBee and Cotterill, 1979); delipidation by organic solvents (Bade and Stegemann, 1984; Horikoshi et al., 1993); lipoprotein precipitation by: a) polyethylene glycol (Polson et al., 1980; Polson, 1990); b) sodium dextran sulfate (Jensenius et al., 1981); c) sodium alginate (Hatta et al., 1990); and d) ultrafiltration (Kim and Nakai, 1998). Akita and Nakai (1992) precipitated and purified IgY from the water-soluble fraction which was separated from six-fold water diluted yolk using ammonium sulphate followed by precipitation with sodium sulphate or ethanol. Anion exchange chromatography was reported as the final step in purification of IgY in many papers. A DEAE-sephacel column was used to separate IgY from a water-soluble fraction of yolk (McCannel and Nakai, 1990). Cation exchange chromatography was used to separate IgY from a water-soluble fraction of egg yolk (Fichtali et al., 1992). An immobilized metal ion (Fe^{3+}) affinity chromatography was also applied to purify egg yolk IgY (Green and Holt, 1997).

4. IMMUNOTHERAPEUTIC POTENTIAL OF CHICKEN IGY

Oral administration of specific antibodies is an attractive approach to establish passive immunity against gastrointestinal pathogens in human and animals. The increasing number of antibiotic resistant bacteria demands the need to develop alternatives to antibiotics (Crabb, 1998). Such immunotherapy can also be used against infections that are difficult to treat with antibiotics, such as viral infections. Several workers have shown that bovine colostrum derived from immunized cows and monoclonal antibodies, are effective agents for an immunotherapeutic approach against infectious gastrointestinal diseases (Carlander et al., 2000). For practical application of passive immunization, a cost-effective and large scale preparation of antibody will be required. The antigen specific IgY is ideal for this purpose. Many papers have been published on the production of IgY to bacterial pathogens and various protein antigens; however, its application for potential passive immunization to prevent animal and human diseases is still under evaluation. However, research in this area has suggested great future opportunities for designing new prophylactic strategies in the gastrointestinal tract for the treatment of infections or perhaps even to allow absorption of IgY for the treatment or prevention of systemic conditions.

We highlight below four successful applications of IgY against human rotavirus, *Yersinia ruckeri, enteritic E. coli* and *Salmonella* pathogenesis.

4.1 Application of IgY for the Prevention of Human Rotavirus Infection

Human rotavirus (HRV) has been identified as the major causative agent of acute infantile gastroenteritis (Yolken et al., 1988), infecting up to 90% of children under the age of three and resulting in more than 1 million infant deaths annually in developing countries (Prasad et al., 1988). Characteristically localized to the epithelial cells of the gastrointestinal tract, HRV causes a shortening and atrophy of the villi of the intestines (Kapikian and Chanock, 1996), resulting in decreased water absorption, leading to severe diarrhea and vomiting, and eventually death due to dehydration. While symptomatic treatment of rotavirus gastroenteritis is available, currently no HRV vaccine exists, as individuals most susceptible to infection are the least responsive to current vaccination techniques.

A member of the *Reoviridae* family, rotaviruses are triple-layered icosahedral particles, with a double-stranded RNA genome containing 11 gene segments. The outer capsid of the virus is comprised of two coat proteins, VP7, a 37 kDa glycoprotein which forms a smooth layer, and VP4, an 88 kDa nonglycosylated protein which forms spikes that extend from the surface of the particle (Ludert et al., 1996). The spike protein, VP4, has been implicated in several important functions, including cell attachment, penetration, hemagglutination, neutralization, host range, and virulence (Nejmeddine et al., 2000). Viral infectivity is enhanced by the proteolytic cleavage of VP4 into VP5 (the 60 kDa C-terminal segment) and VP8 (the 28 kDa N-terminal segment) (Ruggeri and Greenberg, 1991). The VP8 subunit in particular has been found to play a significant role in viral infectivity and neutralization of the virus. Studies have identified eight neutralization epitopes on VP4, with five of them located on the VP8 portion (Mackow et al., 1988), and monoclonal antibodies to VP8 were reported to neutralize rotavirus in vitro and passively protect mice against rotavirus challenge in vivo (Matsui et al., 1989). Ruggeri et al. (Ruggeri and Greenberg, 1991), found that monoclonal antibodies directed against VP8 not only functioned to inhibit viral attachment to cells, but may also mediate a significant release of previously bound virus from the cell surface, indicating that VP8 may be an important target in immunization research.

The importance of HRV in human disease has prompted much research into the development of strategies for the prevention of rotavirus gastroenteritis. It has been found that local intestinal immunity played a key role in preventing rotavirus infection (Greene and Holt, 1997), while no viremic state had been detected, leading to the conclusion that a rotavirus vaccine would be most effective if administered orally. Current strategies involve the oral immunization with attenuated strains of human or antigenically related animal rotavirus, or with rotavirus antigens, in order to elicit an active immune response. There are, however, individuals at high risk for serious rotavirus infection who might not be expected to respond to such active immunization approaches, including newborns, children with congenital or acquired immunodeficiency syndromes, and adults or children rendered immunodeficient by chemotherapy, malnutrition, or aging (Yolken et al., 1988; Hammarstrom, 1999). An alternative method for the prevention of infectious gastroenteritis in individuals unable to mount an active immune response involves the administration of preformed antibodies capable of neutralizing the infectious agent. In the case of rotavirus, this would involve the introduction of antibodies directly into the gastrointestinal tract, the principal site of viral replication. Human immunoglobulins, derived from serum (Losonsky et al., 1985) and breast milk (Ebina, 1996) have both been shown to inhibit the intestinal replication of rotavirus; however, the high cost and limited supply of human products limits their use for widespread applications for HRV prevention. Immunoglobulins derived from other species have also been shown to passively protect against HRV, and the effective use of bovine and chicken immunoglobulins for the passive protection against HRV have both been described by Yolken and coworkers (Yolken et al., 1988). It was found that the oral administration of antibodies isolated from the eggs of chickens immunized with three different serotypes of rotavirus (mouse, human, and monkey) were capable of preventing rotavirus-induced diarrhea in mice infected with murine rotavirus, whereas IgY isolated from the eggs of unimmunized chickens failed to prevent rotavirus infection. Using an HRV infection model in suckling mice, it was reported that anti-HRV IgY decreased the incidence of rotavirus-induced diarrhea both when administered before and after HRV challenge, suggesting its use for both therapeutic and prophylactic applications (Hatta et al., 1993). Furthermore, Kuroki et al. observed that the sustained oral administration of anti-bovine rotavirus IgY, having a neutralization titre of 6400, protected calves against challenge with bovine rotavirus (Kuroki et al., 1994). In addition, it has been observed that hens immunized with HRV continuously laid eggs without any change in the egg laying rate, and demonstrated a sustained high level of neutralizing anti-HRV antibodies (Hatta et al., 1993), reiterating the fact that IgY would provide a plentiful and consistent source of anti-HRV antibodies for passive immunization applications.

Recently, the VP8 cDNA from the WA strain of HRV was prepared by RT-PCR, and cloned into the pUC18 plasmid. The VP8 gene was then inserted into a pGEX-4T-2 glutathione S-transferase (GST) fusion vector, which contains a thrombin protease recognition site. The GST-VP8* fusion protein was expressed in *Escherichia coli*, in a soluble form, and purified using Glutathione Sepharose 4B affinity chromatography (Kovacs-Nolan et al., 2001). Ten 25 week-old Shaver White Leghorn chickens were injected intramuscularly with 0.2 mg of the purified VP8* with Freund's Complete Adjuvant. After the initial injections, significant IgY titres were observed within three weeks, and remained relatively high for approximately three months, upon which time a booster was administered. Upon boosting, the titres rapidly returned to near-maximum levels (Figure 46).

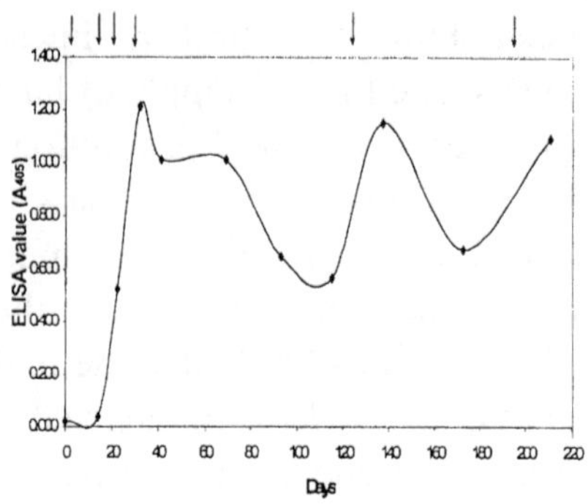

Figure 46. Antibody response in the egg yolks of a hen injected with recombinant VP8*. Antibody level in a 1000 dilution of egg yolk water-soluble fraction was measured by ELISA using VP8* as the antigen. Arrows indicate days of injection.

While antibodies against HRV have been previously produced in chickens, they were raised against the whole virus particle. Injecting chickens with VP8* should produce highly specific antibodies, capable of neutralizing the virus, and preventing HRV infection. All IgY samples exhibited a significant neutralization activity in vitro, with neutralization titres ranging from 8000 to 48000 (Table 13). Thus, the recombinant VP8* subunit protein of HRV has significant implications as a vaccine candidate for humans, and the production of anti-VP8* antibodies presents potential applications for the passive immunization of infants (and others susceptible) against HRV infection. This approach could provide insight into developing novel ways for the administration of IgY, and its possible use in the fortification of food products

Table 13. Anti-human rotavirus VP8* IgY neutralization titres

Chicken	Antibody titre	Neutralization titre
1	8000	12000
2	16000	24000
3	8000	8000
4	32000	48000
5	16000	12000

4.2 Application of IgY for the Prevention of *Yersinia ruckeri* in Rainbow Trout

A trial of passive immunization of rainbow trout, *Oncorhynchus mykiss,* against infection with *Y. ruckeri* using IgY has been studied (Lee et al., 2000). *Y. ruckeri* is the causative agent of enteric redmouth disease, a systemic bacterial septicemia of salmonid fish (Stevenson et al., 1993). *Y. ruckeri* can be isolated from the intestinal lining of fish (Busch and Lingg, 1974) and Evelyn (1996) has suggested that this bacterium may be resistant enough to use the gastrointestinal tract as one of its routes of infection.

Persistence of *Y. ruckeri* for long periods in carrier fish and shedding of bacteria in feces can present a continuing source of infection. If a population of carrier fish could be substantially cleared by oral administration of anti-*Y. ruckeri* antibody treatments, it may be a cost-effective alternative to slaughtering a stock of fish which pose a health risk. The IgY against *Y. ruckeri* was used in attempts to determine if infections could be reduced, particularly when administered orally. *Y. ruckeri* strains RS1154 (serovar 1) and RS 1153 (serovar 2) were injected into chickens to raise IgY. After 30 to 40 days after the first immunization, antibodies to whole cells of both serovars of *Y. ruckeri* were found at high titre in egg yolk. Hens immunized with purified LPS of either serovar (in FCA) showed very poor antibody responses. It is important to protect the IgY from degradation by the gastric enzymes and low pH. The IgY microencapsulation was prepared using a microbial transglutaminase. Groups of 5 rainbow trout which had been fed anti-*Y. ruckeri* IgY 2 h prior to an immersion challenge with *Y. ruckeri* RS 1154 had lower mortalities after 8 days compared with fish fed with normal food before the challenge. The group fed IgY appeared to contain a lower number of infected fish after 8 days, based on organ and intestine culture. In a subsequent trial of the feeding procedures with triplicate replicates of the groups, the IgY-fed group showed lower mortalities than groups receiving normal feed (Lee et al., 2000). The numbers of IgY-fed fish carrying *Y. ruckeri* in intestine samples appeared lower than the normal-feed controls, regardless of whether the feeding was given before or after the challenge. The groups fed IgY three times post-infection gave inconsistent results. Two groups of sampled fish, one in each of these treatments, showed no *Y. ruckeri* in individual intestinal samples, nor in a pool of kidney tissues.

Table 14 shows the effect of passive immunization with anti-*Y. ruckeri* IgY following an immersion challenge (Mine et al., 1999). Four days after a challenge dose of 10^8 cfu, *Y.ruckeri* was injected intraperitoneally (i.p.) into three rainbow trout (about 285 g). All fish had *Y. ruckeri* associated with the intestinal tract tissue, but none in the kidney tissue by direct streak. If the dose was mixed with either 4 mg or 40 mg of the IgY preparation immediately prior to injection, only one fish in each group had *Y. ruckeri* after 4 days, indicating some direct passive immunization effect of IgY, but not by the carrier material.When groups of 5 fish were passively immunized by i.p. injection with IgY and subsequently immersion challenged, no *Y. ruckeri* were found in the intestine or kidney after 7 days, while fish groups injected with a non-specific IgY or saline had the bacteria in both kidney and intestinal tissue.

Therefore, the i.p. injection of anti-*Y. ruckeri* at a dose of 4mg is effective against an immersion challenge of *Y. ruckeri* of 10^8 cfu/mL. Many pathogens of fish have been reported to spread by infection through intestinal mucosa. The naturally oral feeding of specific IgY against fish pathogens with feed will be an alternative to methods using antibiotics and chemotherapy for the prevention of fish diseases in fish farms. Moreover, the oral feeding of active IgY would be a novel approach for preventing viral infection diseases of fish because no medicine has been reported to be effective.

4.3 Application of IgY for the Prevention of Enterotoxigenic *Escherichia coli* Infection

Diarrhea is a major health problem in humans and animals. Enterotoxigenic *Escherichia coli* (ETEC) is the most common enteric colibaccilosis encountered in neonatal calves (Moon et al., 1976), piglets (Morris and Sojka, 1985) and children in developing countries and travelers to these countries (Sack, 1986)

Table 14. Effects on rainbow trout by passive immunization[a] with anti- *Y. ruckeri* IgY following immersion challenge[b]

	CFU g^{-1} in group[c]		No. infected with *Y ruckeri* [d]			*Y. ruckeri* positive	
treatment	intestine	kidney	intestine	kidney	no. of fish	intestine	kidney
challenge					1	-	-
high-dose specific IgY					2	-	-
	5.6 x 10^4	0	0/5	0/5	3	-	-
					4	-	-
					5	-	-
challenge					1	-	-
low-dose IgY	7.2x10^3	0	0/5	0/5	3	-	-
					4	-	-
					5	-	-
challenge					1	-	-
nonspecific IgY					2	+	+
	7.4x10^9	6.2X10^7	4/5	4/5	3	+	+
					4	+	+
					5	+	+
challenge					1	-	-
saline					2	-	-
	3.6x10^5	3.8x10^6	3/5	3/5	3	+	+
					4	+	+
					5	+	+

[a]Passive immunization: fish are injected with 0.1 ml of anti-*Y. ruckeri*
[b]Immersion challenge with viable *Y. ruckeri* (1.8 x 10^8 cfu ml^{-1}).
[c]CFU g^{-1} of group in intestine and kidney 7 days after immersion challenge.
[d]Number of *Y. ruckeri* positive/number of tested fish.

ETEC accounts for one billion diarreal episodes annually and perhaps one million deaths each year (Sack, 1986). One half of all travelers to developing countries also develop diarrhea (Svennerholm et al., 1989). Colonization of the small intestine of the pig by ETEC adhering to the epithelium accounts for most gastrointestinal disorders in both neonatal and post-weaning piglet. It causes economic losses to the pig industry from both mortality and reduced growth rates, and on farms, killing 1.5-2.0% pigs weaned (Hampson, 1994). The strains of ETEC which are associated with an intestinal colonization and cause severe diarrhea are the K88, K99 and 987P fimbrial adhesions (Parry and Rooke, 1985). It has been reported that feeding colostrum from vaccinated cows prevented diarrhea due to infectious *E. coli* in infants (Hilpert et al., 1977). Milk IgG has been used as an effective prophylactic against travelers' diarrhea (Tacket et al., 1988). IgY could be an alternative source of immunoglobulins for the prevention of ETEC infection. IgY raised against ETEC antigen has been administered orally to piglets and has offered a potential prophylactic and therapeutic approach for controlling ETEC induced diarrhea (Wiedemann et al., 1990; Wiedemann et al., 1991; Schmidt et al., 1989; Jungling et al., 1991). Yokomaya and coworkers (Ikemori et al., 1993; Yokomaya et al., 1992) studied the passive protective effect of IgY against ETEC infection in neonatal piglets. The antibodies were prepared by spray drying the water-soluble proteins of egg yolk from chickens immunized with *E. coli* K-88, K-99 and K-987P fimbrial adhesins. All piglets in their studies when challenged with different strains of ETEC developed mild to severe diarrhea within 12 hrs after infection. The orally administered IgY protected in a dose dependent manner against infection with each of the three strains of *E. coli* in passive immunization trials (Table 15). More than 80% of the pigs died in the control group while none of the pigs died when treated with antibodies containing an IgY titre of 625 or 2500. It was also revealed that fewer piglets in the group receiving the IgY excreted *E. coli* and showed resistance to bacterial adhesion. They also demonstrated that E. coli K88, K99 and 987P strains adhered equally to porcine duodenal and epithelial cells but failed to do so in the presence of homologous anti-fimbrial IgY. This suggests that anti-fimbrial antibodies are active components. Marquardt et al. reported that the IgY titre was much greater when E. coli K88 fimbriae were used rather than the whole cell (Marquardt et al., 1999). In the animal feeding study, 21-day old pigs were challenged with a dose of the ETEC (10^{12} cfu). The IgY was administered to the piglets in the milk three times a day for 2 days. Control piglets developed severe diarrhea within 12 hrs and 30% of the pigs died. In contrast, the IgY treated pigs exhibited no sign of diarrhea 24 or 48 hrs after treatment. They also block the binding of *E. coli* K88 to the mucosal receptor. The interaction of IgY with E.coli K88 was fairly rapid, as maximum binding was done within 15 min (Jin et al., 1998). The IgY against *E. coli* F18ac fimbriae was raised and in vitro adhesion tests demonstrated that the IgY inhibited attachment of F18ac positive *E. coli* bacteria to the intestinal mucosa. The F18ab antibodies diminished the cases of diarrhea and death in animals infected with F18ac positive *E, coli* (Imberechts et al., 1997).

Table 15. Rates of isolation of ETEC K88⁺, K99⁺, and 987P⁺strains from newborn piglets after challenge and treatment with antibody powder at various titres

Trial and Strain	Antibody treatment (titre)	Positive rectal swabs/total on day:		
		1	3	5
1. K88	0	7/7	4/4	1/1
	156	7/7	5/5	4/5
	625	7/7[a]	4/7	4/7
	2500	4/7[a]	2/7[b]	0/7[a]
	Absorbed	4/4	2/2	1/1
2. K99	0	4/4	0/0	0/0
	156	4/4	2/2	0/2
	625	4/4	4/4	1/4
	2500	3/4[a]	1/4	0/4
	Absorbed	4/4	1/1	0/0
3. 987P	0	5/5	1/1	1/1
	1456	5/5	3/3	3/3
	625	5/5	3/5	0/5 [a]
	2500	5/5[a]	0/5[a]	0/5 [a]
	Absorbed	4/4	1/1	1/1

[a] $P<0.01$; [b] $P < 0.05$.
From Yokoyama et al. (1992).

The passive protective effect of anti ETEC (B44 strain) for neonatal calves against fatal enteric collibacillosis was also studied (Ikemori et al., 1992). Calves fed milk containing IgY powder with an anti pili agglutinin titre of 1:800 and 1:1600 had transient diarrhea, 100% survival and good body weight gain during the course of the study. Anti ETEC (K88) IgY was produced as a potential food ingredient and showed that the IgY was relatively heat stable without loss in antibody activity by pasteurization (70°C x 15 min)(Shimizu et al., 1988). The IgY against heat labile toxin (LT) produced by ETEC and studied its neutralization activity (Akita et al., 1998; O'Farrelly et al., 1992). The Fab' fraction of IgY was found to be as effective in neutralizing LT. Although oral administration of IgY from ETEC has proven to be successful for treatment of gastrointestinal infections of animals, clinical application of passive immunization of IgY against diarrhea is now encouraged to develop new systems to prevent infant ETEC infection and for its treatment.

4.4 Application of IgY for the Prevention of *Salmonellosis*

In recent years, many cases of food poisoning caused by *Salmonella* have been widespread in the world. *Salmonella enteritidis* (SE) and *Salmonella typhimurium* (ST) are particularly prominent as major agents of the food poisoning (Bell and Kriakides, 1998). The original source of SE and ST are suspected to be poultry products, eggs and meat. *Salmonella* can cause typhoid fever: the symptoms include fever, abdominal pain, headache, malaise, lethargy, skin rash, constipation and change in mental state. The elderly, infants and those with impaired immune systems may experience a more severe illness. In this case, the infection may spread from the intestines to the blood stream and then to other body sites, and can cause death. *Salmonella* is a rod-shaped, motile bacterium, non-spore forming and gram negative. It is estimated that 204 million cases of salmonellosis occur in the U.S. annually (Bell and Kriakides, 1998). *Salmonella* has a variety of surface components which are virulence related. The outer membrane protein (OMP) plays a role as pathogenicity determinants (Galdiero et al., 1990). OMP has been used successfully as a vaccine antigen and a number of workers have shown that they are good immunogens and are protective in both active and passive immunization studies (Ishibasi et al., 1988; Udhatakumar and Muthukkaruppan, 1987). Lipopolysaccharide (LPS) was also shown to induce a strong immunogenic reaction producing a large amount of LPS specific IgY and showed potential application for the prevention of *Salmonella* adhesion and diseases (Sunwoo et al., 1996; Mine, 1997). A novel fimbrial antigen (SEF14) was first described on strain SE (Thorns et al., 1990). SEF14 is produced mainly in SE and *S. dublin* strains and thus can be regarded as specific to SE among *Salmonella* isolates from poultry (Thorns et al., 1992), Chickens were immunized with SEF14 to raise egg yolk IgY for protective trials. Seventy-nine mice were challenged orally with $2x10^{10}$ cfu of SE. The test mice treated with anti SEF14 IgY (titre 128) had a survival rate of 77.8%

compared to 32% survival in control mice fed normal egg yolk IgY (titer<10)(P<0.01)(Peralta et al., 1994). The passive protective efficacy of chicken IgY specific for OMP, LPS or flagella (Fla) in controlling experimental salmonellosis in mice was compared (Table 16).

Table 16. Effect of egg yolk antibody on in vitro adhesion of *S. enteriditis* and *S. typhimurium* to HeLa cells.

Antibody titre		Mean no. of adherent *Salmonella* cells at different antibody dilutions					Minimum inhibitory titre[a]
		x 10	x 20	x 40	x 80	x 160	
S. enteriditis							
OMP	256	6+6[b]**	10+4**	21+7**	23+14**	36+12	3.2
LPS	128	12+8**	17+9**	25+8**	33+12	39+11	3.2
Fla	1024	23+17**	24+8**	33+12	43+15	39+3	51.2
Control	< 2	42+12					
S. typhimurium							
OMP	256	3+4[b]**	9+13**	15+9**	26+10**	36+10	3.2
LPS	128	15+8**	31+12**	38+12**	42+11	46+12	6.4
Fla	2048	12+9**	15+7**	27+7**	39+10	43+15	51.2
Control	< 2	42+15					

[a]Minimum adhesion inhibitory concentration = Antibody titer/Dilution of antibody solution with significant adhesion inhibition

[b]Number of bacteria attached per HeLa cell + Standard deviation

**p < 0.01 relative to the control (Student's *t* test)

Adapted from Yokoyama et al. (1998).

The attachment of SE against HeLa cells was reduced by homologous anti-OMP, -LPS and ÐFla IgY. The minimum antibody titers that significantly inhibited an adherence compared to control was 3.2 for OMP, 3.2 for LPS and 51.2 for Fla in the case of SE IgY. In a mouse challenge test with SE (2 x 10^9 cfu), antibody treatment resulted in a survival rate of 80%, 47% and 60% using OMP, LPS or Fla specifc IgY, respectively in contrast to only 20% in control mice. In the case of ST (2 x 107), the survival rate was 40%, 30% and 20%

using OMP, LPS or Fla specific IgY, while it was 0% in control mice (Yokoyama et al., 1998). *Salmonella* infection in calves is still a worldwide problem and now serovars, ST and *S. dublin* are accounting for most salmonellosis within the first 2 week after birth. IgY specific against ST and *S. dublin* was investigated by oral inoculation with 1011 SE or *S. dublin*. IgY was administered orally 3 times a day for 7 to 10 days after challenge exposure. All control calves died within 7-10 days, whereas low titre IgY treated calves had 60-70% mortality. Only fever and diarrhea, but not death ($p<0.01$) were observed in calves given the higher titre IgY (Yokoyama et al., 1998). These results show IgY specific for ST or *S. dublin* is protective against fatal salmonellosis when given in sufficient quantities, and may be clinically useful during a salmonellosis outbreak.

4.5 Other Applications of IgY

As discussed, the most attractive application of IgY is passive immunization therapy against gastrointestinal infections. Several researchers have reported the potential prevention of other bacterial and viral pathogeneses with specific IgY. Dental caries is perhaps the most prevalent disease affecting humans today. *Streptococcus mutants* serotype c is thought to be the principal causative bacterium of dental caries. Chicken antibodies against *Streptococcus mutans* MT8148 serotype c or cell-associated glucosyltransferase were prepared and tested against dental caries (Otake et al., 1991; Chang et al., 1999; Hamada et al., 1991). Feeding a cariogenic diet containing more than 2% IgY yolk powder showed significantly lower caries scores (Otake et al., 1991) and effective passive protection for the prevention of colonization of mutant *Streptococci* in the oral cavity. It has been reported that a mouth rinse containing IgY specific to *Streptococcus mutants* was effective in preventing dental plaque in humans (Hatta et al., 1997). It has been estimated that 1.7 million people are bitten or stung by venomous snakes, scorpions, jelly fish or spiders each year, 40,000-50,000 fatally. The most widely accepted treatment of envenomation is the use of specific antivenoms to neutralize the toxic, potentially lethal effects of venoms. Chicken anti-venoms IgY has been produced and had a higher bioactivity than antivenoms raised in horse (Thalley and Carroll, 1990; Almeida et al., 1998). The uses of specific IgY to control barley yellow dwarf virus (Hu et al., 1985), influenza virus (Cuceanu et al., 1991), Coccobacilloid of South African Clawed frog, *Xenopus laevis* (Haynes et al., 1992), *Edwardsiellosis tarda* of Japanese eels (Gutierrez et al., 1993) murine rotavirus (Bartz et al.,

1980) and canine distemper virus (Schmidt et al., 1989) have been reported. Chicken IgY also has potential advantages for diagnostic purposes (Mine, 2000; Bauwens et al., 1988; Larsson et al., 1993; Kuronen et al., 1997; Bar-Joseph and Malkinson, 1980) and in immunoaffinity chromatography (Kim et al., 1999; Li-Chan et al., 1998).

5. CONCLUSION AND FUTURE STUDIES

Oral administration of antigen specific IgY is a promising approach for the prevention of gastrointestinal infections. The higher productivity and mass production on an industrial scale strongly suggest the attractiveness of IgY technology as a practical reality in the near future to prevent infectious diseases. To be effective, the IgY must survive the gastrointestinal environment and reach their target local areas with their biological activity intact. Unfortunately, the stability of IgY against lower pH and peptic digestion seems a little lower than other species such as bovine IgG. However, there has been increasing interest in the oral administration of IgY for localized treatment of infections of viral and bacterial pathogenesis in the gastrointestinal tract. Enzymes such as pepsin, trypsin and chymotrypsin initially degrade the antibodies to $F(ab')_2$, Fab and Fc fragments. The $F(ab')_2$ and Fab fragments retain some of their neutralizing activity locally in the gastrointestinal tract. There is still controversy regarding survival of IgY through gastrointestinal tract and its implications for human immunotherapy. Finding an effective way to protect the antibodies from degradation in the gastrointestinal tract would be useful for the economic and systemic approach of IgY technology in future work.

REFERENCES

Akita EM and Nakai S, 1992. Immunoglobulins from egg yolk: Isolation and purification. J Food Sci, 57: 629-634

Almeida CM, Kanashiro MM, Rangel Filho FB, Mata MF and Kipnis TL, 1998. Development of snake antivenom antibodies in chickens and the purification from yolk. Vet Rec, 143:579-584

Bade H and Stegemann H,1984. Rapid method of extraction of antibodies from hen egg yolk. J Immunol Methods, 71:421-426

Bar-Joseph M and Malkinson M,1980. Hen egg yolk as a source of antiviral antibodies in the enzyme-linked immunosorbent assay (ELISA): a comparison of two plant viruses. J Virol Methods, 1:179-183

Bartz CR, Conklin RH, Tunstall CB and Steele JH,1980. Prevention of murine rotavirus infection with chicken egg yolk immunoglobulins. J Infect Dis, 142:439-441

Bauwens RM, Devos MP, Kim JA and Leenheer AP, 1988. Chicken egg yolk and rabbit serum compared as sources of antibody for radioimmunoassay of 1,25-dihydroxyvitamin D in serum or plasma. Clin Chem, 34:2153-2159

Bell C, and Kriakides A, 1998. *Salmonella: A practical approach to the organism and its control in foods.* Blackie, A & P, New York

Bollen LS and Hau J, 1999. Freund's complete adjuvant has a negative impact on egg laying frequency in chickens. In Vivo, 13:107-108

Burley RW and Vadehra DV, 1989. An outline of the physiology of avian eggs. In *The avian egg: Chemistry and biology*, Wiley International, New York, pp.17-24

Burley RW and Cook WH, 1961. Isolation and composition of avian egg yolk granules and their consitituents α and β lipovitellines. Can J Biochem Physiol, 39:1295-1307

Busch RA and Lingg AJ, 1974. Establishment of an asymptomatic carrier state infection of enteric redmouth disease in rainbow trout (*Salmo gairdneri*). J Fish Res Education Canada, 32:2429-2432

Carlander D, Kollberg H, Wejaker PE and Larsson A, 2000. Peroral immunotherapy with yolk antibodies for the prevention and treatment of enteric infections. Immunol Res 21:1-6

Carlander D, Stalberg J and Larsson A, 1999. Chicken antibodies: a clinical chemistry perspective. Ups J Med Sci 104:179-189

Chang HM, Ou-Yang RF, Chen YT and Chen CC, 1999. Productivity and some properties of immunoglubulin specific against *Streptococcus mutans* serotype c in chicken egg yolk (IgY). J Agric Food Chem, 47:61-66

Crabb JH, 1998. Antibody-based immunotherapy of cryptosporidiosis. Adv Parasitol 40: 121-149

Cuceanu N, Constantinescu C and Ionita E. 1991. Isolation and characterization of egg yolk antibody IgY from chickens immunized with different influenza virus strains. Roum Arch Microbiol Immunol, 50:215-222.

Ebina T, 1996. Prophylaxis of rotavirus gastroenteritidis using immunoglobulin. Arch Virology (suppl), 12: 217-223

Epple A and Steson M, 1980. *Avian endocrinology.* Academic Press, New York

Evelyn TPT, 1996. Infection and disease. In *The fish immune system: Organism, pathogen and environment.* Eds. Iwama G and Nakanishi T, Academic Press, San Diego, CA. pp 339-362

Fichtali J, Charter EA, Lo KV and Nakai S, 1992. Separation of egg yolk immunoglobulins using an automated liquid chromatography system. Biotech Bioeng, 40: 1388-1394

Galdiero F, Tufano MA, Masiello S and Rose MD, 1990. Inflammatory effects of *Salmonella typhimurium* porins. Infect Immun, 58:3183-3191

Gardner PS and Kaya, S, 1982. Egg globulin in rapid virus diagnosis. J Virol Methods, 4:257-262

Gassmann M, Thommes P, Weiser T and Hubscher U, 1990. Efficient production of chicken egg yolk antibodies against a conserved mammalian protein. Federation Am Sci Exp Biol J, 4: 2528-2532

Gottstein B and Hemmeler E, 1985. Egg yolk immunoglobulin Y as an alternative antibody in the serology of echinococcosis. Z Parasitenkd, 71:273-276

Greene CR and Holt PS, 1997. An improved chromatographic method for the separation of egg yolk IgG into subpopulations utilizing immobilized metal ion ($Fe3+$) affinity chromatography. J Immunol Methods, 209:155-164

Gutierrez MA, Miyazaki T, Hatta H and Kim M, 1993. Protective properties of egg yolk IgY containing anti-*Edwardsiella tarda* antibody against paracolo disease in the Japanese eel, *Anguilla japanica Temminck & Schlegel*. J Fish Dis, 16:113-122

Hamada S, Horikoshi T, Minami T, Kawabata S, Hiraoka, J, Fujiwara T and Ooshima T, 1991. Oral passive immunization against dental caries in rats by use of hen egg yolk antibodies specific for cell-associated glucosyltransferase of *Streptococcus mutans*. Infect Immun 59:4161-4167

Hammarstrom L, 1999. Passive immunity against rotavirus in infants. Acta Paediatr Suppl, 430: 127-132

Hampson DJ, 1994. Postweaning *Escherichia coli* diarrhoea in pigs. In *Escherichia coli in domestic animals and humans*. Eds. Gyles CL. CAB International, Oxon, UK, pp171-191

Hatta H, Kim M and Yamamoto T, 1990. A novel isolation method for hen egg yolk antibody "IgY". Agric Biol Chem, 54: 2531-2535

Hatta H, Tsuda K, Akachi S, Kim M, Yamamoto T and Ebina T, 1993. Oral passive immunization effect of anti-human rotavirus IgY and its behavior against proteolytic enzymes. Biosci Biotech Biochem, 57:1077-1081

Hatta H, Tsuda K, Akachi S, Kim M and Yamamoto T, 1993. Productivity and some properties of egg yolk antibody (IgY) against human rotavirus compared with rabbit IgG. Biosci Biotech Biochem, 57:450-454

Hatta H, Tsuda K, Ozeki M, Kim M, Yamamoto T, Otake S, Hirosawa M, Katz J, Childers NK and Michalek SM, 1997. Passive immunization against dental plaque formation in humans: Effect of a mouth rinse containing egg yolk antibodies (IgY) specific to *Streptococcus mutans*. Caries Res, 31:268-274

Hatta H, Ozeki M and Tsuda K, 1997. Egg yolk antibody IgG and its application. In *Hen eggs; Their basic and applied science*. Eds. Yamamoto T, Juneja LR, Hatta H and Kim M. CRC Press, New York, pp 151-178

Haynes L, Harding FA, Koniski AD and Cohen N, 1992. Immune system activation associated with a naturally occurring coccobacilloid in *Xenopus laevis*. Dev Comp Immunol, 16:453-462

Hilpert H, Gerber H, Pahud JJ, Ballabriga A, Arcalis L, Farriaux F, de Leyer E and Nussle D, 1977. Bovine milk immunoglobulins (Ig). Their possible utilization in industrially prepared infant's milk formula. In *Food and immunology*. Eds. Hambraeusm L, Hanson L and McFarlene H. Almquist and Wiksell International, Stockholm, pp182-196

Horikoshi T, Hiraoka J, Saito M and Hamada S, 1993. IgG antibody from hen egg yolk: Purification by ethanol fractionation. J Food Sci, 58:739-742

Hu JS, Rochow WF and Dietert RR, 1985. Production and use of antibodies from hen eggs for the SGV isolate of barley yellow dwarf virus. Phytopathology, 75:914-919

Ikemori Y, Peralta RC, Kuroki M, Yokoyama H and Kodama Y, 1993. Avidity of chicken yolk antibodies to enterotoxigenic *Escherichia coli* fimbriae. Poultry Sci, 72:2361-2365

Ikemori Y, Kuroki M, Peralta RC, Yokoyama H and Kodama Y, 1992. Protection of neonatal calves against fatal enteric colibacillosis by administration of egg yolk powder from hens immunized with K99-piliated enterotoxigenic *Escherichia coli*. Am Vet Res, 53:2005-2008

Imberechts H, Deprez P, van Driessche E and Pohl P, 1997. Chicken egg yolk antibodies against F18ab fimbriae of *Escherichia coli* inhibit shedding of F18 positive *E. coli* by experimentally infected pigs. Vet Microbiology, 54:329-341

Janeway Jr CA and Travers P, 1996. Structure of the antibody molecule and immunoglobulin genes. In *Immunobiology: The immune system in health and disease.* Eds. Janeway Jr CA and Travers P, Current Biology Ltd, London

Janson AK, Smith CI and Hammarstrom L, 1995. Biological properties of yolk immunoglobulins. Adv Exp Med Biol, 371:685-690

Jensenius JC, Anderson I, Hau J, Crone M and Kock C, 1981. Eggs: conveniently packaged antibodies. Method for purification of yolk IgG. J Immunol Methods, 46:63-68

Jin LZ, Samuel K, Baidoo K, Marquardt RR and Frohlich AA, 1998. In vitro inhibition of adhesion of enterotoxigenic *Escherichia coli* K88 to piglet intestinal mucus by egg yolk antibodies. FEMS immunol. Med Microbiol, 21:313-321

Jungling A, Wiedemann V, Kuhlmann R, Erhard M, Schmidt P and Losch U, 1991. Chicken egg antibodies for prophylaxis and therapy of infectious intestinal diseases: in vitro studies on protective effects against adhesion of enterotoxogenic *Escherichia coli* to isolated enterocytes. J Vet Med 38:373-381

Kapikian AZ and Chanock RM, 1996. Rotaviruses. In: *Fields virology*, Third Edition. Eds. Fields BN, Knipe DN, Howley PM, Chanock RM, Melnick JL, Monath TP, Rolzman B and Strauss SE. Raven Press. New York, pp 1657-1708

Kim H and Nakai S, 1998. Simple separation of immunoglobulin from egg yolk by ultrafiltration. J Food Sci 63:485-490

Kim H, Durance TD and Li-Chan ECY, 1999. Reusability of avidin-biotinylated immunoglobulin Y columns in immunoaffinity chromatography. Anal Biochem, 268:383-397

Klemperer F. 1983. Uber naturliche immunitat und ihre verwertung fur die im munisierungstherapie. Archiv Exp/ Pathologie und Pharma, 31:356-382

Kovacs-Nolan J, Sasaki E, Mine Y and Yoo D, 2001. Expression of human rotavirus VP8* protein in *Escherichia coli*. Biochem Biophys Res Commun, 282:1183-1188

Kronvall G, Seal US, Sevensson S and Williams RC Jr, 1974. Phylogenetic aspect of *Staphylococcal* protein A-reactive serum globulins in birds and mammals. Acta Path Microbiol Scand Section B, 82:12-18

Kuklmann R, Wiedemann V, Schmidt P, Wanke R, Linckh E and Losch U, 1988. Chicken egg antibodies for prophylaxis and therapy of infectious intestinal disease: immunization and antibody determination. J Vet Med, 35:610-616

Kuroki M, Ohta Y, Ikemori Y, Peralta RC, Yokoyama H and Kodama Y, 1994. Passive protection against bovine rotavirus in calves by specific immunoglobulins from chicken egg yolk. Arch Virol, 138:143-148

Kuronen I, Kokko H, Mononen I and Parviainen M, 1997. Hen egg yolk antibodies purified by antigen affinity under highly alkaline conditions provide new tools for diagnostics. Human intact parathyrin as a model antigen. Eur J Clin Chem Clin Biochem, 35:435-440

Larsson A and Sjoquist J, 1988. Chicken antibodies: A tool to avoid false positive results by rheumatoid factor in latex fixation tests. J Immunol Methods, 108:205-208

Larrson A and Sjoquist J, 1990. Chicken IgY: Utilizing the evolutionary difference. Comp. Immun. Microbiol. Infec Dis, 13:199-201

Larsson A, Balow RM, Lindahl TL and Forsberg PO, 1993. Chicken antibodies: Taking advantage of evolution-A review. Poultry Sci, 72:1807-1812

Lee SB, Mine Y and Stevenson RMW, 2000. Effects of hen egg yolk immunoglobulin in passive protection of rainbow trout against *Yersinia ruckeri*. J Agric Food Chem, 48:110-115

Leslie GA and Clem LW, 1969. Phylogeny of immunoglobulin structure and function. III Immunoglobulins of the chicken. J Exp Med 130:1337-1352

Li X, Nakano T, Sunwoo HH, Paek BH, Chae HS and Sim JS, 1998. Effects of egg and yolk weights on yolk antibody (IgY) production in laying chickens. Poultry Sci, 77: 266-270

Li-Chan ECY, Ler SS, Kummer A and Akita EM, 1998. Isolation of lactoferrin by immunoaffinity chromatography using yolk antibodies. J Food Biochem, 22:179-195

Loeken MR and Roth TF, 1983. Analysis of maternal IgG subpopulations which are transported into the chicken oocyte. Immunology, 49:21-28

Losch U, Schranner I, Wanke R and Jurgens L, 1986. The chicken egg, an antibody source. J Vet Med 33:609-619

Losonsky GA, Johnson JP, Winkelstein JA and Yolken RH, 1985. Oral administration of human serum immunoglobulin in immunodeficient patients with viral gastroenteritidis. A pharmacokinetic and functional analysis. J Clin Invest, 76:2362-2367

Ludert JE, Krishnaney AA, Burns JW, Vo PT and Greenberg HB, 1996. Cleavage of rotavirus VP4 in vivo. J Gen Virol, 77:391-395

Mackow ER, Shaw RD, Matsui SM, Vo PT, Dang M-N and Greenberg HB, 1988. The rhesus rotavirus gene encoding protein VP3: location of amino acids involved in homologous and heterologous rotavirus neutralization and identification of a putative fusion region. Proc Natl Acad Sci USA, 85:645-649

Marquardt, RR, Jin LZ, Kim JW, Fang L, Frohlich AA and Baidoo SK, 1999. Passive protective effect of egg yolk antibodies against enterotoxigenic *Escherichia coli* K88+ infection in neonatal and early weaned piglets. FEMS Immunol. Medical Microbiology. 23: 283-288

Matsui SM, Offit PA, Vo PT, Mackow ER, Benfield DA, Shaw RD, Padilla-Noriega L and Greenberg HB, 1989. Passive protection against rotavirus-induced diarrhea by monoclonal antibodies to the heterotypic neutralization domain VP7 and the VP8 fragment of VP4. J Clin Microbiol 27:780-782

Matsuura F, Ohta M, Murakami K and Matsuki Y, 1993. Structures of asparagine linked oligosaccharides of immunoglobulins (IgY) isolated from egg yolk of Japanese quail. Glycoconjugate J, 10:202-213

McBee LE and Cotterill SJ, 1979. Ion-exchange chromatography and electrophoresis of egg yolk proteins. J Food Sci, 44: 656-660

McCannel AA and Nakai S, 1990. Separation of egg yolk immunoglobulins into subpopulations using DEAE-ion exchange chromatography. Can Inst Food Sci Technol J, 23:42-46

Mine Y, 2000. Production of hens immunoglobulin Y against an enzymatic-crosslinked human insulin for immunodiagnosis using a microbial transglutaminase. Submitted to J. Immun. Methods.

Mine Y, Lee SB and Stevenson RMW, 1999. Prevention of *Yersinia rukeri* infection in rainbow trout with hens egg yolk immunoglobulin. In *Egg nutrition and biotechnology.* Eds. Sim JS, Nakai S and Guenter W. CABI Publishing, Oxon, UK, pp. 341-350

Moon HW, Whipp SC and Skartvedt, SM, 1976. Etiologic diagnosis of diarrheal diseases of calves: frequency and methods for detecting enterotoxic and K99 antigen produced by *Escherichia coli.* Am J Vet Res, 37:1025-1029

Morris JA and Sojka WJ, 1985. *Escherichia coli* as a pathogen in animals. In *The Virulence of Escherichia coli.* Ed. Sussman M. Academic Press, London, pp 47-77

Nakai S, Li-Chan E and Lo KV, 1994. Separation of immunoglobulin from egg yolk. In *Egg uses and processing technologies, new developments.* Eds. Sim JS and Nakai S. CAB International, Oxon, UK, pp 94-105

Nejmeddine M, Trugnan G, Sapin C, Kohli E, Svensson L, Lopez S and Cohen J, 2000. Rotavirus spike protein VP4 is present at the plasma membrane and is associated with microtubules in infected cells. J Virol, 74:3313-3320

O'Farrelly C, Branton D and Wanke CA, 1992. Oral ingestion of egg yolk immunoglobulin from hens immunized with an enterotoxigenic *Escherichia coli* strain prevents diarrhea in rabbits challenged with the same strain. Infect Immun 60:2593-2597

Ohta M, Hamako J, Yamamoto S, Hatta H, Kim M, Yamamoto T, Oka S, Mizuochi T and Matsuura F, 1991. Structure of asparagine-linked oligosaccharides from hen egg yolk antibody (IgY). Occurrence of unusual glucosylated oligo-mannose type oligosaccharides in a mature glycoprotein. Glycoconjugate J, 8:400-413

Otake S, Nishihara Y, Makimura M, Hatta H, Kim M, Yamamoto T and Hirasawa M, 1991. Protection of rats against dental caries by passive immunization with hen egg yolk antibody (IgY). J Dent Res, 70:162-166

Otani H, Matsumoto K, Saeki A and Hosono A, 1991. Comparative studies on properties of hen egg yolk IgY and rabbit serum IgG antibodies. Lebensm Wiss U Technol 24:152-158

Parry SH and Rooke DM, 1985. Adhesins and colonization factors of *Escherichia coli*. In *The virulence of Escherichia coli*. Ed. Sussman M. Academic Press, London, pp79-159

Parvari R, Avivi A, Lentner F, Ziv E, Tel-Or S, Burstein Y and Schechter I, 1988. Chicken immunoglobulin c heavy chains: Limited Vh gene preparation, combinatial diversification by D gene segments and evolution of the heavy chain locus. EMBO J, 7:739-744

Peralta RC, Yokoyama, H, Ikemori Y, Kuroki M and Kodama Y, 1994. Passive immunization against experimental salmonellosis in mice by orally administered hen egg yolk antibodies specific for 14-kDa fimbriae of *Salmonella enteritidis*. J Med Microbiol, 41:29-35

Polson A, Wechmar MBV and Fazakerley G, 1980. Antibodies to proteins from yolk of immunized hens. Immunol Commun, 9:495-514

Polson A, 1990. Isolation of IgY from the yolk of eggs by a chloroform polyethylene glycol procedure. Immun Invest 19:253-258

Prasad BVV, Wang GJ, Clerx JPM, and Chiu W, 1988. Three-dimensional structure of rotavirus. J Mol Biol, 199:269-275

Reilly RM, Domingo R and Sandhu J, 1997. Oral delivery of antibodies; Future pharmacokinetic trends. Clin Pharmacokinet, 4:313-323

Reynaud CA, Dahan A and Weill JC, 1983. Complete sequence of a chicken λ light chain immunoglobulin derived from the nucleotide sequence of its mRNA. Proc Natl Acad Sci USA, 80:4099-4103

Romanoff AL and Romanoff AJ, 1949. *The avian egg*. Wiley, New York

Rose ME, Orlans E and Buttress N, 1974. Immunoglobulin classes in the hen's eggs: Their segregation in yolk and white. Eur J Immunol 4:521-528

Ruggeri RM and Greenberg HB, 1991. Antibodies to the trypsin cleavage peptide VP8* neutralize rotavirus by inhibiting binding of virions to target cells in culture. J Virol, 65:2211-2219

Sack RB, 1986. Antimicrobial prophylaxis of travellers' diarrhea: A selected summary. Rev. Infect. Diseases, 8: S160-166

Schade R, Pfister C, Halatsch R and Henklein P, 1991. Polyclonal IgY antibodies from chicken egg yolk-an alternative to the production of mammalian IgG type antibodies in rabbits. ATLA, 19:403-409

Schade R, Burger W, Schoneberg T, Schniering A, Schwarzkopf C, Hlinak A and Kobilke H, 1994. Avian egg yolk antibodies. The egg laying capacity of hens following immunization with antigens of different kinds, origin and the efficiency of egg yolk antibodies in comparison to mammalian antibodies. Alternativen Zu Tierexperimenten, 11:75-84

Schmidt P, Hafner A, Reubel GH, Wanke R, Franke V, Losch U and Dahme E, 1989. Production of antibodies to canine distemper virus in chicken eggs for immunochemistry. J Vet Med 36:661-668

Schmidt P, Wiedemann V, Kuhlmann R, Wanke R, Linckh E and Losch U, 1989. Chicken egg antibodies for prophylaxis and therapy of infectious intestinal diseases: in vitro studies on gastric and enteric digestion of egg yolk antibodies specific against pathogenic *Escherichia coli* strains. J Vet Med, 36:619-628

Shimizu M, Fitzsimmons RC and Nakai S, 1988. Anti-*E. coli* immunoglobulin Y isolated from egg yolk of immunized chickens as a potential food ingredient. J Food Sci, 53:1360-1366

Shimizu M, Fitzsimmons RC and Nakai S, 1989. Serum and egg antibody responses in chickens to *Escherichia coli.* Agric Biol Chem 53:3233-3238

Shimizu M, Nagashima H, Sano K, Hashimoto K, Ozeki M, Tsuda K and Hatta H, 1992. Molecular stability of chicken and rabbit immunoglobulin G. Biosci Biotech Biochem 56:270-274

Sim JS and Nakai S, 1994. *Egg uses and processing technologies; New developments.* CAB International, Oxon, UK.

Sim JS, Nakai S and Guenter W, 1999. *Egg nutrition and biotechnology*. CAB, Oxon, UK.

Stevenson RMW, Flett D and Raymond BT, 1993. Enteric redmouth (ERM) and other enterobacterial infections of fish. In *Bacterial diseases of fish*. Eds. Inglis V, Roberts RJ and Bromage NR. Blackwell Scientific Publications, Oxford, pp 80-105

Svennerholm AM, Holmgren J and Sack DA, 1989. Development of oral vaccines against enterotoxigenic *Escherichia coli.* Vaccine, 7:196-198

Tacket CO, Losonsky G, Link H, Hoang Y, Guerry P, Hilpert H and Levine MM, 1988. Protection by milk immunoglobulin concentrate against oral challenge with enterotoxigenic *Escherichia coli* . New Eng J Medicine, 318:1240-1243

Thalley BS and Carroll SB, 1990. Rattlesnakes and scorpion antivenoms from the egg yolk of immunized hens. Biotechnol, 8:934-937

Thorns CJ, Sojka, MG and Chasey D, 1990. Detection of a novel fimbrial structure on the surface of *Salmonella enteritidis* by using a monoclonal antibody. J Clin Microbiol, 28:2409-414

Thorns CJ, Sojka MG, McLaren M and Dibb-Fuller M, 1992. Characterization of monoclonal antibodies against a fimbrial structure of *Salmonella enteritidis* and certain other serogroup D salmonelae and other application as serotyping reagents. Res Vet Sci 53:300-308

Wang HY and Imanaka T, 1995. *Antibody expression and engineering*. American Chem Soc, Washington, DC.

Warr GW, Magor KE and Higgins DA, 1995. IgY: clues to the origins of modern antibodies. Immunol Today, 16:392-398

Wiedemann V, Linkh E, Kuhlmann R, Schmidt P and Losch U, 1990. Chicken egg antibodies for prophylaxis and therapy of infectious intestinal diseases: *in vivo* tenacity test in piglets with artificial jejunal fistula. J Vet Med 37:163-172

Wiedemann V, Linckh E, Kuhlmann R, Schmidt P and Losch U, 1991. Chicken egg antibodies for prophylaxis and therapy of infectious intestinal diseases: in vivo studies on protective effects against *Escherichia coli* diarrhea in pigs. J Vet Med, 38:283-291

Yokoyama H, Peralta RC, Diaz R, Sendo S, Ikemori Y and Kodama Y, 1992. Passive protective effect of chicken egg yolk immunoglobulins against experimental enterotoxigenic *Escherichia coli* infection in neonatal piglets. Infect Immun, 60:998-1007

Yokoyama H, Peralta RC, Umeda K, Hashi T, Icatlo FC, Kuroki M, Ikemori Y and Kodama Y, 1998. Prevention of fatal salmonellosis in neonatal calves, using orally administered chicken egg yolk *Salmonella* specific antibodies. Am J Vet Res, 59:416-420

Yokoyama H, Umeda K, Peralta RC, Hashi T, Icatlo F, Kuroki M, Ikemori Y and Kodama Y, 1998. Oral passive immunization against experimental salmonellosis in mice using chicken egg yolk antibodies specific for *Salmonella enteritidis* and *S. typhimurium*. Vaccine, 16:388-393

Yolken F, Leister S, Wee R, Miskuff S and Vodnerfecht S, 1990. Immunoglobulins and other modalities for the prevention and treatment of enteric viral infections. J Clin Immunol 10:80-86

Yolken RH, Leister F, Soik-Bi W, Miskuff R and Vonderfecht S, 1988. Antibodies to rotavirus in chickens eggs: a potential source of antiviral immunoglobulins suitable for human consumption. Pediatrics, 81:291-295

Chapter 14

THE PRODUCTION OF RECOMBINANT CYTOKINES IN PLANTS

R. Menassa[1], A. Jevnikar[2] and J. Brandle[1]
[1]*Southern Crop Protection and Food Research Centre, Agriculture and Agri-Food Canada, 1391 Sandford St., London, Ontario, N5V 4T3, Canada;* [2] *MOTS -Transplantation and Immunobiology Group, London Health Sciences Centre, 339 Windermere Road, London, Ontario, N6A 5A5, Canada*

Keywords: recombinant cytokines, transgenic plants, molecular farming, biopharmaceuticals

Abstract Cytokines are soluble, secreted glycoproteins that, like hormones, regulate the functions of cells and tissues. Many cytokines possess therapeutic properties and attempts to produce them in large quantities are ongoing. Transgenic plants are one system that has the potential to yield large amounts of low cost recombinant proteins. In addition, plants have the ability to introduce post-translational modifications needed for biological activity of many cytokines. In this review, we describe the efforts aimed at producing recombinant cytokines in plants and discuss some of the issues that have arisen from such activities.

1. INTRODUCTION

Cytokine is a term used to describe a broad array of soluble proteins and peptides that, at very low concentrations (10^{-9} to 10^{-12}M), act as biological regulators of the functional activities of cells and tissues. Most cytokines are glycoproteins that are secreted through normal pathways. Cytokines are usually unrelated in terms of amino acid sequence, but they can be grouped according to secondary or tertiary structure or function. Since cytokines are involved in a vast array of physiological processes, including the pathogenesis of many diseases, they have great potential for therapeutic use. There are over 150 cytokines, many of which (e.g. GM-CSF, erythropoietin, interferon (a, b), IL-2

L. Erickson et al. (eds.),
Molecular Farming of Plants and Animals for Human and Veterinary Medicine, 319–338.

and IL-11 etc.) have established or emerging clinical applications (Meager, 1998; Walsh, 2000). All of these cytokines are produced in vivo at extremely low concentrations and because of their size and complexity, conventional synthesis is not practical. Therefore, recombinant DNA technology must be used to produce cytokines in amounts sufficient for clinical experimentation and therapeutic application (Mire-Sluis, 1999). Although effective production systems based on *E. coli*, mammalian, yeast or insect cells exist, they all have limitations related to cost and scale-up. New methods of expression based on transgenic plants address these issues.

The expression of the first mammalian gene in plants was a major advance in plant biotechnology (Lefebvre et al., 1987). Since those early experiments numerous mammalian genes have been expressed and plants have been recognized as efficient, low-cost, non-sterile bioreactors that have enormous potential for the production of proteins valuable to both medicine and industry (Evangelista et al., 1998). Plants offer several distinct advantages over conventional expression systems for the production of recombinant proteins. Plants are higher eukaryotes and are able to properly fold and process proteins that bacterial systems cannot. Clear evidence of this was presented by Julian Ma and co-workers who showed that the four chains of a secretory immunoglobulin were properly expressed and assembled in plants, and that the antibody was fully functional in human clinical applications (Ma et al., 1995; Ma et al., 1998). This ability may be particularly important in producing plant recombinant cytokines that have in vivo activities identical to their native counterparts (Rollwagen and Baqar 1996). In addition, plant systems produce recombinant proteins free of mammalian pathogens. Blood proteins derived from human sources carry potential risks of pathogens, triggering research into heterologous systems for the production of recombinant blood proteins. In response to this need, both serum albumin and human haemoglobin have now been produced in plants (Dieryck et al. 1997; Sijmons et al. 1990). Unlike fermentation-based bacterial and mammalian cell culture systems, protein production in plants is not restricted by physical facilities. Agricultural scale production ensures the availability of recombinant proteins in amounts sufficient for extensive clinical studies and therapeutic use. Although extraction and purification costs may be similar to other systems, the upfront cost of producing the raw plant material will be significantly lower than that of fermentation (Evangelista et al., 1998). The importance of cytokines to modern medicine makes this group of glycoproteins a clear target for the development of low cost production methods.

This review examines progress towards the production of recombinant cytokines in plants and identifies research directions that will support these efforts.

2. EXPERIMENTAL RESULTS FROM PLANTS

Several cytokine genes have been introduced into plant genomes. Expression of these genes and levels of protein accumulation have been variable,but much can be learned from these experiments (Table 17). The properties of each cytokine (or "phytokine") and details of its characterization in plants are described below.

Table 17. Human cytokines expressed in plants

Cytokine	Plant	Expression levels	Biological activity		Reference
			In vitro	In vivo	
Erythropoietin	Tobacco cell suspension	0.0026% total extractable protein	Yes	No	Matsumoto, 1995
	Tomato	0.342 ng/g leaf fresh weight	Nd	Nd	He et al, 1998
Alpha interferon	Tobacco	Nd	Nd	Nd	Smirnov et al., 1990
	Rice	2048 units/g tissue	Yes	Nd	Zhu et al., 1994
	Turnip via CaMV	2 μg/g tissue fresh weight	Yes	Nd	De Zoeten et al., 1989
Beta interferon	Tobacco	$5x10^7$ units/mg IFN-β	Yes	Nd	Edelbaum et al, 1992
GM-CSF	Tobacco cell suspension	250 ug/liter culture medium (0.5% total proteins)	Yes	Nd	James et al., 2000
	Tobacco seeds	258 ng/ml	Yes	Nd	Ganz et al., 1996

Cytokine	Plant	Expression levels	Biological activity		Reference
			In vitro	In vivo	
Interleukin-4	Tobacco cell culture	455 ug/l cell culture medium and 1.1 μg/g callus	Yes	Nd	Magnuson et al., 1998
Interleukin-6	Tobacco plants	Nd	Yes	Nd	Kwon et al., 1995
Interleukin-10	Tobacco	0.005% total extractable protein	Yes	Nd	Menassa et al., in press

Nd: not determined.

2.1 Human Erythropoietin

Erythropoietin (Epo) is a cytokine involved in the regulation and maintenance of circulating erythrocyte mass. Recombinant human Epo (rhEpo) produced in mammalian cell lines is approved for the treatment of anemia both in the USA and in Europe (Walsh, 2000) and represents a huge worldwide market. The Epo gene encodes a 193 amino acid precursor protein containing a 27 residue amino terminal signal peptide. The mature Epo protein contains 166 amino acids with four cysteines, which form internal disulphide bridges, and three N-glycosylation sites (Recny et al., 1987). The protein is heavily glycosylated, doubling its predicted mass of 18 kDa to 36 kDa. Glycosylation of Epo is important for both stability and in vivo biological activity (Wasley et al., 1991).

Recombinant human Epo was expressed both in cultured tobacco cells and in tomato plants (Matsumoto et al., 1993; Matsumoto et al., 1995; He et al., 1998). In both cases, the hEpo cDNA was placed under the control of the CaMV 35S promoter. In tobacco, two constructs were produced, one with and the other without the hEpo native signal peptide. Extremely low concentrations of Epo were produced with the truncated signal peptide construct (1pg/g wet cells). Concentrations of 0.8 ng/g cells (0.0026% of total extractable proteins) were obtained when the hEpo signal peptide was used and the protein targeted through the secretory pathway (Matsumoto et al., 1995; Matsumoto et al., 1993). The hEpo protein accumulated in transgenic tomato leaves to a concentration of 0.342 ng/g of leaf fresh weight (He et al., 1998).

Human Epo produced in cultured tobacco cells was not secreted into the medium, and was thought to remain associated with the cell wall. The glycosylation pattern of Epo from tobacco cells was different from that of rhEpo produced in CHO cells. Deglycosylation revealed that the size difference between tobacco Epo and rhEpo was due to differences in the N-linked oligosaccharides. Also, tobacco Epo bound to Con A, indicating a high mannose content. The rhEpo bound to wheat germ agglutinin, which indicates the presence of sialic acid in the N-linked oligosaccharides (Matsumoto et al., 1995). These differences are typical of those found with other mammalian glycoproteins produced in plants (Wee et al., 1998)

Recombinant Epo produced in tobacco was found to possess higher in vitro biological activity than rhEpo produced in CHO cells, presumably due to the absence of sialic acid in the carbohydrates attached to Epo. Indeed, enzymatic removal of the terminal sialic acids from recombinant human erythropoietin resulted in increased affinity of EPO to its receptor (Tsuda et al., 1990). In vivo activity, though, was not detectable in tobacco Epo, which was attributed to the rapid clearance of the protein in the circulation, again the result of the absence of terminal sialic acid on the glycan chains (Matsumoto et al., 1995).

2.2 Human Interferons

Interferons (IFNs) induce antiviral activity in a wide variety of cell types and are multifunctional and pleiotropic regulators of immune function. They stimulate differentiation in many cells and have anti-proliferative activity in certain types of tumor cells in vitro (Meager, 1998). Recombinant interferon a_{2a} is approved in the USA and in Europe for the treatment of hairy cell leukemia, hepatitis B, C and various cancers. Several variants of recombinant IFN-b_1 are also approved for treatment of multiple sclerosis, and recombinant IFN-g_{1b} is approved for treatment of chronic granulomatous disease (Walsh, 2000). All recombinant interferons used in approved therapies are currently produced in *E. coli*.

Human alpha interferon genes constitute a superfamily of related genes, all of which encode precursor polypeptides with a signal peptide of 23, mainly hydrophobic, amino acids. Upon cleavage of the signal sequence, mature IFN-a subtypes of 166 amino acids are generated (Henco et al., 1985). The calculated molecular weight of IFN-a is 18.5 kDa, although observed weights vary between 17 and 26 kDa because of post-translational modifications. All IFN-a subtypes contain four cysteine residues that form 2 disulphide bridges. No glycosylation sites exist in IFN-a subtypes except for IFN-a_{14}, which contains two N-glycosylation sites, and IFN-a_2 which contains one O-glycosylation site. Both the N-sites and the O-site have glycans attached (Nyman et al., 1998). The crystal structure of recombinant human interferon-alpha 2b (hIFN-a_{2b}) revealed that hIFN-a_{2b} exists in the crystal as a noncovalent dimer, and that extensive interactions in the dimer interface are mediated by a zinc ion (Radhakrishnan et al., 1996).

A single gene encodes human interferon b. The mature peptide contains 166 amino acids that are 30% homologous to the interferon a genes, thought to be due to a common origin for the alpha and beta interferon (De Meyer and De Meyer-Guignard, 1994). IFN-b contains three cysteine residues, two of which form a disulphide bridge. One N-glycosylation site exists to which oligosaccharides are attached in natural IFN-b. The overall molecular weight of glycosylated IFN-b is 21 kDa. Human IFN-b is predicted to contain five alpha helices folded into an alpha helical bundle structure, and to dimerize at a zinc binding site at the interface of the two subunits (Karpusas et al., 1997). The details of the observed carbohydrate structure, together with biochemical data, implicate the glycosylation of HuIFN-b as an important factor in the solubility, stability and, consequently, activity of the protein (Karpusas et al., 1998).

Tobacco, rice and turnip have been used for the expression of alpha interferon (Zhu et al., 1994; Smirnov et al., 1990; De Zoeten et al., 1989), and beta interferon has been produced in tobacco cell suspension (Edelbaum et al., 1992).

2.2.1 Interferon Alpha

The hIFN-a gene was placed under the control of the CaMV 35S promoter, and introduced into tobacco cells by Agrobacterium infection of leaf discs. Whole plants were regenerated, and both transcription and translation were observed (Smirnov et al., 1990). In rice, the bi-directional promoter P1'2' from *Agrobacterium tumefaciens* T- DNA was used to drive the expression of both the *Npt II* gene and the IFN-a_D cDNA (Zhu et al., 1994). Crude extracts from transgenic rice leaf tissue or cell suspension cultures inhibited infection of human amniotic WISH cells by the vesicular stomatitis virus (VSV), thus exhibiting interferon activity. Maximum activity was reported as 2048 units/gram tissue (Zhu et al., 1994).

De Zoeten and coworkers (1989) engineered the cauliflower mosaic virus (CaMV) to carry the hIFN-a_D gene by replacing the ORFII of CaMV DNA with the human IFN-a_D coding sequence minus the signal peptide. The stop codon of ORF II was added to IFN-a_D and gave a stable CaMV strain. The presence of hIFN-a_D was determined by immuno-gold labeling and electron microscopy. IFN-a_D was localized in inclusion bodies produced by CaMV. The highest IFN activity was observed in young systemically infected leaves four weeks after symptom appearance. In those leaves, the level of expression of IFN was of the order of 2 micrograms/gram tissue fresh weight.

An antiviral assay using reduction of vesicular stomatitis virus infection of MDBK cells, and standardized against the WHO international IFN-a standard, was used to evaluate the biological activity of plant recombinant IFN-a. Interestingly, deletion of a cysteine known to be involved in disulphide bridge formation did not affect IFN-a D activity. Instead, plants infected with this construct showed the highest level of biological activity. It was proposed that the antiviral activity of IFN-a_D would inhibit superinfection of turnip plants by turnip yellow mosaic virus, (TYMV), a single stranded plus sense RNA virus. However, such inhibition was not observed upon simultaneous or sequential inoculation of CaMV and TYMV (De Zoeten et al., 1989).

2.2.2 Interferon Beta

Sequences containing the human IFN-b gene, including the signal peptide and 3'UTR, were placed under the control of the 35S promoter and terminator, then introduced into the tobacco genome by Agrobacterium transformation (Edelbaum et al., 1992). IFN-b was transcribed and translated in tobacco cells. A 21 kDa band was observed by immunoblot analysis, indicating that tobacco-produced IFN-b was glycosylated to similar degree as hIFN-b. Purified tobacco IFN-b was shown to inhibit the cytopathic effect of VSV in human WISH cells. The specific activity of plant-derived IFN-b was calculated to be 5×10^7 units/mg, which compares well with the value of 3.2×10^7 unit/mg of the rhIFN-b standard. However, preliminary results suggest that IFN-b was not able to protect plants from TMV infection.

2.3 Human Granulocyte-Macrophage Colony Stimulating Factor (GM-CSF)

GM-CSF is produced by a broad range of normal cell types, and induces the proliferation of murine bone marrow- or spleen-derived haematopoietic cells containing granulocyte and macrophage precursors. Recombinant GM-CSF produced in *E. coli* is approved in the USA for the treatment of chemotherapy-induced neutropenia and to enhance haematopoietic recovery after bone marrow transplantation (Walsh, 2000).

The human GM-CSF cDNA codes for a 144 amino acid precursor protein. Cleavage of the 17 amino acid signal peptide gives rise to a 127 amino acid mature protein containing two internal disulphide bridges and two potential N-linked glycosylation sites. When expressed in mammalian cells, GM-CSF is secreted as a glycoprotein of 18-34 kDa. Non-glycosylated GM-CSF produced by *E. coli*, is fully functional in vitro. Similarly, yeast or baculovirus-derived recombinant GM-CSF, which exhibit different glycosylation patterns from mammalian GM-CSF, are also active. Thus, glycosylation is not essential for GM-CSF activity, but it leads to an increased affinity of GM-CSF to its receptor (Meager, 1998; Rasko and Gough, 1994).

In plants, GM-CSF was produced in tobacco suspension culture (James et al., 2000) and in tobacco seeds (Ganz et al., 1996). James and coworkers (2000) placed the GM-CSF cDNA sequence under the control of the CaMV 35S promoter and introduced the construct into tobacco cells by Agrobacterium-mediated transformation. A 6-His tag was added to the carboxy-terminus of GM-CSF to facilitate purification of the protein, and the tobacco etch virus leader sequence was added at the 5'end of the gene to enhance translation. The mean concentration of GM-CSF protein was increased significantly when the TEV leader sequence was added to the gene, at 32.4 mg GM-CSF/g total soluble protein, more than two fold the GM-CSF production in non-TEV controls. When grown in suspension cell culture, transgenic cells secreted GM-CSF protein into the medium to a level of 250 mg GM-CSF/l culture medium, making up 0.5% of total proteins in the culture medium. Some GM-CSF, however, was retained intracellularly (150 mg GM-CSF/l culture medium). In constructs carrying a His tag, the maximum secreted protein reached 100 mg GM-CSF/l medium. Immunoblot analysis showed that different clones produce different forms of GM-CSF interpreted to be either different glycosylation patterns, differences in the processing, and/or dimerization of the recombinant

GM-CSF. The secreted GM-CSF, however, showed similar biological activity as a recombinant standard in a proliferation assay, based on mass equivalency and a single point measurement.

Ganz and coworkers (1996) took advantage of the ability of plants to secrete proteins into intracellular compartments (endosperm protein bodies), and produced GM-CSF in tobacco seeds. The promoter of the rice glutelin Gt3 gene was used to target GM-CSF to the endosperm of tobacco seeds. The glutelin signal sequence, and eight amino acids from the glutelin coding sequence were fused to the GM-CSF coding sequence. Protein extracts from seeds showed expression of GM-CSF up to a level of 258 ng/ml, and a single 18 kDa polypeptide was detected, which co-migrated with the recombinant GM-CSF protein. Biological activity of tobacco seed recombinant GM-CSF was assessed by a proliferation assay with a GM-CSF dependent cell line, TF-1, based on mass equivalency and a single point measurement. Tobacco seed recombinant GM-CSF was able to stimulate proliferation of TF-1 cells to a greater level than purified recombinant GM-CSF.

2.4 Human Interleukin-2

The principal activity of interleukin-2 (IL-2) is growth stimulation of antigen-activated T cells (Smith, 1988). It is now recognized to be critical in immunological tolerance induction following activation induced cell death (AICD) and so has a diverse role in the immune system (Refaeli et al., 1998). In addition, IL-2 promotes T cells to produce other lymphokines, including interferon gamma and interleukin-4, revealing its capacity to act as a differentiation signal (Goldsmith and Greene, 1994). Because of its central role in the mediation of the immune response, IL-2 has been shown to have important diagnostic and therapeutic value (Yang et al., 1991). Recombinant IL-2 produced in *E. coli* is approved in the USA for the treatment of renal cell carcinoma (Walsh, 2000).

The IL-2 cDNA encodes a polypeptide of 153 amino acids, of which 20 amino acids form the signal sequence. The mature IL-2 protein has three cysteines, two of which form the intra-chain disulphide bond that is required for biological activity (Tsuji et al., 1987). Substitution of the cysteine residue at position 125 (which is not involved in disulphide bond formation) with alanine improves the yield of active IL-2 from *E. coli*, probably by preventing the mismatching of cysteines and the formation of interchain dimers (Wang et al., 1984). Naturally occurring IL-2 is a glycoprotein with varying amounts of glycosylation, and a molecular weight range of 15-18 kDa. The IL-2 protein does not contain N-glycosylation sites, but an O-glycosylation site is present near the N terminus. Recombinant IL-2 produced in *E. coli* lacks the O-linked oligosaccharides, but is still fully functional.

In plants, the human interleukin-2 gene was placed under the control of the 35S promoter, introduced into the tobacco genome via *Agrobacterium*-mediated transformation. Transformed cells were analyzed in cell suspension culture (Magnuson et al., 1998) and as whole regenerated tobacco plants (Park and Hong, 1997). Following northern analysis, a single IL-2 transcript of 900 nucleotides was detected in transgenic tobacco plants, and IL-2 transcript levels were much higher in imbibed seeds than in fully-grown leaves (Park and Hong, 1997). In cell suspension culture, the IL-2 protein was retained intracellularly and a minimal amount of IL-2 was found to have been secreted into the medium. Maximum accumulation levels of 250-350 ng IL-2/g callus, and 90 ng/ml of culture medium were observed. The IL-2 dependent cell line CTLL-2 was used in a thymidine incorporation assay to assess the biological activity of the plant-recombinant IL-2 secreted into the medium or retained inside the cells. Most activity was found in the culture medium and very little in the cell lysates, which contained most of the measured IL-2. However, crude cell lysates were added directly to the mouse indicator cells, and might have been inhibitory in the proliferation assay (Magnuson et al., 1998).

2.5 Human Interleukin-4

The IL-4 protein exhibits multiple immunomodulatory functions on a wide range of cell types, suggesting a central role for IL-4 in the modulation of immune and inflammatory responses (Banchereau and Rybak, 1994). In mice, IL-4 is crucial to TH2-cell responses, some of which are directed against pathogenic organisms, in particular parasites such as flatworms and nematodes (Kopf et al., 1995). Human IL-4 is synthesized as a polypeptide of 153 amino acids. A signal peptide of 24 amino acids is cleaved to give rise to the mature IL-4 protein (Ohara et al., 1987). The primary amino acid sequence of IL-4 contains six cysteines, which are engaged in disulphide bridges within the molecule, and two N-glycosylation sites. In its native form, human IL-4 is a glycoprotein with a molecular weight of about 19 kDa (Le et al., 1988). The biological activity of IL-4 is completely dependent on the proper folding and disulfide bond formation necessary for receptor binding (Tsuji et al., 1987)The human IL-4 gene was expressed in tobacco cell culture. Overall concentrations of IL-4 in transformed calli were 1100 ng/g tissue fresh weight. In suspension culture, the intracellular level of IL-4 reached 275 ng/ml of the culture medium, while the secreted level of IL-4 was 180 ng/ml, for a total of 455 ng/ml of culture medium. Immunoblot analysis of the plant recombinant IL-4 revealed a doublet at approximately 18-19 kDa, while the commercially available recombinant human IL-4 migrates as a single band of 18 kDa. The doublet is thought to represent either two different glycosylated forms of IL-4 or incomplete processing of the IL-4 protein (Magnuson et al., 1998).

A thymidine incorporation assay was used to measure the activity of intracellular or extracellular IL-4 produced by plant cells. The IL-4 dependent cell line CT.h4S was used for this assay. The secreted IL-4 exhibited biological activity, but only about one fourth of that expected from the amount of protein being produced. Crude extract containing intracellular IL-4 exhibited no activity. The lack of, or reduction of activity could be due to inhibitory factors in the culture medium, or to the two forms of IL-4 detected by immunoblotting (Magnuson et al., 1998).

2.6 Human Interleukin-6

Human IL-6 is associated with a wide range of biological activities, including immunological, haematological and anti-tumor effects, which suggests that it has potential as an anti-tumor agent (for a review, see Borden and Chin, 1994). It is both a growth factor and a stimulatory factor; has differentiating effects (Van Snick, 1990), and plays an important role in human defense against various human pathogens (Hirano et al., 1990). IL-6 knock-out mice have both impaired immune and acute-phase responses to tissue damage and infection (Kopf et al., 1994).

The human IL-6 pre-protein consists of 212 amino acids, including a hydrophobic signal sequence of 28 amino acids (Hirano et al., 1986). Mature IL-6 has a molecular mass ranging from 21-28 kDa. The variation in mass is due to post-translational modifications, including N- and O-linked glycosylations and phosphorylations (May et al., 1988). Glycosylation of the protein was found to have no important effect on activity, but is important for stability (Orita et al., 1994)

The human IL-6 coding sequence, including 3 amino acids from the signal peptide, was introduced into tobacco under the control of the 35S promoter (Kwon et al., 1995). A single RNA species was detected upon northern analysis and a single protein band of 21 kDa was observed in immunoblots, which is the expected size of a properly processed, monomeric, unmodified IL-6 protein. The plant recombinant IL-6 was assayed with IL-6-dependent hybridoma B9 cells. Partially purified prIL-6 was able to stimulate the proliferation of B9 cells in a dose dependent manner, starting at about 10 pg of IL-6/ ml of the reaction mixture. The stimulatory activity was saturated at about 200 ng IL-6/ml of the reaction mixture, and was completely neutralized with co-incubation of prIL-6 with goat anti-IL-6 antisera (Kwon et al., 1995), indicating that no plant factor other than IL-6 was involved in the stimulation of the B9 cells.Human Interleukin-10

Interleukin-10 is a contra-inflammatory cytokine that has multiple roles in the regulation of immune responses. It acts by inhibition of T-cell, monocyte and macrophage function, B-cell proliferation and inhibition of TH1-type cytokines like g-interferon and IL-6 (Moore et al., 1993). Indications for IL-10 are developing for the treatment of Crohn's disease, psoriasis, allergic inflammation, septic shock, diabetes and thyroiditis (Batteux et al., 1999; Leach et al., 1999; Pauza et al., 1999; Pretolani et al., 1997; Reich et al., 1998; Zheng et al., 1995). Interleukin-10 is a homodimeric protein produced in the secretory system of activated Th-2 cells, B-cells, macrophages and monocytes (Vieira et al., 1991). Assembly of the dimer begins with cleavage of an 18 amino acid signal peptide, followed by the formation of two disulfide bonds within the monomer (Windsor et al., 1993). A non-covalent association between monomer units leads to the formation of the biologically active homodimer (Zdanov et al., 1995). In contrast to murine IL-10, human IL-10 is not glycosylated although it is biologically active in murine cell assays (Vieira et al., 1991). Recombinant IL-10 can be produced by insect cells or by bacteria when coupled with in vitro re-folding (Moore et al., 1993; Vellekamp et al., 1994, 1998).

We have produced the human interleukin-10 cytokine (hIL-10) in low nicotine tobacco plants (cv. "81V9"). Targeting hIL-10 to the apoplast resulted in very low protein accumulation (0.8 ng/mg protein; 0.00008% TSP). Retention of IL-10 in the lumen of the ER resulted in a 70-fold increase in accumulation (55 ng/mg protein or 0.0055% TSP). The highest expressing plant contained a single copy of the transgene and produced biologically active human IL-10 (Menassa et al., in press). This plant was selfed and homozygous progeny were used as pollen donors in a cross with male-sterile low nicotine tobacco (81V9). The resulting progeny were hemizygous for IL-10 and male sterile. A field test was conducted with homozygous fertile and hemizygous sterile tobacco plants, and plants were examined for nicotine content and IL-10 levels. No differences were observed between the male sterile and fertile plants (Menassa et al., 2001). This system allows containment of the transgene by growing in the field only male sterile plants that cannot spread pollen to other plants, and containment of the recombinant protein by producing it in a non-food crop. The level of nicotine in 81V9 is 10 times lower than commercial tobacco varieties, making the oral administration of plant tissue or crude protein extracts possible while reducing nicotine side effects.

3. PRODUCTION PLATFORM CONSIDERATIONS

Many crops are being investigated for use in the production of recombinant proteins for the prevention and treatment of disease. It is not yet clear which crop platform is best. Many technical issues, such as intra-cellular targeting, optimization of recombinant protein yield, and purification are being addressed. However, as the production of plant recombinant proteins moves from the bench to pilot scale, practical issues such as safety and containment need consideration. The recombinant cytokines produced in molecular farming are biologically active in mammals, and the effects of chronic exposure following escape into the environment are not known. As a result, plants can only compete with existing recombinant protein systems if costly containment procedures and regulatory restrictions do not inflate production costs. From both a regulatory and public safety standpoint, a non-food crop platform with limited capacity for the spread of pollen or seed offers sufficient containment.

Tobacco is a non-food crop that has been the subject of many years of breeding and agronomic research and can be used as a strong base for a field production system for recombinant protein production. We have developed a production system based on low-nicotine tobacco transgenic lines. The low-alkaloid background genotype addresses the issue of unwanted secondary metabolites, it allows for in vivo use, and is optimized for agricultural production. The plants express the transgene uniformly and protein production is based on leaves, not seeds or tubers, which further limits the potential for escape. The plants are grown at a high density to maximize biomass yield and are harvested after 30-40 days, eliminating flower production and allowing the system to be adapted to a broad range of production environments. Multiple staggered plantings ensure the availability of plant material for processing throughout the growing season.

4. QUALITY CONTROL AND STANDARDIZATION ISSUES

Although the biggest challenge facing the production of recombinant proteins in plants is optimizing yields, the quality of therapeutic recombinant proteins needs careful attention. No single test is able to completely characterize a protein, so a range of different tests is needed. Immunological assays and physico-chemical analyses can reveal the composition of the protein, and biological activity assays are essential in determining the functionality of the recombinant proteins. However, these assays need to be standardized to ensure consistency of results between laboratories (Mire-Sluis, 1999). Various methods of cytokine production have been developed that utilize cell sources such as *E. coli*, yeast and Chinese hamster ovary (CHO) cells. This results in a situation where identical amounts of a given cytokine from different sources may have very different activities. It is therefore not possible to correctly calibrate such proteins by mass alone. It is thus essential that the biological activity of a cytokine be monitored with a stable, well-characterized reference standard calibrated in terms of biological units (Mire-Sluis et al., 1998). To allow a valid comparison of the standard and the sample, in vitro biological activity experiments should be conducted using parallel-line analysis. Such analysis involves the production of dose-response curves by a dilution series of both calibrant and unknown. (Mire-Sluis and Thorpe, 1998).

Establishing biological activity in vitro is the first step toward the development of a recombinant therapeutic protein. However, it is not always an indication of the in vivo activity of that protein. Because most cytokines are glycoproteins, differences in the glycosylation patterns of the recombinant versus the native protein can lead to problems of rapid clearance or of immunogenicity in the host (Matsumoto et al., 1995; van Ree et al., 2000). The identification of the N-glycosylation pattern is a key step in the quality control of these proteins. Mammalian glycoproteins produced in transgenic plants are glycosylated, and usually retain biological activity (Cabanes-Macheteau et al., 1999). However, no available heterologous system can produce mammalian glycoproteins with their native glycans (Altmann, 1997). Indeed, plants, insects and yeast do not introduce sialic acid on their glycan side chains (Cabanes-Macheteau et al., 1999). Non-native glycosylation patterns can reduce the in vivo half-life of recombinant plasma glycoproteins such as erythropoietin (Matsumoto et al., 1995). Moreover, plant glycans contain b (1,2)-xylose and a(1,3)fucose which are absent from mammalian glycans. Although the differences in glycosylation patterns do not affect biological activity in vitro, they could potentially cause an allergic reaction when administered in vivo (Lerouge et al., 1998, also ch. 3, this volume). ER-resident proteins such as calreticulins were recently shown to lack complex glycans containing b (1,2)-xylose and a(1,3)fucose (Pagny et al. 2000). The tetrapeptide KDEL or HDEL was previously shown to retain heterologous proteins in the lumen of the ER (Munro and Pelham, 1987). However, it does not prevent protein recycling from the Golgi to the ER. Such a retrieval mechanism modifies the structure of the glycan side chains and adds the immunogenic fucose and xylose residues (Pagny et al., 2000). Other signals, as yet unidentified, exclude some ER proteins from retrieval to the Golgi apparatus. Once these signals are defined, recombinant proteins could be engineered to accumulate in the ER compartment. Engineering the plant glycosylation machinery by inactivating target enzymes such as the fucosyl transferase or xylosyl transferase or by introducing new enzymes such as the sialyl transferase or galactosyl transferase into plants is another approach toward the humanization of recombinant glycoproteins produced in plants (Palacpac et al., 1999; Wee et al., 1998; Zablackis et al., 1996; Lerouge et al., ch. 3, this volume).

5. CONCLUSION

The body of literature on the expression of cytokines in plants is developing quickly, and in every case examined so far, plants were able to produce biologically active cytokines. However, the standardization of biological activity is still variable despite its importance for the comparison of plants to other cytokine production systems. Issues such as the choice of a crop platform or the development of scalable purification systems have not been addressed. Most of the cytokines produced in plants have not been analyzed with regard to glycosylation and immunogenicity. Thus far, no cytokine has been expressed in plants at economic concentrations. Translational enhancers, tissue-specific promoters, intracellular targeting and optimized 3'UTRs are examples of elements that have potential for improving recombinant protein yields. If production yields and glycosylation issues can be addressed, plant-based cytokine production will offer huge advantages over conventional systems.

REFERENCES

Altmann F, 1997. More than silk and honey--or, can insect cells serve in the production of therapeutic glycoproteins? Glycoconj J, 5:643-646

Banchereau J and Rybak ME, 1994. Interleukin-4. In *The Cytokine handbook.* Academic Press. London, pp 99-126

Batteux F, Trebeden H, Charreire J and Chiocchia G, 1999. Curative treatment of experimental autoimmune thyroiditis by in vivo administration of plasmid DNA coding for interleukin-10. Eur J Immunol, 29:958-963

Borden EC and Chin P, 1994. Interleukin-6: a cytokine with potential diagnostic and therapeutic roles. J Lab Clin Med, 123:824-829

Cabanes-Macheteau M, Fitchette-Laine AC, Loutelier-Bourhis C, Lange C, Vine ND, Ma JK, Lerouge P and Faye L, 1999. N-Glycosylation of a mouse IgG expressed in transgenic tobacco plants. Glycobiology, 9:365-372

De Meyer E and De Meyer-Guignard J, 1994. Interferons. In *The Cytokine handbook.* Academic Press. London, pp 265-288

De Zoeten GA, Penswick JR, Horisberger MA, Ahl P, Schultze M and Hohn T, 1989. The expression, localization, and effect of a human interferon in plants. Virology, 172:213-222

Dieryck W, Pagnier P, Poyart C, Marden MC, Gruber V, Bournat P, Baudino S and Merot B, 1997. Human haemoglobin from transgenic tobacco. Nature 386:29-30

Edelbaum O, Stein D, Holland N, Gafni Y, Livneh Q, Novick D, Rubinstein M and Sela I, 1992. Expression of active human interferon-beta in transgenic plants. J Interferon Res 12:449-453

Evangelista RL, Kusnadi AR, Howard JA and Nikolov ZL, 1998. Process and economic evaluation of the extraction and purification of recombinant beta-glucuronidase from transgenic corn. Biotechnol Prog 14:607-614

Ganz PR, Dudani AK, Tackaberry ES, Sardana R, Sauder S, Cheng XY and Altosaar I, 1996. Expression of human blood proteins in transgenic plants: the cytokine GM-CSF as a model protein. In *Transgenic plants: A production system for industrial and pharmaceutical proteins,* Eds. Owen MRL and Pen J, Wiley, New York. pp 281-297

Goldsmith MA and Greene WC, 1994. Interleukin-2 and the Interleukin-2 receptor, In *The Cytokine handbook.* Academic Press. London, pp 57-80

He ZM, Cao J, Li BJ and Huang XQ, 1998. Expression of erythropoietin gene in tomato. Acta Genetica Sinica, 25:155-159

Hirano T, Akira S, Taga T and Kishimoto T, 1990. Biological and clinical aspects of interleukin 6. Immunol Today, 11:443-449

Hirano T, Yasukawa K, Harada H, Taga T, Watanabe Y, Matsuda T, Kashiwamura S, Nakajima K, Koyama K, Iwamatsu A, et al., 1986. Complementary DNA for a novel human interleukin (BSF-2) that induces B lymphocytes to produce immunoglobulin. Nature, 324:73-76

James EA, Wang C, Wang Z, Reeves R, Shin JH, Magnuson NS and Lee JM, 2000. Production and characterization of biologically active human GM-CSF secreted by genetically modified plant cells. Protein Expr Purif, 19:131-138

Karpusas M, Nolte M, Benton CB, Meier M, Lipscomb WN and Goelz S, 1997. The crystal structure of human interferon beta at 2.2-A resolution. Proc Natl Acad Sci USA,94:11813-11818

Karpusas M, Whitty A, Runke L and Hochman P, 1998. The structure of human interferon-beta: implications for activity. Cell Mol Life Sci, 54:1203-1216

Kopf M, Baumann M, Freer M, Freudenberg M, Lamers M, Kishimoto T, Zinkernagel R, Bluethmann H and Kohler H, 1994. Impaired immune and acute-phase responses in interleukin-6-deficient mice. Nature, 368:339-342

Kopf M, Le Gros G, Coyle AJ, Kosco-Vilbois M and Brombacher F, 1995. Immune responses of IL-4, IL-5, IL-6 deficient mice. Immunol Rev, 148:45-69

Kwon SY, Yang Y,. Hong CB and Pyun KH, 1995. Expression of active human interleukin-6 in transgenic tobacco. Mol Cells, 5:486-492

Le HV, Ramanathan L, Labdon JE, Mays-Ichinco CA, Syto R, Arai N, Hoy P, Takebe Y, Nagabhushan TL and Trotta PP, 1988. Isolation and characterization of multiple variants of recombinant human interleukin 4 expressed in mammalian cells. J Biol Chem, 263:10817-10823

Leach MW, Davidson NJ, Fort MM, Powrie F and Rennick DM, 1999. The role of IL-10 in inflammatory bowel disease: "Of mice and men". Toxicol Pathol, 27:123-133

Lefebvre DD, Miki BL and Laliberte JF, 1987. Mammalian metallothionein functions in plants. Bio/Technology, 5:1053-1056

Lerouge P, Cabanes-Macheteau M, Rayon C, Fischette-Laine CM, Gomord V and Faye L, 1998. N-glycoprotein biosynthesis in plants: recent developments and future trends. Plant Mol Biol, 38:31-48

Ma JK, Hiatt A, Hein M, Vine ND, Wang F, Stabila P, van Dolleweerd C, Mostov K and Lehner T, 1995. Generation and assembly of secretory antibodies in plants. Science, 268:716-719

Ma JK, Hikmat BY, Wycoff K, Vine ND, Chargelegue D, Yu L, Hein MB and Lehner T, 1998. Characterization of a recombinant plant monoclonal secretory antibody and preventive immunotherapy in humans. Nat Med, 4:601-606

Magnuson NS, Linzmaier PM, Reeves R, An G, HayGlass K and Lee JM, 1998. Secretion of biologically active human interleukin-2 and interleukin-4 from genetically modified tobacco cells in suspension culture. Protein Expr Purif, 13:45-52

Matsumoto S, Ikura K, Ueda M and Sasaki R, 1995. Characterization of a human glycoprotein (erythropoietin) produced in cultured tobacco cells. Plant Mol Biol, 27:1163-1172

Matsumoto S, Ishii A, Ikura K, Ueda M and Sasaki R, 1993. Expression of human erythropoietin in cultured tobacco cells. Biosci Biotechnol Biochem, 57:1249-1252

May LT, Ghrayeb J, Santhanam U, Tatter SB, Sthoeger Z, Helfgott DC, Chiorazzi N, Grieninger G and Sehgal PB, 1988. Synthesis and secretion of multiple forms of beta 2-interferon/B-cell differentiation factor 2/hepatocyte-stimulating factor by human fibroblasts and monocytes. J Biol Chem, 263:7760-7766

Meager, Tony. 1998. The molecular biology of cytokines. *Molecular Medical Science Series*. Eds. James K and Morris A. John Wiley & Sons, London

Menassa R, Jevnikar A and Brandle J, 2001. A contained system for the field production of plant recombinant proteins. In *Molecular farming*. Eds. Toutant JP and Balazs E. INRA Editions, Versailles, France

Menassa R, Nguyen V, Jevnikar A and Brandle J, 2001. A self contained system for the field production of plant recombinant interleukin-10. Mol Breed, (in press)

Mire-Sluis AR, 1999. Cytokines: from technology to therapeutics. Trends Biotechnol, 17:319-325

Mire-Sluis AR, Das RG and Padilla A, 1998. WHO cytokine standardization: facilitating the development of cytokines in research, diagnosis and as therapeutic agents. J Immunol Methods, 216:103-116

Mire-Sluis AR and Thorpe R, 1998. Laboratory protocols for the quantitation of cytokines by bioassay using cytokine responsive cell lines. J Immunol Methods, 211:199-210

Moore KW, O'Garra A, de Waal Malefyt R, Vieira P and Mosmann TR, 1993. Interleukin-10. Annu Rev Immunol, 11:165-190

Munro S and Pelham HR, 1987. A C-terminal signal prevents secretion of luminal ER proteins. Cell, 48:899-907

Nyman TA, Kalkkinen N, Tolo H and Helin J, 1998. Structural characterisation of N-linked and O-linked oligosaccharides derived from interferon-alpha2b and interferon-alpha14c produced by Sendai-virus-induced human peripheral blood leukocytes. Eur J Biochem, 253:485-493

Ohara J, Coligan JE, Zoon K, Maloy WL and Paul WE, 1987. High-efficiency purification and chemical characterization of B cell stimulatory factor-1/interleukin 4. J Immunol, 139:1127-1134

Orita T, Oh-eda M, Hasegawa M, Kuboniwa H, Esaki K and Ochi N, 1994. Polypeptide and carbohydrate structure of recombinant human interleukin-6 produced in Chinese hamster ovary cells. J Biochem (Tokyo), 115:345-350

Pagny S, Cabanes-Macheteau M, Gillikin JW, Leborgne-Castel N, Lerouge P, Boston RS, Faye L and Gomord V, 2000. Protein recycling from the Golgi apparatus to the endoplasmic reticulum in plants and its minor contribution to calreticulin retention. Plant Cell, 12:739-756

Palacpac NQ, Yoshida S, Sakai H, Kimura Y, Fujiyama K, Yoshida T and Seki T, 1999. Stable expression of human beta1,4-galactosyltransferase in plant cells modifies N-linked glycosylation patterns. Proc Natl Acad Sci USA, 96:4692-4697

Park SM and Choo Bong Hong, 1997. Transcription of human interleukin-2 gene in *Nicotiana tabacum* driven by CaMV35S promoter. J Plant Biology, 40:110-114

Pauza ME, Neal H, Hagenbaugh A, CheroutreH and Lo D, 1999. T-cell production of an inducible interleukin-10 transgene provides limited protection from autoimmune diabetes. Diabetes, 48:1948-1953

Pretolani M and Goldman M, 1997. IL-10 a potential therapy for allergic inflammation? Immunol Today, 18:277-280

Radhakrishnan R, Walter LJ, Hruza A, Reichert P, Trotta PP, Nagabhushan TL and Walter MR, 1996. Zinc mediated dimer of human interferon-alpha 2b revealed by X-ray crystallography. Structure, 4:1453-1463

Rasko JEJ and Gough NM, 1994. Granulocyte macrophage colony stimulating factor. In *The Cytokine handbook*. Academic Press. London, pp 343-70

Recny MA, Scoble HA and Kim Y, 1987. Structural characterization of natural human urinary and recombinant DNA-derived erythropoietin: Identification of des-arginine 166 erythropoietin. J Biol Chem, 262:17156-17163

Refaeli Y, Van Parijs L, London CA, Tschopp J and Abbas AK, 1998. Biochemical mechanisms of IL-2-regulated Fas-mediated T cell apoptosis. Immunity, 8:615-623

Reich K, Bruck M, Grafe A, Vente C, Neumann C and Garbe C, 1998. Treatment of psoriasis with interleukin-10. J Invest Dermatol, 111:1235-1236

Rollwagen FM and Baqar S, 1996. Oral cytokine administration. Immunol Today, 17:548-550

Sijmons PC, Dekker BM, Schrammeijer B, Verwoerd TC, van den Elzen PJ and Hoekema A, 1990. Production of correctly processed human serum albumin in transgenic plants. Biotechnology (NY), 8:217-221

Smirnov SP, Teverovskaia EKH, Krasheninnikova LV and Pukhal'ski VA, 1990. Design of an expression integrative vector and its application for introducing the human recombinant alfa interferon gene into plants. Genetika, 26:2111-2121

Smith KA, 1988. Interleukin-2: inception, impact, and implications. Science, 240:1169-1176

Tsuda E, Kawanishi G, Ueda M, Masuda S and Sasaki R, 1990. The role of carbohydrate in recombinant human erythropoietin. Eur J Biochem,188:405-411

Tsuji T, Nakagawa R, Sugimoto N and Fukuhara K, 1987. Characterization of disulfide bonds in recombinant proteins: reduced human interleukin 2 in inclusion bodies and its oxidative refolding. Biochemistry, 26:3129-3134

van Ree R, Cabanes-Macheteau M, Akkerdaas J, JP Milazzo, Loutelier-Bourhis C, Rayon C, Villalba M, Koppelman S, Aalberse R, Rodriguez R, Faye L and Lerouge P, 2000. Beta (1,2)-xylose and alpha(1,3)-fucose residues have a strong contribution in IgE binding to plant glycoallergens. J Biol Chem, 275:11451-11458

Van Snick J, 1990. Interleukin-6: an overview. Annu Rev Immunol, 8:253-278

Vellekamp G, Cannon-Carlson S and Tang J, 1994. Purification of human interleukin-10 from a cell culture medium. US Patent # 5,328,989

Vellekamp G, Cannon-Carlson S and Tang J, 1998. Purification of bacterially expressed human interleukin-10. US Patent # 5,710,251

Vieira P, de Waal-Malefyt R, Dang MN, Johnson KE, Kastelein R and Fiorentino DF, 1991. Isolation and expression of human cytokine synthesis inhibitory factor cDNA clones: homology to Epstein-Barr virus open reading frame BCRFI. Proc Natl Acad Sci USA, 88:1172-1176

Walsh G, 2000. Biopharmaceutical benchmarks. Nat Biotechnol, 18:831-833

Wang A, Lu SD and Mark DF, 1984. Site-specific mutagenesis of the human interleukin-2 gene: structure-function analysis of the cysteine residues. Science, 224:1431-1433

Wasley LC, Timony G, Murtha P, Stoudemire J, Dorner AJ, Caro J, Krieger M and Kaufman RJ, 1991. The importance of N- and O-linked oligosaccharides for the biosynthesis and in vitro and in vivo biologic activities of erythropoietin. Blood, 77:2624-2632

Wee EG, Sherrier DJ, Prime TA and Dupree P, 1998. Targeting of active sialyltransferase to the plant Golgi apparatus. Plant Cell, 10:1759-1768

Windsor WT, Syto R, Tsarbopoulos A, Zhang R, Durkin J and Baldwin S, 1993. Disulfide bond assignments and secondary structure analysis of human and murine interleukin-10. Biochemistry, 32:8807-8815

Yang SC, Grimm EA, Parkinson DR, Carinhas J,. Fry KD, Mendiguren-Rodriguez A, Licciardello J, Owen-Schaub B, Hong WK and Roth JA, 1991. Clinical and immunomodulatory effects of combination immunotherapy with low-dose interleukin 2 and tumor necrosis factor alpha in patients with advanced non-small cell lung cancer: a phase I trial. Cancer Res, 51:3669-3676

Zablackis E, York WS, Pauly M, Hantus S, Reiter WD, Chapple CC, Albersheim P and Darvill A, 1996. Substitution of L-fucose by L-galactose in cell walls of Arabidopsis mur1. Science, 272:1808-1810

Zdanov A, Schalk-Hihi C, Gustchina A, Tsang M, Weatherbee J and Wlodawer A, 1995. Crystal structure of interleukin-10 reveals the functional dimer with an unexpected topological similarity to interferon gamma. Structure, 3:591-601

Zheng XX, Steele AW, Nickerson PW, Steurer W, Steiger J and Strom TB, 1995. Administration of noncytolytic IL-10/Fc in murine models of lipopolysaccharide-induced septic shock and allogeneic islet transplantation. J Immunol, 154:5590-5600.

Zhu Z, Hughes KW, Huang L, Sun B, Liu C, Li Y, Hou Y and Li X, 1994. Expression of human alpha-interferon cDNA in transgenic rice plants. Plant Cell Tissue and Organ Culture, 36:197-204

Chapter 15

EDIBLE VACCINES IN PLANTS FOR LIVESTOCK PATHOGENS

L. Erickson, W-J.Yu, T. Tuboly[1,2], E. Nagy[1], A. Bailey, J. Zhang, D. Yoo[1] and S. Du

Department of Plant Agriculture, Ontario Agricultural College, and [1]Department of Pathobiology, Ontario Veterinary College, University of Guelph, Guelph, Ontario N1G 2W1.
[2]Current address: Department of Microbiology and Infectious Diseases, University of Veterinary Sciences, Budapest, Hungary

Keywords: oral vaccines, transgenic plants, synthetic genes, antigens, porcine, swine, bovine, mucosal immunity, virus, bacteria, intestinal disease

Abstract Edible vaccines derived from plants offer clear advantages to the livestock industry with respect to cost and ease of delivery. From a consumer perspective, the widespread adoption by the industry of efficient, low-cost, edible plant-based vaccines offers the prospect of reduced exposure to animal pathogens in food products, as well as in the water supply. A major obstacle encountered in the initial development of plant-based vaccines has been the very low levels of antigens expressed in transgenic plants. Some of the strategies adopted to meet this obstacle include synthetic forms of the antigen genes, fusion proteins and inducible expression systems. However, even with augmented levels of antigen in plant tissue, a successful edible vaccine in plants will probably require in addition strategies for reducing degradation in the gut and enhancing immunogenicity. It may even be possible to develop DNA-based vaccines in plants.

1. INTRODUCTION

Animal health is a major concern in any livestock operation regardless of species or the size and nature of the operation. The recent outbreak of foot and mouth disease in Europe demonstrates how vulnerable the livestock industries are to pathogens and how costly it can be when existing practices fail. Public

L. Erickson et al. (eds.),
Molecular Farming of Plants and Animals for Human and Veterinary Medicine, 339–367.

concern is even more heightened when farm animal health problems become human health problems. The negative impact of "mad cow" disease on public perceptions of the industry, its products and practices is worldwide, deep and probably long-term. There is also increased awareness of the involvement of livestock operations in human illness related to toxigenic *E. coli* and cryptosporidium. The intensification of animal agriculture, characterized by very large operations and crowded, stressful conditions, is seen by many to increase the probability of disease in these animals, as well as in adjacent human populations. Finally, the use of antibiotics by the industry to enhance animal health and productivity is facing mounting criticism for its potential role in developing resistance to antibiotics of significance to human health.

It is apparent to many that if the livestock industries are to survive, increased emphasis on animal health will be necessary. This will involve, among other things, the development of new approaches to enhancing the immunity of animals to pathogens. As many of these pathogens infect the host via mucosal tissues, such as those in the lungs and intestines, strategies for generating a protective mucosal immune response are very attractive. Conventional approaches to vaccines, which utilize attenuated or killed strains of a viral pathogen, for example, will likely continue to be a mainstay of the industry. Problems with such vaccines, such as reversion of attenuated strains to virulence, has prompted development of alternative systems, such as DNA vaccines or those based on adenovirus (Babiuk and Tikoo, 2000). Another option is the development of edible vaccines in plants, to be consumed either directly or in some processed form.

2. ADVANTAGES OF EDIBLE PLANT-BASED VACCINES TO INDUSTRY AND THE CONSUMER

The commercial advantages of plant-based production systems for therapeutic proteins, such as high volume/low cost, ease of scale-up and storage, and absence of animal pathogens, have been well-publicized. However, there are particular advantages to using edible plant material as vaccines for livestock. There is a growing desire to eliminate, where possible, the use of needles for vaccines. Injections are time-consuming, labour-intensive and unpleasant to administer, especially with large animals. Producers also refer to problems in the administration of such injections by their labour force, which is generally unskilled in veterinary techniques. Such problems are highlighted by reports of broken needles in meat purchased by consumers. Although infrequent, these incidents are widely broadcast in the media, much to the detriment of the industry. Producers and processors also refer to the losses incurred due to lower carcass quality because of needle marks in the meat.

The oral consumption of antigens as a component of plant tissue may have considerable advantages from an efficacy standpoint, as well. The kind of strong mucosal immune response needed for resistance to many of the pathogens which infect animals through such tissues is most effectively generated by presentation of antigens to those same tissues through what is generally referred to as a common mucosal response (McGhee et al. 1999). The generation of such immunity by the presentation to intestinal tissues of large amounts of antigen in edible plant material is a very attractive strategy for achieving this. Crop plants and existing cropping/processing technologies associated with the various crop species offer very cost-efficient systems for production and delivery of large mounts of antigen. The potential advantage of orally consuming an antigen as part of intact plant tissue, rather than in a purified form, was demonstrated by Modelska et al. (1998); in this study, a rabies antigen expressed as part of a plant virus particle induced a much stronger immune response when fed to mice in spinach leaves than when fed as virus particles purified from leaf tissue.

From a consumer perspective, the widespread adoption by the industry of efficient, low-cost, edible plant-based vaccines offers the prospect of reduced exposure to animal pathogens in food products, as well as in the water supply. In addition, there is the benefit of the perception by the public of enhanced animal health and welfare by means of a less intrusive, more "natural" approach of "medicinal plants". Vaccines produced in crop plants and incorporated as a feed component represent a combination of plants as a traditional source of nutrients and modern pharmaceuticals for prevention of disease.

3. OBSTACLES TO THE DEVELOPMENT OF EDIBLE PLANT-BASED VACCINES

3.1 Expression Levels of Antigens in Plants

The first obstacle faced by those attempting to develop this technology is achieving an adequate level of antigen protein in plants. There is a growing list of antigens which have been expressed in plants, and a striking feature of all of the data is the low levels of protein reported. There is still only a handful of antigens expressed in plants for the purpose of developing vaccines for livestock, and levels are either not reported (na) or very low also (Table 18).

Table 18. Expression of antigen genes from livestock pathogens in plants

Pathogen	Gene	Plant Species	Expr. Lev[a] (max)	Biological Activity[b]	Reference
foot and mouth virus	epitope of VP1	cowpea	na	na	Usha et al. 1993
	VP1	arabidopsis	na	I,P(inj mice)	Carrillo et al. 1998
		alfalfa	na	I,P(inj/or mice)	Wigdorovitz et al. 1999
transm. gastroent. virus(swine)	S(spike)	arabidopsis	0.06	I(inj mice)	Gomez et al. 1998
	S(trunc.)	tobacco	0.1	I(inj swine)	Tuboly et al. 2000
	S(trunc. synthetic)	alfalfa, tobacco	0.1	I(inj mice)	Yu et al. 1999
	D epitope of S gene fused to plant prot	alfalfa, tobacco	0.2	I(inj mice)	Bailey, 2000
	S(synthetic)	maize	na	P(or swine)	Jilka(ch. 10, this vol)
porcine resp/reprod virus	ORF5	alfalfa, tobacco	0.2	na	Zhang, 1999
shipping fever (Mann. hemolytica)	Lkt50 (leucotoxin) (trunc/fused to GFP))	white clover	1.0	I(inj rabbits)	Lee et al., 2000

[a] Percent soluble protein. [b]I = immunogenic, P = protective, inj = injected, or = orally delivered.

Although many feel that for commercial purposes the antigen should be present in plant tissue in the 1.0% range of total soluble protein, such a figure is fairly meaningless as to the final efficacy of an oral vaccine produced from such tissue. First, it is very difficult to translate concentrations of an antigen in plant tissue to effective concentrations at the mucosal surface; for example, an antigen at a level of 1% of soluble protein in starting plant material will be present at a much lower level in the total biomass passing through the intestine, even if the diet consists entirely of the plant material containing the vaccine, which may not practical. Ultimately, however, the concentration required in the starting plant tissue is determined by the amount needed at the mucosal surface to induce an optimum immune response, and the amount lost prior to delivery to those surfaces. Antigens vary a great deal in their immunogenicity. The B subunit of the heat labile toxin (LTB) of E coli, for example, is known to be a potent oral immunogen (Haq et al., 1995), and therefore protective immunization may not require the levels of such an antigen that are needed for another antigen. Furthermore, the structure of the antigen protein in plant tissue may affect levels reached, as well as immunogenicity. The formation of viral particles, e.g. Norwalk virus (Mason et al., 1996), or other polymeric structures, such as the pentameric ring of LTB (Haq et al., 1995), may reduce susceptibility to degradation and interfere less with cell structure and functioning than other antigen proteins which may associated freely with other cellular components. Undoubtedly, the formation of stable, inert Bt protein crystals in chloroplasts was a major factor enabling the very high accumulation of this protein in leaf tissue (DeCosa et al., 2001). In addition, the ability to target expression to a specific plant tissue which would allow subsequent processing or concentration, has implications for the levels needed for commercial production. Finally, the amount of antigen needed in the plant tissue is very much dependent on the degree of degradation experienced by that antigen in the digestive tract prior to delivery to the target tissues in the animal.

3.2 Degradation of Antigens in the Gut

The emphasis to date on enhancing expression of antigens in plants for the purpose of edible vaccines reflects the difficulties associated with what is only the first step in developing this technology. It is surprising how little attention has been paid to what may be an even greater obstacle to successful oral immunization - break-down of the antigen proteins in the stomach and intestine. The pH of the stomach of carnivores and omnivores is quite low, and this acidity, in conjunction with the presence of proteolytic and hydrolytic enzymes, results in the reduction of many food constituents to smaller components for absorption by tissues lining the gut (Sanderson and Walker, 1999).The complexity of the digestive environment, as well as the variability between animals and even within animals over time, make this obstacle difficult to predict reliably and overcome. For example, there are many significant differences between the structure and function of the digestive tract of a young piglet vs that of a lactating sow regarding pH, the presence of proteinases, other digestive enzymes, mucins, antimicrobial peptides and antibodies. Other factors such as diet, microbial populations, housing conditions, use of antibiotics can also have significant impact on gut function, and especially on the break-down and utilization of feed components. Given the variations that can exist between animals in a livestock operation regarding the physiological condition and function of the gastrointestinal tract, devising a feed component and feeding strategy which will effectively vaccinate all animals in that operation has to be considered a major hurdle for the development of this technology.

Perhaps because of the complexity of the variables and interactions involved in gastrointestinal degradation, as well as the difficulties inherent in addressing this obstacle, those working in this field have chosen to attempt feeding trials despite these unknowns and the low levels of antigen present in their plant material. Although the results of some of these experiments are surprisingly positive (Wigdorovitz et al. 1999; Richter et al., 2000;), it must be conceded that we are still some distance from the desired level of protection from pathogens necessary for commercialization. Even if the levels of antigen in plants can be significantly increased over current levels, it is unlikely that simply producing an antigen in plants and feeding the plant material will ultimately succeed without developing strategies for reducing degradation in the gut and for enhancing immunogenicity.

3.3 Immunogenicity of Antigens Produced in Plants and Delivered Orally

The levels of antigen in the starting plant tissue, losses due to length and conditions of storage and processing of plant tissue, and degradation in the gut will determine how much antigen reaches the mucosal surfaces in the animal. However, even if those obstacles are adequately addressed, there is no guarantee of the kind or extent of the immune response in the animal ingesting the antigen. The immunogenicity of an antigen protein in plant tissue will depend on many factors. First, a protein which is immunogenic as a component of a live, replicating virus, for example, may not be as immunogenic when expressed as a separate protein in plant tissue. The response of the intestinal mucosa to orally ingested antigen may range from absence of a specific immune response, such as when a pathogen is cleared by non-specific innate mechanisms, to more specific B and T cell responses in the mucosal and systemic branches (Strobel and Mowat, 1998). The specific response depends on the nature of the antigen, how it is administered and what cells are involved in uptake.

Generally, antigen entering the body through the intestinal mucosa does so via epithelial cells or M cells (Hershberg and Meyer, 2000; Neutra, 1998). Following entry, the antigen will encounter a range of immune cells, and, depending on how it is processed during internalization, will be taken up by professional antigen presenting cells (Neutra, 1998) or presented directly to T cells (Mayer and Blumberg, 1999). The nature of the immune response will be determined by the specific cytokines produced by the CD4+ helper cells to which the antigen is presented (Kelsall and Strober, 1999). The response may be largely localized to the common mucosal immune system, characterized mainly by secretion of sIgA at immune effector sites (Strobel and Mowat, 1998). Alternately, the primary mucosally-induced response will consist of IgG antibodies in the systemic compartment (Strober et al. 1998). In other cases, the major immune response may be the generation of IgE antibodies, characteristic of allergic reactions (Lorentz et al., 2000). There is currently insufficient understanding of the immune response to orally-administered antigens to enable us to manipulate the process, but clearly, strategies to meet this final challenge are needed to develop edible vaccines in plants. Such strategies will have to address the difference in the presentation to the immune system between an antigen as a component of plant tissue versus an antigen as part of a replicating pathogen during an infection. Another hurdle is delivering the appropriate dosage of an antigen to absorption sites in the digestive tract using current feeding practices in the industry. The importance of this consideration cannot be overlooked in light of the evidence for generating oral tolerance by feeding an antigen expressed in plants (Ma et al., 1997). Finally, plants are known to contain many compounds with immunomodulator properties, which could inhibit a mucosal response to associated compounds in the diet or dominate the immune response by virtue of their strongly immunogenic properties (Mason et al., 1998).

4. STRATEGIES FOR PRODUCTION AND DELIVERY OF EDIBLE VACCINES IN PLANTS

4.1 Expression of Antigens Using Native Gene Constructs

Most antigens expressed to date in plants have been encoded by the native gene, i.e. a cDNA or bacterial clone, modified appropriately for expression in plants by the addition of regulatory and targeting sequences. Almost all of the antigens produced to this time in plants for oral immunization of farm animals are viral in origin (Table 1). Expression levels in plants containing the native cDNA clones of the foot and mouth viral gene VP1 were very low (Carrillo et al. 1998; Wigdorovitz et al. 1999), as was that reported by Gomez et al. (1998) for swine transmissible gastroenteritis virus (TGEV).

The initial work with TGEV in our labs involved two constructs containing the full-length cDNA coding for the spike (S) protein on the surface of the virus particle. Specifically, the full length cDNA clone (4300 bp) was released from the recombinant plasmid pTS (Tuboly et al., 1994) by *Eco*RI digestion, rendered blunt-ended with Klenow fragment, and digested with *Bam*HI. The resulting fragment was cloned into the binary vector pBI121 (Jefferson, 1987) following digestion of the vector with *Sac*I (adjacent to the 5' end of the nos terminator), end-filling with Klenow and digestion with *Bam*HI, creating vector pS1 (Figure 47). In pS3, an enhanced 35S promoter and 35S terminator from pFF19G (Timmermans et al., 1990) were fused to the full-length S gene to which an endoplasmic reticulum targeting sequence had been added. This form of the gene was created by fusing the 3.8 kb *Bam/Xba* fragment from the 5' end of the cDNA to a 0.6 kb *Xba/Pst* fragment containing the remainder of the coding region of the S gene and a SEKDEL tail. This terminal portion of the coding region had been amplified by PCR from the cDNA using a primer starting at nt 3826 containing an *Xba*I site (5'-GCTTCAGATG GTGATCG-3') and a 49 nt primer (5'-TGCACTGCAG TCATAGCTCA TCTTTCTCAG AATGGAGGTG CACTTTTTC-3'), containing sequentially the end of the coding region of the S gene, the SEKDEL coding region, a stop codon and a *Pst*I site. This construct was transferred to pBI121 which had been digested with *Hind* III/*Eco*RI.

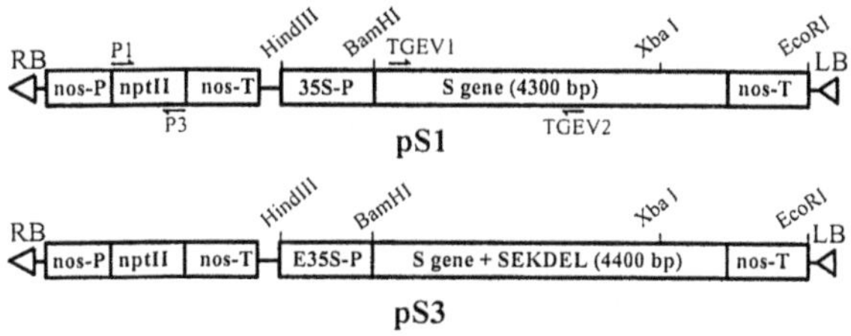

Figure 47. Constructs containing the native cDNA for the S gene. nos-P: nopaline synthase promoter; npt II: neomycinphosphotransferase II; nos-T: nos terminator; 35S-P: CaMV 35S promoter; E35S-P: enhanced CaMV 35S promoter; 35S-T: CaMV 35S terminator. Positions of primers for amplifying the NPT-II gene (P1, P3) and the S gene (TGEV1, TGEV2) are indicated by small arrows parallel to the vector sequence.

The native S gene contains a putative signal peptide-coding region, and such signals of genes from mammals and mammalian viruses have been shown to be correctly recognized in transgenic plants, resulting in targeting to the endoplasmic reticulum (e.g. Khoudi et al. 1999). The above two vectors represent two common and alternate strategies for expression of xenoproteins in plants, i.e. secretion from the cell (pS1) and retention in the endomembrane system (pS3).

Of the approximately 150 tobacco and alfalfa plants containing the full length native cDNA for the S gene controlled by the 35S promoter (pS1), only 3 contained mRNA detectable by Northern blots and this consisted mainly of a faint band of about 1200 nt in size (Figure 48B, lane 2). Addition of an enhanced 35S promoter to this cDNA resulted in higher levels of S mRNA than with the 35S promoter and in a greater proportion of transgenic plants (12 out of 12), but still of the 1200 nt size. Southern blot analysis of these transgenic plants indicated that all plants contained full-length copies of the gene (Figure 48A). A Western blot of protein from plants containing these two constructs did not detect any S protein (Figure 48C, lanes 2,3 and 4).

Analysis of transcripts in transgenic plants by RT-PCR produced bands of expected size for the NPT-II-gene with the use of an internal primer (P3) or oligo-dT primer. However, when RNA extracts from the same plants containing the native cDNA clone of the S gene (vector pS3) were used as the template for reverse transcriptase, a band was produced only with an internal primer (TGEV2) and not with an oligo-dT primer.

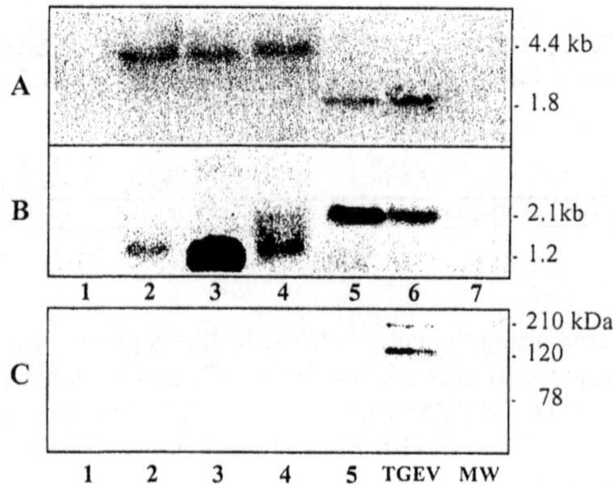

Figure 48. Analysis of transgenic plants containing the S gene by Southern (A), Northern (B) and Western (C) blots. Transgenic tobacco (lanes 2-5) and alfalfa (lane 6) contain the cDNA of the native S gene: pS1 (35S promoter, lane 2), pS2 (pollen-specific promoter, lane 3), pS3 (enhanced 35S promoter, lane 4) and the synthetic S gene, pSS1 (enhanced 35S promoter, lane 5, 6); DNA from plants containing vectors pS1, pS2 and pS3 was digested with BamHI and EcoRI and plant DNA containing the synthetic gene was digested with XbaI and SacI.. Lanes 1 and 7 are the control non-transgenic tobacco and alfalfa, respectively. TGEV: TGEV-infected swine testicle cell lysate; MW: protein molecular weight standards.

Reports of truncated transcripts present as distinct bands on a Northern blot are uncommon in plant literature. Analysis of two short mRNA fragments of 600 and 900 nt in transgenic plants containing a wild-type insecticidal gene from bacteria gene revealed that they were polyadenylated, and probably resulted from premature termination at an upstream cryptic poly(A+)-like signal present in the gene (Diehn et al., 1998). The native S gene contains 8 potential premature polyadenylation signals (Figure 49), five of which are concentrated in the region from nt 838 to 1615, and most of the form AATAAT, reported to be a very strong substitute of the more typical polyadenylation signal AATAAA (Rothnie, 1996). These AT-rich near-upstream-elements, NUEs, occur typically 10 to 30 bases upstream of the polyadenylation site in plants, and for efficient formation of 3' ends are accompanied by TG-rich far-upstream-elements, FUEs, located 40 to 150 bp upstream of the polyadenylation site (Hunt, 1994; Rothnie, 1996). Five FUEs were found in the S gene, but most occur downstream of the main cluster of NUEs (Figure3). However, several less typical NUEs were positioned downstream of some of the these FUEs: AATATA(2085, 2182, 2657,2917), AATAGA (2688), AATACA (2948). Despite the presence and possible effects of premature poly(A+) signals in the S gene, we were unable to detect polyadenylated S gene mRNA of any size in our extracts using RT-PCR and oligo-dT primers, although full-length polyadenylated mRNA for the NPT-II gene was detected in parallel experiments.

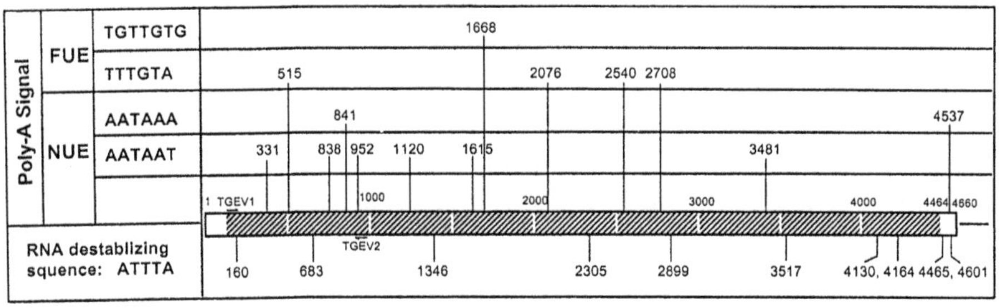

Figure 49. Potential mRNA processing signals in the native S gene. Poly-A signals are divided into near upstream elements (NUEs) and far upstream elements (FUEs). Arrows indicate positions of primers used for PCR analysis: TGEV1(138-155 nt) and TGEV2(977-994 nt). Hatched area represents protein coding region of the gene.

The above data suggest that the mRNA observed from the native S gene is not polyadenylated. There are very few reports with examples of discrete bands of non-polyadenylated truncated transcripts in the plant literature. Truncated mRNA from the soybean small subunit gene SRS4 of Rubisco have been detected by Northern blots by Tanzer and Meagher (1995). Analysis of these fragments demonstrated that all contained the intact 5' end of the transcript and none were polyadenylated. The authors concluded that the fragments resulted from endonuclease digestion of the full-length transcript and that the location of the cleavage site was determined more by secondary structure than specific sequences in the transcript. Such discrete mRNA degradation intermediates are not the normal products of mRNA degradation in plants, nor in the better-characterized systems of mammals and yeast (Gutierrez et al., 1999).

It is possible that the predominant 1200 nt band of S gene mRNA is also the result of endonuclease activity, but unlike the SRS4 fragments, which are derived from the degradation of full-length message, the S mRNA is probably not derived from full-length transcripts, as these could not be detected. We propose rather that the truncated and non-polyadenylated S gene mRNA results from the failure of the elongation stage of transcription. Such failure may involve endonuclease activity, such as that associated with Pol II during transcriptional arrest (Shilatifard, 1998), as well as subsequent 3' to 5' exonuclease activity presumed to be part of the mRNA degradation pathway for plants (Gutierrez et al., 1999). However, extensive exonuclease activity would likely result in a smear of mRNA which was not evident in the Northern blots (Figure 48B).

There is very little information on the factors affecting transcription elongation in plants. Based on the information available from other eukaryotes such as yeast, there are many genes encoding unique factors which control this process (Shilatifard, 1998; Reines et al., 1999). As truncated transcripts were not evident in plants containing the synthetic S gene, we assume that elongation failure with the native S gene was due to DNA-dependent arrest. In yeast and mammalian systems, there are elongation factors such as SII which assist Pol II in overcoming arrest sites by means of a reiterative process of cleavage and re-extension (Shilatifard, 1998). It may be that the kind and/or degree of pausing at a particular site in the S gene caused destabilization of the elongation complex and subsequent cleavage with no re-extension. In a similar experiment involving the expression of a malarial gene in yeast, Milek et al. (2000) also reported truncated transcripts, but did not mention whether the transcripts were polyadenylated ; when the malarial gene was re-synthesized to optimize codon usage and reduce AT content, only full-length mRNA was detected.

The specific features of the native S gene which might cause transcriptional arrest are unknown, but must include those altered in the synthetic form of the gene (see below), which did not result in truncated transcripts. The first of these is the high AT content (63%) of the native gene, which increases the probability of cryptic mRNA processing sites that are AT rich, such as poly(A+) sites. Aside from the numerous poly(A+) sites in the first half of the gene, there are also in the same region some T-rich sequences which could block elongation by Pol II (Shilatifard, 1998). For example, T occurs in 31 of the 65 nt starting at position 906, in 26 of the 54 nt starting at position 1045, 7 of the 10 nt starting at position 1173, and 31 of the 60 nt starting at position 1385 (Figure 49). The S gene also contains 10 sequences of the form ATTTA, which has been shown to target mRNA for rapid decay when located in multiple copies in the 3' untranslated region of mammalian genes, and has produced similar effects when introduced into plant genes (Ohme-Takagi et al.,1993; Gutierrez et al., 1999); 2 of the 10 copies of this sequence occur in the 3' untranslated region of the S gene (Figure 49).

Given that we could not detect polyadenylated mRNA for the native S gene in any transgenic plants containing that gene, it is not surprising that S protein was not detectable on Western blots with protein extracts from any of these plants, which prompted us to adopt alternative strategies with different gene constructs. It is interesting that expression of the full-length S gene in insect cells employing baculovirus vectors has also been at very low levels (Tuboly et al.1994)

4.2 Synthetic Genes

The very low levels of expression frequently observed for foreign proteins in plants may be due to a variety of factors affecting transcription, translation, and stability of the protein. Some of these factors can be predicted from the DNA sequence, and steps taken to eliminate them. An example of this is the 100-fold increase in expression of a bacterial insecticidal protein gene in maize following modification of the gene to eliminate features such as premature poly A signals, high AT content and unusual codons (Perlak et al., 1991).

In our labs, we adopted a similar strategy for expressing the S gene of TGEV when it became clear that the native form of the gene would not provide the levels of antigen needed for oral immunization. As it was our intention to use alfalfa as our crop production vehicle, codon usage in the synthetic S gene fragment was optimized for dicot species (Ikemura, 1993; Murray et al., 1989), and GC base content was increased to 49% from the original 37%. Possible mRNA destabilizing segments, such as ATTTA, were removed, and unstable codons such as xCG and xTA (Grantham and Perrin, 1986; Murray et al., 1989) were avoided. The translational initiation site was optimized for expression in plants using the sequence ACCATGG (Koziel et al., 1996). Although non-plant signal sequences appear to be generally recognized in plants, the putative 16-aa signal peptide of the native gene was replaced by a 25-aa tobacco pathogenesis-related protein signal peptide, MNFLKSFPFLOFGQYFVAVTHA, to ensure secretion of the S protein from the cell. This strategy and particular signal sequence has been utilized to enhance the synthesis of foreign proteins in plants (Sijmons et al., 1990; Firek et al., 1993; Verwoerd et al., 1995). The remainder of the coding region of the synthetic gene encoded 579 aa of the N-terminal part of the native mature S protein to which was added a six-histidine tag for purification from plant extracts (Figure 50).

The resulting 1856 bp synthetic S gene DNA was 83% identical to the same coding region in the native S gene cDNA. It should be noted that this synthetic form of the gene codes for much less than the 1447 aa of the entire native protein, but does include the major immunogenic epitopes of the full-length protein.

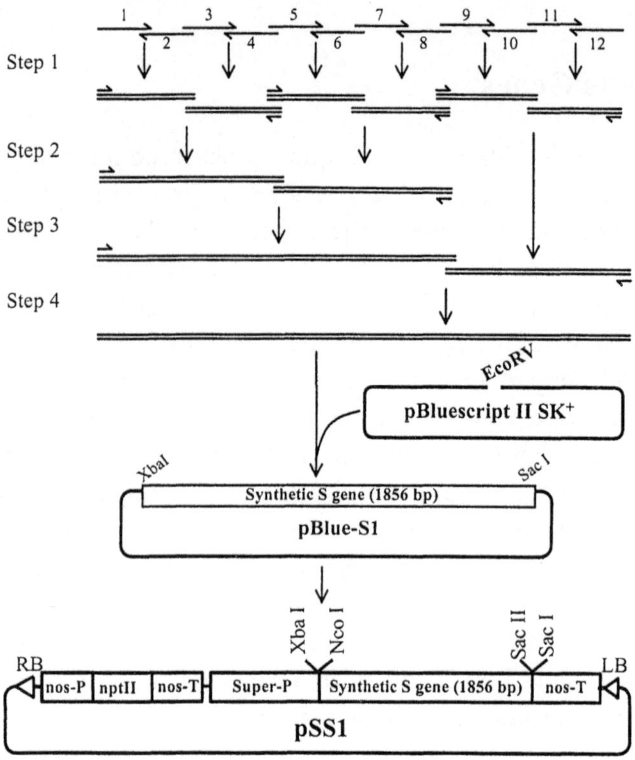

Figure 50. Synthesis of the redesigned 1856 bp S gene by overlap extension PCR and construction of the expression vector pSS1. The starting synthetic oligos for Step 1 are numbered 1-12.

The modified and truncated version of the *S* gene was synthesized by overlap PCR (Ho et al., 1989; Rouwendal et al., 1997), using twelve oligomers, 175-180 nt in length, synthesized by ACGT Corp., Toronto, Canada, and purified by polyacrylamide gel electrophoresis. The oligomers were designed such that six, numbered 1, 3, 5, 7, 9, 11 comprised the sense strand of the designed gene and six, numbered 2, 4, 6, 8, 10, 12, comprised the antisense strand, with the ends of adjacent oligomers overlapping by 23 to 25 nt (Figure

50). Four restriction enzyme sites, *Xba*I and *Nco*I at the 5'-end, *Sac*II and *Sac*I at the 3'-end, were added for cloning into different expression vectors.

Step 1: Six amplification reactions were carried out using pair-wise combinations of the overlapping oligomers: 1+2 (nt 1-323), 3+4 (nt 301-628), 5+6 (nt 605-934), 7+8 (nt 908-1236), 9+10 (nt 1213-1540), and 11+12 (nt 1515-1850). Each reaction (50 μL) contained 0.5 μM of both oligomers, 0.2 mM of dNTP mixture, and 2.5 U of *Pfu* DNA poymerase (Stratagene). The reaction was performed by denaturing the DNA at 94°C for 5 min, followed by 20 cycles of: 94°C for 45 sec, 50°C for 45 sec and 72°C for 1 min and a final extension at 72°C for 10 min. The amplified products of ~ 330 bp each were purified from an agarose gel following electrophoresis.

Step 2: Three overlap PCR reactions were carried out, each containing two contiguous and overlapping products from step 1: bp 1-323 + bp 301-628, bp 605-934 + bp 908-1236, and bp1213-1540 + bp 1515-1850. An initial amplification step consisted of three cycles at 94°C for 3 min, 50°C for 1 min, and 72°C for 8 min. This was followed by the addition of two primers, 18 to 24 nt in length, to a final concentration of 0.5 μM in each reaction. The primers were designed to amplify the entire length of the combined parent fragments in that reaction, i.e. primers P1 and P628c for the first fragment reaction, P605 and P1236c for the middle fragment , and P1213 and P1850c for the end fragment, the number of each primer indicating the position of the nucleotide at the 5' end of the primer. A denaturation step at 94°C for 3 min was followed by 20 cycles of: 94°C for 45 sec, 50°C for 45 sec and 72°C for 1 min, and the reaction was terminated with a step at 72°C for 10 min. The amplification products were separated by gel electrophoresis and the bands of expected size were cut out and purified as templates for further amplifications.

Step 3: The PCR products, bp 1-628 plus bp 605-1236 were combined and amplified as in step 2, using primers P1 and P1236c.

Step 4: The purified 1245 bp PCR fragment from step 3 and 628 bp product from step 2 were coupled, using primers P1 and P1850c to amplify the entire S gene fragment under the same conditions as in steps 2 and 3.

The synthetic *S* gene fragment was purified from an agarose gel and inserted into pBluescript II SK$^+$ at an *Eco*RV site. Seventeen clones were sequenced, and, as all contained one or more sequence errors, the final correct version of the S gene fragment was created by assembly of fragments from these clones and site-directed mutagenesis (Gene Editor, Promega, USA). The binary vector pSS1 was constructed by excising the synthetic S gene from pBluescript using

*Xba*I and *Sac*I and inserting it into pBSN1 (Ni et al., 1995), cut with the same enzymes. The promoter for the synthetic gene in this vector, composed of elements of the mannopine and octopine synthase gene promoters from Agrobacterium, is reported to result in high expression levels in various plant tissues and is frequently referred to as the "super" promoter.

A single mRNA transcript of predicted size was evident in most transgenic tobacco and alfalfa plants carrying the synthetic S gene (Figure 48B). The elimination of the poly(A+) sites and the ATTTA sequences, as well as a reduction of AT (principally T) content from 63% to 51% in the synthetic gene were the major alterations made to the native S gene, and probably account for the presence of full-length transcripts from the synthesized gene. The alterations in codon usage would not likely reduce DNA-dependent arrest of transcript elongation, aside from the effects on AT content.

Most of the S protein detected by immunolocalization in tobacco leaf tissue containing the synthetic S gene was extracellular (data not shown). Interestingly, the labelled protein was located predominantly in the cell wall rather than in the extracellular space, where other foreign proteins have been reported to be primarily deposited when fused to the same PR signal peptide (Sijmons et al., 1990; Firek et al., 1993; Verwoerd et al., 1995). We can only speculate on the features of the S protein which might account for this, but the sequestering of an antigen in the cell wall matrix may provide certain advantages for reducing degradation and enhancing immunogenicity. The levels of S protein in these plants was higher than in those with the native gene to the extent that it was detectable on a Western blot (Figure 48C), reaching a maximum of about 0.01% of total soluble protein. Similar increases in protein levels of xenoproteins in plants following reconstruction of the coding region have been reported for another antigen (Mason et al., 1998) and for the insecticidal Bt protein (Koziel et al., 1993). However, such levels may be inadequate for some applications requiring purification or direct feeding, especially where large amounts of antigen are required and degradation of the foreign protein occurs in the plant tissue/extract (Khoudi et al., 1999) or in the gut.

It is clear from these results that there are as-yet-unexplained features of genes which may limit the synthesis and/or accumulation of transgenic proteins in plants. It is also becoming clear that synthetic genes and codon optimization may not adequately address these features. In contrast to our experience with the S gene of TGEV, we have expressed an unmodified cDNA clone coding for the ORF5 antigen of the porcine respiratory and reproductive syndrome (PRRS) virus in tobacco and alfalfa at levels considerably in excess of those attained with the synthetic S gene (J. Zhang, 2001). Fortunately, there are many examples from microbial expression systems of alternate or additional strategies which can increase foreign gene expression, such as fusion proteins and inducible promoters.

4.3 Fusion Proteins

The strategy of translational fusions to increase levels of a foreign protein or peptide has not been adopted as widely in plant production systems as in those which are microbially-based. This is due at least in part to the long history of research in microbial systems, particularly for the purpose of producing heterologous proteins. The fusion of antigens to other proteins for expression in plants has taken one of three forms: fusion to a non-plant protein, fusion to a protein of plant origin, and fusion to a protein of a plant virus.

Lee et al. (2001) fused a truncated form of a bacterial leucotoxin to the amino or carboxyl terminus of an ER-targeted green fluorescent protein (GFP) in two separate constructs. Expression of the leucotoxin was detected in clover plants containing the construct in which the antigen was inserted between the signal peptide and the amino terminus of GFP and not in the other construct. The maximum level of the fusion protein reached with this construct containing the 35S promoter was about 1% of soluble protein. Whether such a fusion was necessary to reach such expression levels in plants cannot be determined as no attempt was made to express an unfused form of the protein.

In our lab, we have been experimenting with fusing portions of the S gene, the major antigen of transmissible gastroenteritis virus (TGEV) of swine, to plant proteins to enhance expression. Attempts to express the full-length cDNA clone of this gene using various promoters, including the 35S CaMV promoter did not result in levels of the S protein detectable on Western blots, although antibodies to the viral protein were induced by injecting plant tissue extracts into pigs and mice. As an alternative, portions of the S gene encoding a linear epitope (D) responsible for generating neutralizing antibodies to the virus, were fused to two plant genes truncated at the 3' end.

One of these proteins was a beta amylase to which an extended D epitope of 390 bp was attached (Tuboly et al., 2000) (Figure 51). The other plant gene used as a fusion partner codes for a pollen-specific protein from alfalfa (Qiu et al., 1997). PO2 is a highy glycosylated protein with properties typical of an arabinogalactan, and possibly plays a role in pollen hydration. It has no homologies to any allergen, enzyme or structural protein and was thought to be a relatively inert protein which might accumulate to high levels when secreted out of the cell. The coding region for the carboxy terminal 21 amino acids of PO2 was replaced with one encoding a nine-amino acid epitope D of TGEV, and an additional nine-amino acid peptide thought to enhance antibody binding (Posthumus et al., 1990); codon usage was optimized in this fragment for expression in dicots.

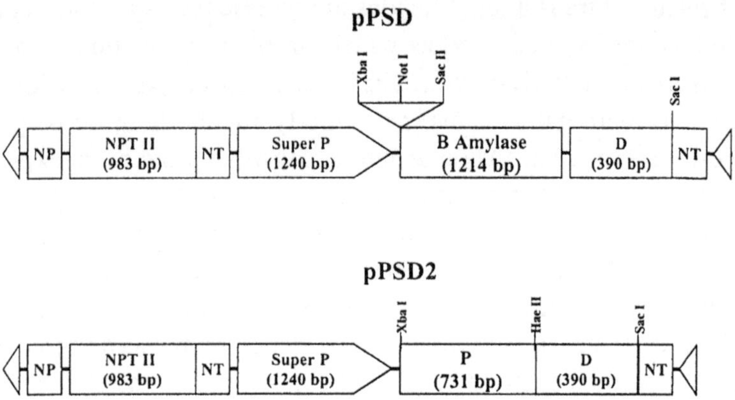

Figure 51. Binary vectors containing a swine viral epitope fused to plant protein genes. NP: nopaline synthase promoter; NPT II: neomycinphosphotransferase II; NT: nos terminator; Super P: mannopine/octopine synthase promoter; D: D epitope region of the S gene; P: truncated PO2 gene from alfalfa.

Both fusion proteins were driven by a chimeric mannopine/octopine synthase promoter (Ni et al., 1995), referred to by some as the "super-promoter". In the case of the fusion to PO2, the fusion product was detected with antibodies to PO2, and the presence of the TGEV epitope was confirmed by the injection of plant extracts into mice, resulting in the production of neutralizing antibodies to the virus (Bailey, 2000). The highest expression levels achieved with this construct was in the 0.2% range in alfalfa plants, but were much lower in tobacco plants. This is in marked contrast to plants containing the full-length cDNA clone for the S antigen of the virus, wherein the S protein could not be detected, and to plants containing a truncated, synthetic version of the S gene in which the S protein constituted only 0.01% of total soluble protein.

Following the initial example of fusing a foot and mouth viral epitope to a protein of a plant virus (Usha et al., 1993), there have been several reports of this approach to expressing viral antigens in plants (rev. by Walmsley and

Arntzen, 2000). Although the particulate structure of antigens produced by this method is very desirable for inducing an immune response, the ability of plant viral particles to survive the gastrointestinal tract is questionable and, as yet untested. However, intranasal administration of such particles has elicited a protective immune response in a mouse model of hepatitis (Koo et al, 1999).

4.4 Inducible Systems

The very low levels of expression frequently observed for foreign proteins in plants may be due to a variety of gene-specific factors which affect transcription, translation, and stability of the protein. Alternatively, the foreign protein may have inhibitory effects on the growth and development of the plant, resulting in the selection of low-expressing plants during regeneration. Mason et al. (1998) enhanced the levels of the B subunit of the E coli heat labile toxin (LTB) in potato shoots and tubers using a synthetic form of the gene driven by the 35S promoter. However, they also noted that as LTB levels increased, rate of shoot growth and tuber yield dropped in plants grown in greenhouse conditions. When LTB synthesis was controlled by a tuber-specific promoter, shoot growth was normal, but tuber development was impaired. Similarly, Richter et al. (2000) noted poor shoot growth and low tuber yield in potato as they increased levels of hepatitis surface antigen B by adding plant terminator sequences to their construct driven by an enhanced 35S promoter. .

The negative physiological impact of foreign genes in transgenic plants was not widely anticipated by those working in this field, as proteins from organisms as distant as mammalian pathogens have little homology with structural or enzymatic proteins of plants. However, with the accumulation of sequence data from many organisms, it is becoming clear that many protein motifs are widely shared across organisms. An example is the particular amino acid pattern of six cysteines in a protein, first noted in genes for epidermal growth factor (EGF) in mammals, and subsequently determined to be essential for the characteristic three-dimensional structure and biological activity of this molecule. There is an abundant and growing number of DNA sequences from many organisms which contain EGF motifs. EGF motifs have been detected in the extracellular domain of a wall-associated kinase gene from arabidopsis (He et al.1996.), as well as in a putative vacuolar sorting receptor (Miller et al. 1999). EGF-binding proteins have been detected in plant extracts (Komatsu et al.1996) and recently membrane-associated proteins have been discovered in plants with homology to EGF receptor proteins (Ahmed et al.1997).These observations raise the possibility that plant proteins with EGF-like domains may have some physiological function in plants, and that expression of mammalian forms of EGF in plants may interfere with that function. This is perhaps the reason for the lack of success recorded by Higo et al. (1993) in expressing a human EGF gene in tobacco and for similar problems in our lab with such a protein in transgenic plants. In addition, there are many other types of non-specific and unpredictable interactions possible between a foreign protein and: other plant proteins, such as enzymes and receptors; plant membranes, such as those of the endoplasmic reticulum, Golgi apparatus, vacuole and plasmalemma; or the host of other molecules critical to the growth and development of the plant

The toxic or inhibitory effects of a foreign protein are particularly problematic when expression of the gene is controlled by constitutive promoters such as the 35S promoter from cauliflower mosaic virus (Mason et al.,1998; Richter et al., 2000). Although tissue-specific promoters may reduce such effects generally throughout the plant, the toxic effect of the protein on the development and productivity of the target tissue, such as tubers (e.g. Mason et al. 1998) or seeds, could probably not be avoided and would likely be more serious as such promoters usually provide higher expression levels in the targeted tissue than do constitutive promoters . For example, the levels of Bt protein in maize leaves were much higher on average across a sample of transgenic plants when the Bt gene was controlled by the PEP carboxylase promoter than the 35S promoter (Koziel et al., 1993). Another disadvantage of a constitutive promoter is the metabolic cost of synthesizing the transgenic protein in all tissues at all stages of growth. If the only tissue to be harvested is the leaves, for example, it is inefficient and wasteful for the plant to produce the foreign protein in other tissues. In this case, leaf-specific promoters or promoters induced in the leaves by some treatment would restrict synthesis to only the harvested tissue.

For these reasons there is growing interest in developing inducible expression systems suitable for practical application to crop plants cultivated in a conventional field setting (Zuo and Chua, 2000). Ideally, synthesis of the foreign protein could be induced by simple, low-cost application of a treatment to plants in the field, resulting in high levels of the protein within hours up to a day or so. Leaf tissue is the most obvious target for such rapid induction, as it would present the largest biomass available for induction and probably the most amenable to rapid metabolic change.

We have used subtraction libraries to clone genes that are induced in alfalfa grown in field conditions, following application of specific treatments. We are concentrating on three clones for which we cannot detect mRNA in above-ground plant tissue from plants growing in the field under normal conditions. All three clones are induced by a specific treatment applied to plants in the field; one of the clones is also induced by a heat treatment and another is induced by wounding. Induction of the third gene appears to be exclusive to a particular treatment.

There are a number of advantages to such an approach. The potential deleterious effects of expressing a foreign protein on the growth and development of a plant are avoided. The synthesis of the transgenic product can be timed to coincide with optimal conditions of growth and stage of development of the crop plant. Induction may be targeted to specific harvestable tissues, thereby avoiding synthesis in other tissues which entails additional metabolic costs and perhaps regulatory issues, as well. It may be, as well, that the possibility of gene silencing could be reduced if expression of the transgene is delayed to a late stage of growth in the plant rather than in all tissues at all times. Finally, inducible transgene systems offer a method of biological containment, i.e. the foreign protein is not present in the crop until the application of the inducing treatment, at which time the crop is harvested.

4.5 Reducing Degradation and Enhancing Immunogenicity

To address concerns regarding degradation of the protein, one might select proteins which can resist such degradation, either because they are adapted to that environment or because of their intrinsic properties. Examples of the former are the various surface antigens of intestinal pathogens, and an example of the latter is the beta-conglycinins of soybean (Astwood et al., 1996), perhaps a candidate for a carrier protein.

There are other strategies one can adopt, as well, to reduce degradation, such as localization in cell walls. Plant cell walls, depending on species, tissue and environment can produce a limitless array of physicochemical structures which can protect a protein in planta, as well as in the digestive tract of animals that ingest the plant tissue. For the purpose of oral vaccination, the gradual breakdown of the cell wall during digestion could result in a slow continual release of antigen as the tissue passes through the gut. The association of the antigen with resulting particulate structures could also enhance the immune response to the antigen.

Dietary context is another very significant factor affecting both degradation and antigenicity. It represents the entirety of biochemical components of the feed, many of which have varied effects on the immunophysiological functions of the intestinal mucosa, as well as on the digestion process itself. An edible vaccine strategy must include considerations not only of the effects of the plant tissue containing the antigen, but also the effects of other components of the diet before, during and after oral vaccination. Experiments by Modelska et al. (1998) demonstrate the effects of feeding an antigen as part of plant tissue. Mice were fed spinach leaves containing a alfalfa mosaic virus modified to express two rabies virus epitopes. The levels of secreted IgA in such mice were twice those in mice that had been orally intubated with purified virus particles at a dose ten-fold that received by those eating the spinach leaves directly. Either the virus particles were protected physically by the associated cell walls and membranes, or biochemically by various protein and non-protein compounds of plant origin, or by both mechanisms. Equally as important may be the immunological, adjuvant-type effects of these physical and biochemical components provided by the leaf tissue.

The amount and type of processing can also affect the amount and condition of peptides in a plant-derived product that reach the target tissues of the gastrointestinal tract. Some seed-derived feed ingredients, such as soya meal, are heat-treated during processing at temperatures ranging from 60° to 80° C for periods ranging from a few to several minutes, depending on the process. It has been shown in many nutritional studies that such heating is necessary to inactivate enzyme inhibitors and other antinutritional factors to enable utilization of these components in animal diets. The potential for denaturation of a therapeutic protein using such a production and delivery system is obvious, but it not may be as extensive as anticipated since incorporation of the peptide within plant tissues may reduce the extent of denaturation compared to subjecting such proteins to heat as isolated molecules.

In addition, there is extensive experimental literature documenting the beneficial effects of heating plant-derived feedstuffs to enhance utilization of the protein component especially (Conrad and Klopfenstin, 1988). This process of rendering proteins less digestible is particularly attractive for ruminant utilization of the high protein content of alfalfa, much of which is rapidly degraded microbially in the rumen and lost as methane and urea. Controlled heating converts much of the readily soluble and digestible plant protein to "bypass protein" which escapes rapid degradation in the rumen, and is broken down more slowly in passage through the intestinal tract. This approach could be conceivably be utilized to reduce degradation of antigens expressed in plant tissue for oral delivery not only to reduce degradation, but also to enhance immunogenicity.

The delivery of oral vaccines in plant tissue or in some form derived from plants presents unique opportunities for mucosal immunization, as well as obstacles. As mentioned above, a problem faced by any subunit vaccine is the extent and kind of immune response induced by an antigen stripped of its association with a particulate, replicating structure typical of a pathogen. The possibility of inducing oral tolerance to an antigen delivered as part of a potato, for example, has been demonstrated by Ma et al. (1997). The ability to control this response has not yet been developed, although a number of strategies have been adopted to produce the desired "vaccinated" response. For example, the selection of antigens which in plants assemble into particulate structures, such as the pentameric ring of LTB (Haq et al., 1995) or the viral particle of Norwalk virus (Mason et al. 1996) should enhance the oral immunogenicity of those antigens (Florence and Jani, 1993). Incorporation of mucosal adjuvants along with an antigen produced in plant material could also increase the immune response. For example, Richter et al. (2000) added cholera toxin B (CTB) to transgenic potato tissue containing hepatitis B surface antigen, but a wide variety of other protein and non-protein compounds could serve a similar function, including polycations, lipid conjugates, streptomycin, bacterial adhesins and lectins of diverse origins (Mahon et al. 1998; De Aizpurura and Russell-Jones, 1998). Plants are especially rich in lectins, some of which are potent mucosal adjuvants while others are not. Intranasal co-administration of mistletoe lectin with ovalbumen (OVA) enhanced secretion of IgA to OVA from mucosal surfaces of rat, whereas phytohemagglutinin did not increase levels over those elicited by the antigen alone (Lavelle et al. 2001). An example of potent non-protein adjuvants from plants is the class of compounds referred to as saponins, which are found in many plant species, but are especially prominent in *Quillaia saponaria* (Sjolander et al., 1998). In this connection, the attachment of antigens to oil-bodies produced in oilseeds (Parmenter et al., 1996) may provide another mechanism for stimulating immunity, depending on the plant source of the oil-bodies.

It is also possible to draw on the extensive literature dealing with mammalian genes affecting the mucosal immune response to engineer plants for edible vaccines. For example, the role of cytokines in modulating T-cell response towards tolerance to an orallly-administered antigen might be exploited by co-producing specific cytokines in plant tissue. Recently, a number of neuropeptides endogenous to the gut have been identified which can affect T-cell response and levels of antibody secretion when orally co-administered with an antigen (Pascual et al.1998).

5. EDIBLE DNA VACCINES IN PLANTS

Recombinant DNA vaccines have shown great promise in animal models for inducing long-lived cell- and humorally-mediated immunity to pathogens (Apostolopoulos and Plebanski, 2000). A particularly attractive feature of this approach is the possibility of creating multivalent vaccines by incorporating multiple genes for antigens in one or more plasmid molecules. An extension of this principle could involve , as well, inclusion of genes for immunomodulatory proteins, such as cytokines, in the DNA construct. The potential of this approach to immunizing livestock against diseases is under active investigation, including its application to enteric diseases (van Drunen Littel-van den Hurk et al.,2001). However, there has been little mention in the literature of plants as vehicles for orally-delivered DNA vaccines. There are some features of plants which may make such an approach more feasible than would be apparent at first glance. First, with the advent of chloroplast transformation, it is now possible to grow plant tissues which contain many thousands of copies of a transgene per cell (DeCosa et al. 2001). Second, as mentioned above, plant cell walls provide a form of encapsulation, the specific features of which can vary, depending on the species, tissue and environment. It is also possible to modify cell wall structures by various processes, such as heating. An example of organism-mediated delivery of DNA to intestinal target tissues in the intestine was provided by Darji et al.(2000) who utilized attenuated strains of *Salmonella*. The possibility of using plant tissue in a similar fashion is strengthened by the observations of Schubbert et al.(1998) who demonstrated that oral administration of naked DNA alone was sufficient to result in the uptake of such DNA in cells of the spleen and liver, as well as macrophages. Histological analyses and in situ hybridization assays provided evidence of incorporation of the foreign DNA into the chromosomal DNA of the mouse and replication of this DNA during cell division resulting in transformed tissue sectors. It may be that such tissues are amenable to oral vaccination with DNA in plant cells containing not only many copies of mammalian genes and the appropriate regulatory sequences, but also other mammalian genes to enhance a mucosal immune response.

ACKNOWLEDGEMENTS

This work was supported by grants from the Natural Sciences and Engineering Research Council of Canada program for Strategic Projects and the Ontario Ministry of Agriculture, Food and Rural Affairs. We are grateful to Dr. Yongzhong Wu, Akwasi Agyare-Tabbi, Ning Chen for their contributions to this research, and to Jennifer Kingswell for preparation of the manuscript.

REFERENCES

Ahmed SU, Bar-Peled M and Raikhel NV, 1997. Cloning and sub-cellular location of an Arabidopsis receptor-like protein that shares common features with protein-sorting receptors of eukaryotic cells. Plant Physiol, 114:325-336

Apostolopoulos V and Plebanski M, 2000. The evolution of DNA vaccines Curr Opin Mol Ther, 2:441-447

Astwood JD, Leach JN and Fuchs RL, 1996. Stability of food allergens to digestion in vitro. Nat Biotechnol, 14:1269-1273

Babiuk LA and Tikoo SK, 2000. Adenoviruses as vectors for delivering vaccines to mucosal surfaces. J Biotechnol, 83:105-111

Bailey A, 2000. Expression in plants and immunogenicity of a fusion protein containing an epitope from the swine transmissible gastroenteritis virus s protein. M.Sc. Thesis, Dept of Plant Agriculture, University of Guelph, Guelph, Ontario, Canada

Carillo C, Wigdorovitz A, Oliveros JC,, Zamorano PI, Sadir AM, Gomez N,Salinas J, Escribano JM and Borca MV, 1998. Protective immune response to foot-and-mouth disease virus with VP1 expressed in transgenic plants J Virol, 72:1688-1690

Conrad HR and Klopfenstin TJ, 1988. Role in livestock feeding– greenchop, silage, hay, and dehy. In *Alfalfa and alfalfa improvement.* Eds. Hanson AA, Barnes DK and Hill RR. American Society of Agronomy, Madison, Wisconsin, USA, pp 539-551

Darji A, zur Lage S, Garbe AI, Chakraborty T and Weiss S, 2000. Oral delivery of DNA vaccines using attenuated *Salmonella typhimurimum* as carrier. FEMS Immunol Med Microbiol, 27:341-349

De Cosa B, Moar W, Lee SB, Miller M and Daniell H, 2001. Hyperexpression of Bt Cry 2Aa2 operon in chloroplasts leads to formation of insecticidal crystals. Nat Biotech, 19:71-74

De Aizpurura HJ and Russell-Jones GJ 1988. Identification of classes of proteins that provoke an immune response upon oral feeding. J Exp Med, 167:440-451

Diehn SH, Chiu WL, De Rocher EJ and Green PJ, 1998. Premature polyadenylation at multiple sites within a Bacillus thuringiensis toxin gene-coding region. Plant Physiol, 117: 433-443.

Firek S, Draper J, Owen MR, Gandecha A, Cockburn B and Whitelam GC, 1993.Secretion of a functional single-chain Fv protein in transgenic tobacco plants and cell suspension cultures. Plant Mol Biol, 23:861-870

Florence AT and Jani PU, 1993. Particulate delivery: the challenge of the oral route. In *Pharmaceutical particulate carriers: therapeutic applications.* Ed. Rolland A, Marcel Dekker Inc NY, pp 65-107

Gomez N, Carrillo C, Salinas J, Parra F, Borca MV and Escribano JM, 1998. Expression of immunogenic glycoprotein S polypeptides from transmissible gastroenteritis coronavirus in transgenic plants. Virology, 249: 352-358

Grantham P and Perrin P, 1986. AIDS virus and HTLV-I differ in codon choices. Nature, 319: 727-728

Gutierrez RA, MacIntosh GC and Green PJ, 1999. Current perspectives on mRNA stability in plants: multiple levels and mechanisms of control. Trends Plant Sci, 4:429-438

Haq TA, Mason HS, Clements JD and Arntzen CJ, 1995. Oral immunization with a recombinant bacterial antigen produced in transgenic plants. Science, 268:714-716

He Z-H, Fujiki M and Kohorn BD, 1996. A cell-wall associated receptor-like protein kinase.J Biol Chem, 271:19789-19793

Hershberg RM and Meyer LF, 2000 Antigen processing and presentation by intestinal epithelial cells-polarity and complexity. Immunology Today, 21:123-129

Higo K, Saito Y and Higo H, 1993. Expression of a chemically synthesized gene for human epidermal growth factor under the control of cauliflower mosaic virus 35S promoter in transgenic tobacco. Biosci Biotechnol Biochem, 57:1477-1481

Ho SN, Hunt HD, Horton RM, Pullen JK and Pease LR, 1989. Site-directed mutagenesis by overlap extension using the polymerase chain reaction. Gene, 77:51-59

Hunt AG and MacDonald MH, 1994. Deletion analysis of the polyadenylation signal of a pea ribulose-1,5-bisphosphate carboxylase small-subunit gene. Plant Mol Biol, 13:125-138

Ikemura T, 1993. Plant nucleic acids codon usage In *Plant molecular biology labfax*. Ed. Croy RDD, BIOS Scientific Publishers Ltd, Oxford, pp 37-48

Jefferson RA, Kavanagh TA and Bevan MW, 1987. GUS fusions: beta-glucuronidase as a sensitive and versatile gene fusion marker in higher plants. EMBO J, 6:3901-3907

Kelsall B and Strober W, 1999 Gut-associated lymphoid tissue: antigen handling and T lymphocyte responses. In *Mucosal Immunology* (2nd ed). Eds. Ogra PL, Mestecky J, Lamm ME, Strober W, Bienenstock J and McGhee JK. Academic Press, San Diego, pp 293-317

Khoudi H, Laberge S, Ferullo J-M, Bazin R, Darveau A, Castonguay Y, Allard G, Lemieux R and Vezina L-P, 1999. Production of a diagnostic monoclonal antibody in perennial alfalfa plants. Biotech and Bioeng, 64:135-143

Komatsu S, Abe K and Karibe H, 1996. Epidermal growth factor binding proteins in rice (*Oryza sativa* L.) leaves. Biol Pharm Bull, 19:905-908

Koziel MG, Carozzi NB and Desai N, 1996. Optimizing expression of transgenes with an emphasis on post-transcriptional events. Plant Mol Biol, 32:393-405.

Koziel MG, Beland GL, Bowman C, Carozzi NB, Crenshaw R, Crossland L, Dawson J, Desai N, Hill M, Kadwell S, Launis K, Maddox D, McPherson K, Meghji MR, Merlin E, Rhodes R, Warren GW, Wright M and Evola SV, 1993. Field performance of elite transgenic maize plants expressing an insecticidal protein derived from *Bacillus thuringiensis*. Bio/Technology, 11:194-200.

Lavelle EC, Grant G, Pusztai A, Pfuller U and O'Hagan DT, 200. The identification of plant lectins with mucosal adjuvant activity. Immunology, 102:77-86

Lee RWH, Strommer J, Hodgins D, Shewen PE, Niu Y and Lo RYC, 2001. Towards development of an edible vaccine against bovine pneumonic pasteurellosis using transgenic white clover expressing a *Mannheimia haemolytica* A1 Lkt50 fusion protein.

Lorentz A, Schwengberg S, Sellge G, Manns MP and Bischoff SC, 2000. Human intestinal mast cells are capable of producing different cytokine profiles: role of IgE receptor cross-linking and IL-4. J Immunology 164:43-48

Ma SW, Zhao D-L, Yin, Z-Q and Mukherjee K, 1997. Transgenic plants expressing autoantigens fed to mice to induce oral immune tolerance. Nat Medicine, 3:793-796

Mahon BP, Moore A, Johnson PA and Mills KHG, 1998. Approaches to new vaccines. Crit Rev in Biotech, 18:257-282

Mason HS, Ball JM, Shi J-J, Jiang X, Estes MK and Arntzen CJ, 1996. Expression of Norwalk virus capsid protein in transgenic tobacco and potato and its oral immunogenicity in mice. Proc Natl Acad Sci USA, 93:5340-5353

Mason HS, Haq TA, Clements JD and Arntzen CJ, 1998. Edible vaccine protects mice against *Escherichia coli* heat-labile enterotoxin (LT):potatoes expressing a synthetic LT-B gene. Vaccine, 16:1336-1343

Sanderson IR and Walker WA, 1999. Mucosal barrier: An overview. In: *Mucosal immunology* (2nd ed). Eds. Ogra PL, Mestecky J, Lamm ME, Strober W, Bienenstock J and McGhee JK. Academic Press, San Diego, pp 5-17.

Mayer L and Blumberg RS, 1999. Antigen-presenting cells: epithelial cells. In *Mucosal immunology* (2nd ed). Eds. Ogra PL, Mestecky J, Lamm ME, Strober W, Bienenstock J and McGhee JK. Academic Press, San Diego, pp 365-379

McGhee JR, Lamm ME and Strober W, 1999. Mucosal immune responses: an overview. In *Mucosal immunology* (2nd ed). Eds. Ogra PL, Mestecky J, Lamm ME, Strober W, Bienenstock J and McGhee JK. Academic Press, San Diego, pp 485-506

Milek RL, Stunnenberg HG and Konings RN, 2000. Assembly and expression of a synthetic gene encoding the antigen Pfs48/45 of the human malaria parasite *Plasmodium falciparum* in yeast. Vaccine, 18:1402-1411.

Miller EA, Lee MCS and Anderson MA, 1999. Identification and characterization of a prevacuolar compartment in stigmas of *Nicotiana alata*. Plant Cell 11:1499-1508

Modelska A, Dietzschold B, Sleysh N, Fu ZF, Steplewski K, Hooper DC, Kiprowski H and Yusibov V, 1998. Immunization against rabies with a plant-derived antigen. Proc Natl Acad Sci USA, 95:2481-2485.

Murray EE, Lotzer J and Eberle M, 1989. Codon usage in plant genes. Nucleic Acids Res, 17:477-498.

Neutra MR, 1998. Current concepts in mucosal immunity V. role of M cells in transepithelial transport of antigens and pathogens to the mucosal immune system. Am J Pysiol, 274:G785-91

Ni M, Cui D, Einstein J, Narasimhulu S, Vergara CE and Gelvin SB, 1995. Strength and tissue specificity of chimeric promoters derived from the octopine and mannopine synthase genes. Plant J, 7: 661-676

Ohme-Takagi, M, Taylor CB, Newman TC and Green PJ, 1993. The effect of sequences with high AU content on mRNA stability in tobacco. Proc Natl Acad Sci USA, 90:11811-11815

Parmenter DL, Booth JG and Moloney MM, 1996. Production and purification of recombinant hirudin from plant seeds. In *Transgenic plants: a production system for industrial and pharmaceutical proteins*. Eds. Owen MRL and Pen J. John Wiley and Sons, New York, pp. 261-280

Pascual DW, Kiyono H, McGhee JR, 1998. Mucosal vaccine design for enteric infectious agents. In *Designer vaccines: Principles for successful prophylaxis*. Eds. Hughes T, Huw PA and Campos M. CRC Press, Boca Raton, Florida, pp.120-136

Perlak FJ, Fuchs RL, Dean DA, McPherson SL and Fischoff DA, 1991. Modification of the coding sequence enhances plant expression of insect control protein genes. Proc Natl Acad Sci USA, 88:3324-3328

Qiu X, Wu Y, Du S and Erickson L, 1997. A new arabinogalactan protein-like gene expressed in pollen of alfalfa. Plant Sci 124:41-47

Reines D, Conaway RC and Conaway JW, 1999. Mechanism and regulation of transcriptional elongation by RNA polymerase II. Curr Opin Cell Biol, 11:342-346

Richter LJ, Thanavala Y, Arntzen CJ and Mason HS, 2000. Production of hepatitis B surface antigen in transgenic plants for oral immunization. Nat Biotech, 18:1167-1171

Rothnie HM, 1996. Plant mRNA 3'-end formation. Plant Mol Biol 32:43-61

Rouwendal GJ, Mendes O, Wolbert EJ and de Boer A, 1997. Enhanced expression in tobacco of the gene encoding green fluorescent protein by modification of its codon usage. Plant Mol Biol 33:989-999

Schubbert R, Hohlweg U, Renz D and Doerfler W, 1998. On the fate of orally ingested foreign DNA in mice: chromosomal association and placental transmission to the fetus. Mol Gen Genet, 259:569-576

Shilatifard A, 1998. Factors regulating the transcriptional elongation activity of RNA polymerase II. FASEB J, 12:1437-1446

Sijmons PC, Dekker BM, Schrammeijer B, Verwoerd TC, van den Elzen PJ and Hoekema A, 1990. Production of correctly processed human serum albumin in transgenic plants. Bio/Technology, 8:217-221

Sjolander A and Cox JC, 1998. Uptake and adjuvant activity of orally delivered saponin and ISCOM vaccines. Advanced Drug Delivery Rev. 34:321-338

Strober W, Kelsall B and Marth T, 1998. Oral tolerance. J Clin Immun, 18:1-34

Strobel S and Mowat AM, 1998. Immune responses to dietary antigens: oral tolerance. Immunology Today 19:173-181

Tanzer MM and Meagher RB, 1995. Degradation of the soybean ribulose-1,5-bisphosphate carboxylase small-subunit mRNA, SRS4, initiates with endonucleolytic cleavage. Mol Cell Biol, 5: 6641-6652

Timmermans MC, Maliga P, Vieira J and Messing J, 1990. The pFF plasmids: cassettes utilising CaMV sequences for expression of foreign genes in plants. J Biotechnol, 14:333-344

Tuboly T, Nagy E, Dennis JR and Derbyshire JB, 1994. Immunogenicity of the S protein of transmissible gastroenteritis virus expressed in baculovirus. Arch Virol.,137:55-67

Tuboly T, Yu W, Bailey A, Degrandis S, Du S, Erickson L and Nagy E, 2000. Immunogenicity of porcine transmissible gastroenteritis virus spike protein expressed in plants. Vaccine, 18: 2023-2028

Walmsley AM and Arntzen CJ, 2000. Plants for delivery of edible vaccines. Curr Opin Biotech, 1:126-129

Usha R, Rohl JB, Spall VE, Shanks M, Maule AJ, Johnson JE and Lomonossoff GP, 1993. Expression of a animal virus antigenic site on the surface of a plant virus particle. Virology, 202:949-955

van Drunen Littel-van den Hurk S, Loehr BI and Babiuk LA, 2001. Immunization of livestock with DNA vaccines: current studies and future prospects. Vaccine 19:2474-249

Verwoerd TC, van Paridon PA, van Ooyen AJ, van Lent JW, Hoekema A and Pen J, 1995. Stable accumulation of *Aspergillus niger* phytase in transgenic tobacco leaves. Plant Physiol, 109:1199-1205.

Zhang J, 2002. Expression of foreign proteins in plants using inducible gene systems. Ph.D Thesis, Dept of Plant Agriculture, University of Guelph, Ontario Canada

Zuo J and Chua NH, 2000. Chemical-inducible systems for regulated expression of plant genes. Curr Opin Biotechnol, 11(2):146-151

Index

Lightning Source UK Ltd.
Milton Keynes UK
UKHW020103170719
346308UK00011B/517/P

9 781402 008351